I0015706

Modern Network Observability

A hands-on approach using open source tools such as Telegraf, Prometheus, and Grafana

David Flores

Christian Adell

Josh VanDeraa

Modern Network Observability

Copyright © 2024 Packt Publishing

All rights reserved. No part of this book may be reproduced, stored in a retrieval system, or transmitted in any form or by any means, without the prior written permission of the publisher, except in the case of brief quotations embedded in critical articles or reviews.

Every effort has been made in the preparation of this book to ensure the accuracy of the information presented. However, the information contained in this book is sold without warranty, either express or implied. Neither the authors, nor Packt Publishing or its dealers and distributors, will be held liable for any damages caused or alleged to have been caused directly or indirectly by this book.

Packt Publishing has endeavored to provide trademark information about all of the companies and products mentioned in this book by the appropriate use of capitals. However, Packt Publishing cannot guarantee the accuracy of this information.

Group Product Manager: Pavan Ramchandani
Publishing Product Manager: Prachi Sawant
Book Project Manager: Ashwin Kharwa
Senior Editor: Runcil Rebello
Technical Editor: Yash Bhanushali
Copy Editor: Safis Editing
Proofreader: Runcil Rebello
Indexer: Rekha Nair
Production Designer: Gokul Raj S.T
DevRel Marketing Coordinator: Marylou De Mello

First published: October 2024

Production reference:1120924

Published by Packt Publishing Ltd.
Grosvenor House
11 St Paul's Square
Birmingham
B3 1RB, UK

ISBN 978-1-83508-106-8

www.packtpub.com

To my wife, Yaquelin, for your unwavering support and patience, especially during the writing of this book. To my family, for your endless love and encouragement. And to my friends and colleagues, for inspiring me and pushing me to finally discuss networks and observability—thank you!

- David Flores

To everyone who I have worked or collaborated with during my career. You have been instrumental in allowing me to satisfy my neverending curiosity.

- Christian Adell

To my terrific family, thank you for your support, love, and connection. This book shows my kids that you can do anything that you set your mind to. To those who have been part of my growth from youth through various stages of my life and career, I know I could not be here and do things like this without the path I have taken with you.

- Josh VanDeraa

Foreword 1

Imagine cruising down a winding road at midnight without turning on your headlights. Sure, you might enjoy the excitement and the adrenaline rush, but it is only a matter of time before you end up in an ambulance or worse. In the world of networking, operating a network without monitoring and observability is the equivalent of driving at night without headlights. We all understand the importance of visibility and insights into our networks, but for many years, we have relied on rudimentary tools such as ping and traceroute. Those tools were invented over 30 years ago when the network was much, much simpler. We desperately need to gain visibility into our networks, but we face the challenges of data collection, data processing, storage, scalability, vendor incompatibilities, visualization, and the list goes on. Many advancements have been made in recent years to address those challenges, such as streaming telemetry, accurate sampling of flow data, collecting and storing data at scale, and accurate visualization. However, we lack a comprehensive, non-biased form to illustrate how the individual pieces fit into the architecture and how they can be applied in a modern network.

I first met David and Christian while working at Network to Code and have known Josh for years before then. They were all brilliant engineers, but what impressed me the most beyond their technical abilities was their passion for helping others and empowering the community to succeed. In this book, they combine their deep knowledge of the world of modern network observability with years of practical, battle-tested experience. When I say operational experience, I mean the type that is wake-up-at-2 AM mission critical, leaving the scars that are beer-discussion and conference-talk worthy that normally require admission fees to hear about.

The book addresses the complex world of network observability challenges from architectural concepts to practical configuration examples with a focus on vendor-neutral, open source tools. You will gain the necessary knowledge, from collecting data in various forms, normalizing and enriching it, to visualizing and gaining the necessary insights to help you do your job, all with open source tools. If that sounds too good to be true, it is not. You can even take the architecture and tools introduced in the book and swap out the elements with other tools. That is the beauty of the vendor-neutral, open source tools introduced in this book. I believe the benefits you will gain from the first few chapters will outweigh the investment you've made. Better yet, the knowledge will pay dividends for years to come. I have been waiting for a book of this kind for years, and David, Christian, and Josh have done a masterful job in the delivery.

What are you waiting for? Let's dive into the modern world of network observability!

Eric Chou

Network Automation Advocate, Network to Code

Host, Network Automation Nerds Podcast

Founder, Network Automation Nerds, LLC.

Foreword 2

Observability has revolutionized how we manage infrastructure in the DevOps and cloud era, but its application to networking is still catching up. While network observability shares many similarities with broader infrastructure observability, there are specific challenges that have slowed its adoption in the networking field.

In this book, David, Christian, and Josh have done an outstanding job of covering everything network engineers and network reliability engineers need to know about network observability. They guide you from basic concepts to advanced examples, providing a comprehensive framework that has been successfully used in various production environments.

What makes this book special, in my view, is how network observability is presented not as an isolated component but as part of a larger automation stack. In observability, the quality of the data is everything. One of the benefits of modern observability is the ability to capture additional context, allowing us to slice and dice the data until we extract the most valuable information.

One of the key differences between monitoring and observability is the shift from a vertically integrated stack with limited flexibility to an extensible stack where users are encouraged to explore data and extract valuable insights to improve infrastructure reliability. This shift is important because it requires network engineers to develop new skills in data analysis.

At first, the amount of information may seem overwhelming, but it's important to remember that in a production environment, multiple roles contribute to defining and implementing observability solutions. Network reliability engineers are responsible for building the platform, while network engineers are the main users. However, network engineers should also take the time to understand how the platform works, as the quality of the data collected is crucial to the effectiveness of the observability stack.

A network engineer with a deep understanding of the types of data that can be collected from the infrastructure, and the context in which it is collected, will be an invaluable partner to the network reliability team in building the best possible platform. Whether you are a network engineer or a network reliability engineer, this book covers all the skills and technologies needed for both roles.

As you dive into this book and begin your journey, remember that the tools you use are just one part of the equation. It's more important to focus on understanding the core concepts and functional building blocks that form the foundation of effective observability practices. Tools may change and evolve, but the principles you'll learn here will remain relevant.

Don't be intimidated by the sheer volume of new information. It's natural to feel overwhelmed at first, but with time and practice, these concepts will become second nature. Approach this learning process with curiosity and an open mind, and you'll soon find yourself confidently navigating the exciting world of network observability.

For those new to this field, I recommend starting with a high-level understanding, which will help you identify which topics require deeper exploration and which ones are less critical for your role.

Over the years, I've had the opportunity to help many people discover and learn about network observability, and I've yet to see anyone who has invested time into it who wasn't genuinely excited by the new capabilities that come with these platforms.

Welcome to the exciting field of network observability.

Damien Garros

Infrastructure automation and observability architect

Co-Founder and CEO, OpsMill

Contributors

About the authors

David Flores is passionate about solving complex problems in network infrastructure, software architectures, automation, and observability. With experience with service providers, cloud providers, and system integrators, David has gained expertise in managing, automating, and building observability stacks for network infrastructure. Currently at CoreWeave, he focuses on enhancing automation and observability. David has also contributed to open source projects such as gns3fy, and actively shares his knowledge through blogs, workshops, and technical events. David is always curious and eager to keep himself updated and open to new ideas in the field.

Christian Adell is a principal architect at Network to Code He is focused on building network automation solutions for diverse use cases, with great emphasis on open source software. He is passionate about learning and helping others to grow, but also has more hobbies than hours in the day, so working remotely from Barcelona gives him the time and the space to achieve his dreams. Christian is a co-author of O'Reilly's *Network Programmability and Automation* book and a co-author of *Network Automation with Nautobot* by Packt. Also in relation to sharing knowledge, he is the organizer of the NetBCN community in Barcelona and has been collaborating with several universities for almost 20 years.

Josh VanDeraa is a network engineer and automation leader. Currently, he is a services director at Network to Code, driving value from network automation solutions. Josh has experience in automation and networking across retail, transportation, and managed services. In his free time, he enjoys being with his family or the Minnesota seasons. Josh co-authored *Network Automation with Nautobot* and self-published *Open Source Network Management*.

About the reviewers

Brad Haas is a seasoned professional in network automation, serving as vice president of professional services at Network to Code. With over two decades of experience in the field, Brad leads high-performing teams in complex, customer-focused projects, emphasizing a blend of technical skill and strategic insight. He advocates a data-informed approach to ensure technology aligns with business goals. His career features numerous technical certifications, including multiple CCIEs and cloud credentials. Previously, Brad contributed significantly to network operations and the adaptation of cloud-native applications and microservices, using technology to drive organizational transformation.

Thank you to my wife and children for your patience and love as I ventured into this technical review adventure. While it was fun for me reviewing chapters and testing labs, I know it meant more solo puzzles and one less player for game night at home. Thanks for not hiding my laptop, and for all the laughter and support that kept me going! Love you guys - Always.

Suhaib Saeed is a cloud network engineer with many years of experience in network design, automation, and observability and is currently focused on all things AWS. He has spent most of his career at various ISPs such as BT, working on large-scale projects for FTSE 100 clients. His current role is at Samsara, an IoT company on a mission to improve the safety, efficiency, and sustainability of the operations that power the global economy. Suhaib holds a BSc in Computer Networks and has authored multiple blogs on network automation.

Table of Contents

Preface   xv

Part 1: Understanding Monitoring and Observability

1

Introduction to Monitoring and Observability   3

Defining network observability   4
Network monitoring evolution   5
What has worked so far   5
Trends and requirements   7

Network observability pillars   7
Data quality   8

Scalability and interoperability   9
Actionable data   9
Assisted analysis   10
Benefits   11

Summary   11

2

Role of Monitoring and Observability in Network Infrastructure   13

Networking in the 2020s   14
Technological changes   14
Cultural changes   18

Transforming data into information   19
The importance of using business terms   19
Defining KPIs   20
From data to information   24

Expectations for
network observability   25
Heterogeneous and enriched data   25
Proactive role in network automation   26
Full visibility of network state   26
Faster, more accurate, and at scale   27

Summary   28

3

Data's Role in Network Observability   29

Network monitoring and telemetry   30
Challenges of traditional network monitoring   30
Network telemetry   31
Network observability framework   31

Collecting data, in practice   32

Agent-based versus Agentless approach   33
Network data collection methods   34
Setting up the lab environment   35

Summary   69

Part 2: Building an Effective Observability Stack

4

Observability Stack Architecture   73

**The components of an
observability platform**   73

**The importance of a well-designed
observability stack**   76
Why does an observability stack need to be
well designed?   76
What does it mean to be a
well-designed platform?   79

**Understanding data pipelines
for observability**   80
The versatility of data pipelines   80

Unpacking ETL in data pipelines   81

Challenges and best practices   82
Scalability   82
Reliability   83
Flexibility, extensibility, and customization   84
Cost management   85
Other tips and best practices   86

Setting up a lab environment   88
Lab scenarios   90

Summary   90

5

Data Collectors   91

A deep dive into data collectors   91
Key characteristics   93

A look into Telegraf   95
Telegraf architecture   95
Telegraf configuration   96

Telegraf SNMP input plugin   102
Telegraf synthetic monitoring input plugins   107
Telegraf gNMI input plugin   109
Telegraf exec input plugins   112

A look into Logstash   119

Logstash architecture   119   **Summary**   124
Logstash syslog input   120

6

Data Distribution and Processing   125

Understanding data normalization   126   Data enrichment at query time   157
Observability data models   128   **The scale of the observability**
Breaking down metrics and the data model 130   **data pipeline**   160
                                                  Why message brokers/buses matter in
Enhancing insights with   observability   160
data enrichment   143
                                                  Summary   168
Data enrichment injection   145

7

Data Storage Solutions for Network Observability   171

Databases for observability   172   Grafana Loki architecture   211
Time series databases   173   Writing to Loki   215
Matching databases with observability needs 174 Reading from Loki (LogQL)   216
                                                  Loki rules   228
A look into Prometheus TSDB   177
Prometheus architecture   178   **Persistence tips and best practices** 233
Writing to Prometheus TSDB   185   Performance and scale   234
Reading from Prometheus TSDB (PromQL)   192   Automation is your best friend   235
Prometheus rules   208   **Summary**   236

A look at Grafana Loki   211

8

Visualization – Bringing Network Observability to Life   237

Data visualization principles   238   Creating your first Grafana dashboard   249

A look into Grafana   240   **Visualization tips and best practices** 283
Architecture   242   **Summary**   284
Setting up the lab environment   243

9

Alerting – Network Monitoring and Incident Management   285

Incident management and alerts   288
Challenges and considerations on alerting   289
Alert aggregation and correlation   289
Alert engine architecture   293

A look into rulers and Alertmanager   295
Architecture   295
Creating your first alerts   298
Grafana for alerts   306

External integrations   308

Alerting tips and best practices   311
Addressing common alert challenges   311
Build on top of communication
and transparency   312
Healthy incident management process   313
The role of AI in alerting   314

Summary   314

10

Real-World Observability Architectures   317

Observability stack options   318
All-in-one open source tools   319
Commercial off-the-shelf tools   319
Controller-based systems   320
Time series versus snapshot observability   320

Comparing build versus buy
decision points   321
Defining requirements   321
Evaluating in-house capabilities
and resources   322

Cost analysis   323
Assessing risks   324
Comparing features and flexibility   324
Making a decision   325

Orchestrating an
observability platform   325
Deployment methodologies
and orchestration   326

Summary   329

Part 3: Using Your Network Observability Data

11

Applications of Your Observability Data – Driving Business Success   333

The business value of
observability data   333
Capacity planning   334

Percentiles   335
Forecasting   336
Defining health status   337

Treating your network as a service   338

Monitoring SLIs, SLOs, and SLAs for optimal
network performance   338

How to treat a network as a service   340

Architecting dashboards   341

Network-related personas   341

Dashboard types   343

Summary   357

12

Automation Powered by Observability Data – Streamlining Network Operations   359

Setting up the lab environment   360

Advanced automation techniques
with event-driven automation   385

Event-driven automation   385

Closed-loop automation   388

Event-driven automation with Prefect   389

Summary   401

13

Leveraging Artificial Intelligence for Enhanced Network Observability   403

AI and ML fundamentals   404

ML algorithms   406

Neural networks and language models   413

Real-world AIOps   418

Lab requirements   419

Validating operational changes   420

Assisted root cause analysis   432

Summary   442

Appendix A   445

A lab environment   446

Hardware requirements   447

Software requirements   447

Step 0 – Git repository setup   449

Step 1 – VM provisioning   450

Step 2 – interacting with the lab scenarios   461

Step 3 – removing the lab environment   462

Step 4 – managing lab scenarios   463

Summary   464

Index   465

Index   465

Other Books You May Enjoy   478

Other Books You May Enjoy   478

Preface

Modern Network Observability explores the current technology landscape for understanding the network operational state. Network monitoring tools have been around for many years, providing insights into how the network is performing. But, in today's world, the need for combining information (such as model-driven telemetry and with traditional SNMP monitoring) has brought about new tooling for observing networks that have not been seen before.

In the book, we will take you through some of the reasons for using the tooling, how to implement the tooling, and what you can do to integrate with other systems, such as modern AI and ML techniques.

Who this book is for

This book is for network analysts, network administrators, network architects, support professionals, engineers, and other IT professionals focused on networking. Leaders of networking organizations will find value in understanding capabilities and how these new capabilities may help to observe their networks.

What this book covers

Chapter 1, Introduction to Monitoring and Observability, introduces what monitoring and observability are for networking nowadays.

Chapter 2, Role of Monitoring and Observability in Network Infrastructure, introduces what the role of the data is and covers how data can provide service levels.

Chapter 3, Data's Role in Network Observability, covers the various types of data within the observability stack and how the data is gathered.

Chapter 4, Observability Stack Architecture, introduces a modern observability stack for network observability and how to work with data for a multi-vendor solution in a flexible way.

Chapter 5, Data Collectors, covers how to collect data for metrics and logs.

Chapter 6, Data Distribution and Processing, covers how to enrich data collection with metadata in the metrics pipeline and a method of scaling out data collection.

Chapter 7, Data Storage Solutions for Network Observability, covers the database and storage mechanisms for persisting metrics and logs.

Chapter 8, Visualization – Bringing Network Observability to Life, walks us through using visualization tools to bring dashboards and data out of the database and to your eyes.

Chapter 9, Alerting – Network Monitoring and Incident Management, covers the basics of alerting: getting actionable alerts based on the telemetry data that is collected.

Chapter 10, Real-World Observability Architectures, explains how to use the new information to set you up for success with the data.

Chapter 11, Applications of Your Observability Data – Driving Business Success, introduces how to start to use the data and how to tailor the solutions that you would build to those who need to see the data.

Chapter 12, Automation Powered by Observability Data – Streamlining Network Operations, brings to light the capabilities of integrating the tooling with other systems, with explanations of accessing data programmatically and how to build closed-loop and event-driven automation.

Chapter 13, Leveraging Artificial Intelligence for Enhanced Network Observability, completes the story by connecting the data within the tooling introduced to machine learning and artificial intelligence to supercharge your automation.

Appendix A sets up the lab environment and contains elaboration on how to use its components.

To get the most out of this book

This book assumes that you have a basic understanding of networking and of Linux shell fundamentals. There will be installation of applications on a Linux command line. Most of the command-line examples provided in the book are executed within a containerized environment, or native to most Linux shells. Some basic Python understanding is recommended as well, but not required.

Software/hardware covered in the book	Operating system requirements
Python	Windows Subsystem for Linux (WSL2), macOS, or Linux
Docker	WSL2, macOS, or Linux
Telegraf	WSL2, macOS, or Linux
Kafka	WSL2, macOS, or Linux
Prometheus	WSL2, macOS, or Linux
Grafana	WSL2, macOS, or Linux
ContainerLab	WSL2, macOS, or Linux

There may be some tools that are referenced using a cloud service for ease of gaining experience with the tool. There are no financial requirement to complete the examples provided throughout the book.

Another option is to execute the environments free within the open source community installations. These will be covered in more depth within the relevant chapters.

If you are using the digital version of this book, we advise you to type the code yourself or access the code from the book's GitHub repository (a link is available in the next section). Doing so will help you avoid any potential errors related to the copying and pasting of code.

Download the example code files

You can download the example code files for this book from GitHub at https://github.com/PacktPublishing/Modern-Network-Observability. If there's an update to the code, it will be updated in the GitHub repository.

We also have other code bundles from our rich catalog of books and videos available at https://github.com/PacktPublishing/. Check them out!

Conventions used

There are a number of text conventions used throughout this book.

Code in text: Indicates code words in text, database table names, folder names, filenames, file extensions, pathnames, dummy URLs, user input, and Twitter handles. Here is an example: "The anomaly column gets the value 1 if it's outside of the yhat boundary."

A block of code is set as follows:

```
combined_results = pd.merge(
    current_metric,
    forecast[['ds', 'yhat', 'yhat_lower', 'yhat_upper']],
    on='ds'
)
```

When we wish to draw your attention to a particular part of a code block, the relevant lines or items are set in bold:

```
# Convert the `seconds` representation into datetime object
data_dict['ds'] = pd.to_datetime(data[0], unit='s')
# Save the interface counters value retrieved in a `float`
format
data_dict['y'] = float(data[1])
```

```
        metric_list.append(data_dict)
    df_metric = pd.DataFrame(metric_list)
```

Any command-line input or output is written as follows:

```
>>> pkts = sniff(filter="icmp and host 1.1.1.1", count=2)
```

Bold: Indicates a new term, an important word, or words that you see onscreen. For instance, words in menus or dialog boxes appear in **bold**. Here is an example: "Click the **Settings** menu and then **Security**."

> Tips or important notes
> Appear like this.

Get in touch

Feedback from our readers is always welcome.

General feedback: If you have questions about any aspect of this book, email us at customercare@packtpub.com and mention the book title in the subject of your message.

Errata: Although we have taken every care to ensure the accuracy of our content, mistakes do happen. If you have found a mistake in this book, we would be grateful if you would report this to us. Please visit www.packtpub.com/support/errata and fill in the form.

Piracy: If you come across any illegal copies of our works in any form on the internet, we would be grateful if you would provide us with the location address or website name. Please contact us at copyright@packt.com with a link to the material.

If you are interested in becoming an author: If there is a topic that you have expertise in and you are interested in either writing or contributing to a book, please visit authors.packtpub.com.

Share Your Thoughts

Once you've read *Modern Network Observability*, we'd love to hear your thoughts! Scan the QR code below to go straight to the Amazon review page for this book and share your feedback.

https://packt.link/r/1835081061

Your review is important to us and the tech community and will help us make sure we're delivering excellent quality content.

Download a free PDF copy of this book

Thanks for purchasing this book!

Do you like to read on the go but are unable to carry your print books everywhere?

Is your eBook purchase not compatible with the device of your choice?

Don't worry, now with every Packt book you get a DRM-free PDF version of that book at no cost.

Read anywhere, any place, on any device. Search, copy, and paste code from your favorite technical books directly into your application.

The perks don't stop there, you can get exclusive access to discounts, newsletters, and great free content in your inbox daily

Follow these simple steps to get the benefits:

1. Scan the QR code or visit the link below

https://packt.link/free-ebook/978-1-83508-106-8

2. Submit your proof of purchase
3. That's it! We'll send your free PDF and other benefits to your email directly

Part 1: Understanding Monitoring and Observability

The first part of the book introduces you to monitoring and observability, taking us from where network observability began to where it is now. After reviewing where we have been and where monitoring and observability are going, we dive into the role of monitoring in organizations. The first part of the book wraps up with the different data types, such as logs, metrics, and traces, and how those types fit into the modern observability stack.

This part contains the following chapters:

- *Chapter 1, Introduction to Monitoring and Observability*
- *Chapter 2, Role of Monitoring and Observability in Network Infrastructure*
- *Chapter 3, Data's Role in Network Observability*

1

Introduction to Monitoring and Observability

Since the early days of computer networks, we have needed to detect failures on the different network components (e.g., hardware interface issues, cable cuts, or web service down) to determine outages that require corrective actions. This field has been known as **network monitoring**.

Interestingly, the last decade has witnessed numerous innovations in the field, especially related to new tools and practices around the DevOps culture. This culture emphasizes merging development and operations responsibilities requiring a better understanding of the operational state. Moreover, there has been a significant adoption of network automation. This advancement drives network operations, transforming monitoring from a passive component to an enabler of closed-loop processes. These changes have been the main drivers behind the evolution from network monitoring to **network observability**, and this book wants to help you understand and apply it to improve your network operations.

> **Note**
> Network observability is a broader topic, especially since the rise of running network applications directly in the host with technologies such as **extended Berkeley Packet Filter (eBPF)** and **Data Plane Development Kit (DPDK)**. This kind of observability is not covered in detail in the book, even though most of the concepts are applicable too.

In this book, you will begin understanding the basics concepts related to network observability, and then, for the majority of it, we will explain how to build a modern network observability stack, with a practical, but not limited, emphasis on the **Telegraf** (`https://github.com/influxdata/telegraf`)/**Prometheus** (`https://github.com/prometheus/prometheus`)/**Grafana** (`https://github.com/grafana/grafana`) (**TPG**) stack (details about how to spin up a development environment are in *Appendix A*). Finally, you will learn how to solve real network operations challenges using the flexible observability stack presented.

In this first chapter, we will cover the following topics:

- Defining network observability
- Describing network monitoring evolution
- Exposing the key aspects of network observability

Defining network observability

Let's go straight to the point: what is network observability about?

To answer this, it's convenient to understand first what network monitoring is because network observability supersedes it. Network monitoring is part of the wide IT operations monitoring focused on the network infrastructure.

Even though you are likely used to the network monitoring term, there is no academic definition of it, and everyone understands it slightly differently. We define network monitoring as *measuring the performance and availability of the network infrastructure.*

Related to this goal, you may be familiar with some of the technologies that have provided information about the operational state of the network:

- **Simple Network Management Protocol (SNMP)** polls and traps
- **Internet Control Message Protocol (ICMP)** requests (e.g., ping)
- Flow analysis (e.g., NetFlow)
- Packet capture (e.g., tcpdump)
- Logs (e.g., Syslog)

These technologies make up network monitoring, which provides support for diagnostics and service monitoring, with state visualization and alert generation. Network operation teams leverage network monitoring to detect when something is wrong in the network, but this is not enough anymore.

Nowadays, IT operations have raised the bar, and the focus is not only on the infrastructure status but on translating it to the business level. Therefore, observability is about the *end user's experience*, and this encompasses many layers, from infrastructure to applications.

This convergence of responsibilities materialized in the DevOps culture (i.e., bringing together Development and Operations) that coordinates all the IT efforts around the same business outcome. One basic practice is to consolidate different monitoring systems to enable data correlation. The DevOps movement has broken long-time silos in IT departments, and this new collaboration has produced a lot of innovations, which we will explore in this book.

Moreover, it has transformed the reactive approach of traditional monitoring into a proactive one that helps answer handling issues before impacting the services. Ironically, this leads to simpler (but more effective) systems, capable of getting the data to provide the insights that help solve these issues. This is what IT observability is about, helping to identify the unknown unknowns and having a holistic view.

Within this observability realm, network observability encompasses all the technological trends that support the overall IT observability in the network realm.

In networking, this trend toward adopting network observability has been translated to more flexibility in different aspects:

- Interoperable specialized solutions (e.g., open source solutions provide more flexibility)
- More efficient data retrieval methods (e.g., network streaming telemetry)
- More scalable and advanced data processing (e.g., artificial intelligence)
- Richer context and analysis via data integrations (e.g., source of truth integration)

> **Note**
> That being said, we will use both terms (i.e., monitoring and observability) interchangeably in this book, with the same meaning.

This is what this book is about. We want you to understand how to evolve from traditional network monitoring systems to the new network observability approach, tightly connected with the DevOps culture, and how it connects with the other big revolution in network operations: network automation.

Network monitoring evolution

As already mentioned, modern network observability has evolved from network monitoring, a practice that has been in place for several decades. Before delving into the new approach it introduces, it's important to review what has been effective so far and to understand the trends and requirements that have driven its transformation.

What has worked so far

Networks have been monitored to understand their status since the beginning. **ARPANET** (which stands for **Advanced Research Projects Agency Network**), the first packet-switched network started in 1966, had the **Interface Message Processor** (**IMP**) protocol, which provided a few monitoring features. Fast-forwarding some years to the rise of TCP/IP networks, in 1988, the **SNMP** was defined by the IETF (its last version is SNMPv3) to address this need.

SNMP provides a mechanism to manage networks, but it has been mostly used to *monitor* networks, and not to *manage* configuration changes (which have been mostly done via CLIs, until the rise of newer management interfaces). The main characteristics of SNMP can be summarized in a few aspects:

- The UDP transport protocol is stateless, which is useful for state and status polling
- **Management information bases** (**MIBs**) provide structured data to access specific content
- Massive adoption in all network devices, supporting standard and proprietary MIBs

However, not all that glitters is gold, and SNMP has some limitations such as the performance to retrieve large amounts of data and limited coverage for push mechanisms (i.e., SNMP traps).

> **Note**
> This book doesn't cover SNMP in detail (there are many books dedicated to the topic). We will reference it as one of the available methods to retrieve operational data within a holistic network observability strategy in *Chapter 3*.

Similarly to SNMP, event logs using **Syslog** have been widely used, not only for network monitoring but also for applications. Logs are generated when a specific event is seen by the device, and it brings together several pieces of information such as the generation time, the source, the level, and some meaningful message related to the event. This grouping of data is what we refer to as **multidomain data**. This contrasts with the simple SNMP metrics (integers or strings).

And also, pretty common in network analysis are the flow exporters mechanisms such as **NetFlow**, **sFlow**, and **IPFIX**. With some small differences between them, they represent the basic information to define what a packet flow is about, including the source and destination IP addresses and ports, and some other information. Again, like logs, this is multidomain data.

An important benefit of all these methods is their ubiquitous adoption. It's more than likely that any network device you have supports them. However, the implementation of the monitoring solutions, usually in the form of monolithic platforms, makes it harder to combine and relate this data that may be related.

Also, the initial technologies to manage and persist this data, such as **RRDtool**, came with limitations that modern options such as **Time Series Databases** (**TSDBs**) have overcome. Understanding what a modern network observability stack looks like and how to design and build one is the main goal of *Part 2* of this book.

These methods, together with others such as packet capturing or synthetic monitoring (e.g., ping), have been, and still are, solid pillars to build upon a network monitoring strategy. However, the expectations for observability have disrupted the status quo and, in many environments, traditional network monitoring is no longer enough.

Trends and requirements

Here, we summarize the main trends that have influenced and motivated the evolution of network observability:

- Networks are heterogeneous and abstract. Today, networking takes many forms: campus networks, hyper-scale data centers, cloud-based network services, or service mesh. This variety implies supporting different protocols and interfaces and being able to correlate many data types.

- Network operations, following the DevOps approach, have adopted automation to transform how networks are managed. For most network automation tasks, operational data insights are key and require common data models between developers and operation engineers to get a mutual understanding.

- Focus on application performance has become more predominant, and network monitoring needs to contribute to the common view with all the related information. Moreover, microservice architectures increase the complexity of correlating the data.

- Better visibility requires more data, so we need more efficient retrieval methods and data reusability.

- The volume of the data aggregate can be huge, making it impossible to analyze without the aid of artificial intelligence for IT operations (AIOps).

In this book, we will explain the basic concepts to architect solutions to address these challenges (in *Part 2*), and practical examples to implement them (in *Parts 2* and *3*).

Network observability pillars

With all these expectations, there are four pillars that sustain the network observability solutions (depicted in *Figure 1.1*):

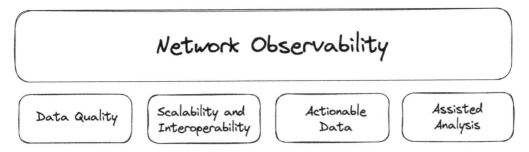

Figure 1.1 – Network observability pillars

We summarize these as follows:

- **Data quality**: Any observation is going to be as good as the data it is based on. We must ensure that the data we collect adheres to some principles that grant its quality.

- **Scalability and interoperability**: To address complex questions, the architecture needs to incorporate specialized tooling that works together in a distributed manner that can scale out as needed.

- **Actionable data**: The active role of network observability within a network automation strategy requires providing data that other components can leverage to act on it.

- **Assisted analysis**: To improve the insights generated, the data needs to be analyzed by machines that can process a large amount of data and applying *intelligence* at scale.

Data quality

It may seem obvious but in any process (even more if it's automated), the quality of the output will be directly proportional to the quality of the input. In network monitoring, operational data is king. Everything depends on the collected data, so it's necessary to carefully select and manage the data that will be used to generate the insights.

But what is data quality exactly? There are many definitions of data quality, but all of them pivot around a few dimensions:

- **Relevance**: Is the data useful (and used) for the purpose it was collected?

- **Accuracy**: Is the data precise enough to give insights into it?

- **Timeliness**: Is the data current enough to provide almost real-time conclusions?

- **Comparability**: Can the data be compared with other datasets?

- **Completeness**: Are there any records missing?

In the context of network observability, we could summarize **quality data** as *the data that is fit for answering the relevant questions about your network.*

This book doesn't cover the data quality topic in depth (there is a myriad of books only focused on this topic), but through this book, we will implicitly refer to data quality characteristics through many sections of the book, as in these instances:

- In *Chapter 3*, we will introduce the relevance of **streaming telemetry** to get almost real-time metrics from network devices. We will be tackling the *timeliness* dimension.

- In *Chapter 5*, we will explain the importance of **normalizing data** from different sources to make it *comparable*.

- In *Chapter 6*, the concept of **data enrichment** increases the data's relevance, adding more context so data can be easily correlated (not only by its timestamp).

- In *Chapter 7*, when explaining the **persistency layer**, we will tackle how some databases can fill missing data records with probable data to help run processing on top, helping with *completeness*.

Scalability and interoperability

Traditional network monitoring systems have been implemented as monolithic ones performing all the necessary functionalities (e.g., collection, storage, and visualization). This all-inclusive approach may seem convenient to get started. However, as more specialized features are required (e.g., a new database type or a new collector agent), it becomes evident that it's unlikely that one tool would be capable of addressing all your needs.

On the contrary, we need to acknowledge that evolving the network observability stack requires plugging in different components, specialized on some of the functionalities. These distributed architectures require clear interfaces to connect the different components, and a solid orchestration to deploy them in a repeatable manner at scale.

Moreover, to make this composable approach efficient, we should avoid duplicating roles. For instance, we have seen many cases where, to leverage the analysis from two different monitoring tools, each one has to run SNMP collectors and collect the very same data. Why not *collect it once, and reuse it many times*?

Related to architecture requirements, the new stack has to allow high scalability to handle the increase in the amount of data and processing required to provide the expected insights. This scalability should allow per-component upgrades (i.e., scale out) instead of the whole stack required by monolithic systems (i.e., scale up).

We will cover these topics in *Part 2*, from a general introduction of the recommended stack in *Chapter 4* to per-component details in the other chapters.

Actionable data

The adoption of network automation has revolutionized how networks are managed. This movement started with the software-defined networking hype (do you remember OpenFlow?) around 2010, and, from there, it evolved in different ways, strongly influenced by DevOps practices (some people refer to it as **NetDevOps**). When we talk about network automation, we refer to replacing the manual changes on the network infrastructure (e.g., the CLI in network devices or the GUI on controller-based ones) by using a repeatable approach that could replace (most) human intervention.

Network automation is a big topic that we don't pretend to cover in this book. However, in every chapter, we will highlight the role of network observability in the network automation space.

One key requirement for network observability within automated network operations is the need to provide actionable data to feed into closed-loop systems (i.e., systems that automatically adjust depending on their output). As we will see later, network automation's heart is about defining the *intended* state of the network that drives the whole system, from defining the configuration and the operational state. Using this as a reference, the network observability will collect the *actual* state and check whether some kind of mitigation action is needed or not.

Part 2 of the book covers how to design and implement the functionalities that support automated operations, and in *Part 3*, we explain how to leverage the recommended stack to solve real problems.

Also, you should not forget that the network automation solutions need to be observed themselves to understand how they behave and influence the network state.

Assisted analysis

Network monitoring has always supported humans with data to understand patterns or provide insights about how the network services were running. Since the early stages, network monitoring has also aspired to provide capable insights on top of the data collected. However, until recently, the analysis of monitoring data had limited (but still useful) use cases. For instance, you have likely defined threshold alerts in your monitoring systems to raise an alarm when the CPU level exceeds some threshold.

Nowadays, with the massive scale of IT systems and their complexity, the need for assisted analysis is more relevant. Answering this call, the advent of **AI/ML** (**Artificial Intelligence and Machine Learning**) technologies has transformed the game, and the promise of implementing **Artificial Intelligence for IT Operations** (**AIOps**) is becoming a reality.

AI/ML provides solutions to various problems, including the identification of data clusters with shared characteristics, such as the number of **access control list** (**ACL**) hits according to device role, and the detection of anomalies by comparing current metrics against historical data while considering seasonality. Even more popular, **large language models** (**LLMs**), used by chatbots such as OpenAI's **ChatGPT** (`https://openai.com/chatgpt`), allow reproducing human language processing to provide complex answers based on all the previous training. For instance, you can ask for the potential impact of a log message that has not been classified before in your system to get an educated insight into the related implications.

The potential of these new tools to contribute to the analysis of network operations is immense, but so are the challenges. There is a learning curve to understand when to use one or another technique, and more importantly, how to select and manage the data (including anonymizing it before sharing it with an external system) to achieve the desired results. We won't go deep into the foundation of ML/AI, but in *Chapter 13*, we will provide some inspirational examples of leveraging them to improve the insights produced by network observability.

Benefits

After this brief introduction to network observability, we want to finish it by highlighting some of the key benefits it brings to the table:

- **Reduced time to solve incidents**: More and better data is available for deeper analysis of multi-dimensional issues that affect the services running on top of the network infrastructure, providing educated suggestions to resolve the issues.

- **Better end user experience**: Due to the shorter time to identify users' issues, and also including the user's perspective in the analysis.

- **More accurate capacity planning**: By combining all the data generated, it is possible to reduce the over-provisioning of network services tailoring to the actual needs.

- **Accelerated network operations**: Being able to validate the state of the network against its intended state enables faster configuration deployments that are validated by observability. It also supports canary deployments, where a big network change is only rolled out to a small subset of the network to reduce the blast radius effect, and incrementally rolled out to the rest once the state is validated.

Understanding the key network observability pillars and the benefits they provide will help you navigate through this book. When we present our proposed architecture and stack in *Part 2*, you will notice the influence of these pillars behind every recommendation.

Summary

In this chapter, we presented the network observability topic and how it evolved from traditional network monitoring. We looked at its evolution and the main trends that have influenced its transformation. Finally, we introduced the four pillars on top of which we will build the observability stack, and some of the expected benefits of this approach.

In the next chapter, we expand on the importance of providing business-level insights and the role of observability in modern network operations.

2

Role of Monitoring and Observability in Network Infrastructure

As previously introduced in *Chapter 1*, the relevance and scope of observability in IT infrastructure goes beyond networking, and most of the ideas and topics proposed in this book are easily portable to other infrastructure realms. However, we want to provide a closer look at networking and its own needs.

For this reason, before going into the pure observability topics, we think it's important to give you a high-level overview of how networks have evolved in the last decade. With this context, it will be easier to later understand the requirements that we expect from a modern network observability solution.

Also in this chapter, we present a shift in how you understand your network. We encourage you to see it as a *product* – either it has a direct impact on your company's revenue or it's supporting it. This mindset will guide how we approach the transformation of the raw data we get from the network, in the form of metrics, logs, and other types, into actual information that can help others set expectations about how to consume the network services.

The chapter covers the following topics:

- The state of networking in the 2020s
- How to transform data into information
- Recapping the expectations for network observability

Let's start doing a quick recap of what networks look like nowadays.

Networking in the 2020s

We have to admit that networking is not the spearhead of rapid innovation in IT. There are good reasons for that. Networks require proven interoperability that has been usually sustained in standards that may take a long time to implement, and the blast radius of a network failure is worse than an electric power outage (no *battery mode* exists). Thus, networking has been resistant to changes until it has become totally necessary to support the evolution of applications running on top.

Changes in networking in the last 20 years have taken many flavors, both technological and cultural, and it has impacted the expectations in network observability. Even though this book is not about network architectures or solutions, we believe that having a 10,000-foot view of these factors will help you better understand the impact of the observability solutions around.

Technological changes

Everything depends on the perspective. If we use the 1990s as our reference, we could say that the current networks forming the internet are more homogeneous than before because the network protocols converged into a few of them, such as IP, TCP/UDP, or HTTP. However, it's also true that today's networks are no longer built only around closed boxes by a few vendors and are adopting open Linux **operating systems (OS)** and virtual network services running on different platforms, making these networks more heterogeneous.

These networks have evolved in different directions, depending on which was the main purpose. We can find network service providers that connect many networks, campus networks that provide access to end users, or data center networks that support backend applications connectivity. Moreover, different nuances are depending on the application nature they are supporting, such as the fintech use case, where a stable and low latency is crucial, to video content delivery, where multicast support allows scalability.

Without going into low-level details, we are going to analyze a few of the most relevant technological changes, starting with how the architecture of these networks has changed.

Network architectures

The authors of this book have been managing networks since the 2000s. We can still remember the days when the networks were connecting a *bunch* of clients (i.e., PCs/personal computers) to a *few* servers (in the order of hundreds at most), and finally providing internet access.

In that context, a popular architecture stood out, the **three-tier architecture**. The name comes from the three layers that compose it: the *core*, *distribution*, and *access* layers. It has been, and still is, the most common architecture (with different variances) for multi-purpose **Local Area Networks (LANs)** because it works well for the north-south traffic pattern (i.e., when most of the traffic goes from one access layer into another domain up in the architecture):

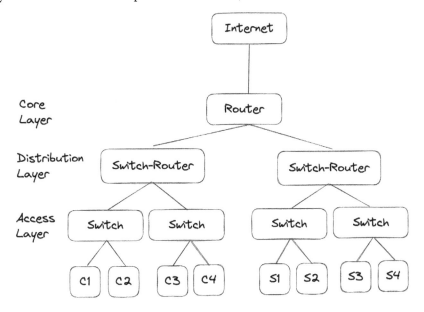

Figure 2.1 – Three-tier architecture

In large scale companies, such as Google or Meta, it became evident that this architecture had scalability limitations to support new application traffic patterns. The new network architecture had to provide a predictable network latency, a higher connectivity capacity, and an easier way to scale it via scaling out (i.e., adding new devices to the network) instead of scaling up (i.e., replacing devices with ones with more capacity). This architecture is known as **leaf and spine** (though, in some cases, it may have different names), but the actual concept comes from the old Clos network design, which provides a consistent flattened network design.

Many articles and books have been written about these topics, but a blog from Meta (formerly Facebook) in 2014 (`https://engineering.fb.com/2014/11/14/production-engineering/ introducing-data-center-fabric-the-next-generation-facebook-data-center-network/`), about the data center fabric network topology, was one of the first explaining it to the public. In the blog, you can notice how the design follows a regular pattern replicated at different levels, and there is always the same distance from one server to another.

On top of this standard architecture, different *virtual* networks can be constructed. The previous is just the underlay network, but on top, many different overlays can be built to provide custom networks. This separation of concerns (i.e., underlay and overlay) allows more flexibility in creating networks dynamically, but it also increments the dimensions to consider about the network state, as there is not only one network but a combination of many.

> **Note**
>
> A common implementation of this architecture in the data center is based on **Ethernet VPN (EVPN)** and **Virtual Extensible LAN (VXLAN)** protocols (i.e., EVPN controls how the overlay VXLAN tunnels are created). In the **Wide Area Network (WAN)** area, the **Software-Defined WAN (SD-WAN)** solutions use a *controller* to orchestrate overlay connections between branches connected over the internet (the underlay).

Both network designs are still in use because they serve different purposes, and in many environments, the network is a combination of both. One way or another, what is a reality is the higher number of connected devices (think about IoT devices, for example), thus, a higher number of network devices (i.e., switches and access points) to connect them (which need to be monitored).

Virtualization in networking

Twenty years ago, servers and network devices were directly mapped to physical boxes. Every box was running proprietary applications running on top of closed OS, consuming the hardware resources directly.

In the late 2000s, the virtualization of servers became mainstream leading to a much flexible way to provision new servers on the same physical box. Running different OSs on top of a hypervisor (abstracting the physical resource) allows OSs to think that they run alone while they are sharing the same physical host.

This transformation on the server side motivated the appearance of similar solutions for networking, and other trends, such as **Network Function Virtualization (NFV)**, which pushed toward taking network functions traditionally implemented in hardware into software.

In addition, containerized solutions came into play as an interesting option to create development environments, which also require observing the resulting state before changes are accepted.

> **Note**
>
> In this book, we use containerized network OSs to create the lab environment. More details are in the *Appendix A*.

You can already notice how all this flexibility requires dynamic network monitoring that is not statically set once and kept for the whole life of the device. It has to change dynamically as the network changes.

Network automation

If you have been networking for a while, likely, you are mostly managing your network via the famous **Command-Line Interfaces (CLIs)**, which provide a human-readable language to define how the network should behave. This worked well for small-scale networks, but as we already mentioned, with the increase in size, and heterogeneity, of modern networks, manually operating network devices doesn't scale and has other limitations.

In this context, around 2010, we witnessed two new paradigms that eventually converged into what we understand as network automation:

- Rise of the **Software-Defined Networking** (**SDN**) movement. It proposed that the network behavior should be managed via software

- DevOps culture, where development and operations are deeply connected, so the software and infrastructure evolve closer

In the present day, most of the new networks are managed in an automated way. Instead of using human language constructs via CLI commands, the network devices have adopted new interfaces for software management **Application Programming Interfaces** (**APIs**), such as NETCONF, RESTCONF, **gRPC Network Management Interface** (**gNMI**), or common-purpose REST APIs. These interfaces also came with new features to support advanced observability use cases. The data (configuration and operational) come modeled, usually using **Yet Another Next Generation** (**YANG**), which enables more effective data processing and also provides model-driven telemetry (i.e., a continuous flow of data) to get data at a higher rate (more details in *Chapter 3*).

Network observability has a key role in this context because it's no longer a passive component that only monitors the network's operational state. It is taking a step further and being the catalyst of the automation tasks to mitigate the network when issues (i.e., divergence between reality and expectations) are detected.

> **Note**
> The *Network Programmability and Automation, 2nd Edition* book by authors Matt Oswalt, Christian Adell, Scott S. Lowe, and Jason Edelman provides a high-level architecture of the role of network monitoring in a complete network automation solution.

Linux networking

Traditional network boxes have been running proprietary OS with null or very limited access to all their capacities. This is the reason most of the network monitoring has been done off the box instead of running a software agent within the box to collect operational data, such as has been done with servers.

In the early 2010s, new incumbent players transformed this rule (e.g., Cumulus Networks and Arista). They started to run network functions on top of a Linux OS while allowing access to running processes directly in the box – in some cases (e.g., Cumulus), without any strong coupling with the hardware platform.

This openness has allowed running the network OS on many different platforms, and the appearance of different network OSs disconnected from specific hardware vendor such as SONiC or VyOS.

In terms of observability, this means that we are no longer only interested in the network control state but also in the state of the OS, and the process running in the box.

Cloud networking

Around the same time, the IT industry was radically transformed by cloud services where the IT infrastructure services (e.g., compute, storage, and networking) were abstracted and provisioned on-demand.

This shift in the way of interacting with IT infrastructure allowed a more rapid go-to-market of new applications as you no longer need to provision and manage your infrastructure directly. Instead, you can leverage APIs to embrace the **Infrastructure as a Service** paradigm and start paying as you grow without worrying about capacity management.

The adoption of cloud services (private and public) is massive today. And, as you may infer, it also involves cloud networking services that need to interact seamlessly with the physical network infrastructure. Most of the network environments are hybrid. This heterogeneity increases the complexity of what to observe and correlate to understand the actual state.

On top of this, if your company is actually *running* the cloud, you must be ready to manage the underlay and overlay networks to support it. For example, if your company manages its own Kubernetes cluster, the network observability has to cover it to provide a complete network state coverage.

All these technological challenges have transformed the network with all the new requirements. However, these changes came along with another important transformation in terms of cultural changes.

Cultural changes

The biggest cultural shift in IT infrastructure has been the change from a slow and static provisioning process to a fast-changing and dynamic approach. Everything in IT is consumed *as a service* (via APIs). We have many flavors, but we can simplify in two:

- **Software as a Service** (**SaaS**), where you simply use an application without caring about how it is built or operated
- **Infrastructure as a Service** (**IaaS**), where the computing, storage, and networking resources are consumed without having to buy, transport, rack, and connect

> **Note**
> IaaS is not magic. There are teams behind these services who do the *real* work to allow others (the users of the IaaS) to consume it with this paradigm.

Regarding networking, the one that applies to networking is the IaaS, which we could name as **Networking as a Service**. Aside from the technical challenges (which we introduced earlier), the key points of the cultural transformation are as follows:

- Network users want to get their requests fulfilled in the order of seconds or minutes, not days or weeks as we used to.

- No one (i.e., users) cares about the network heterogeneity (i.e., physical and cloud network services). People just want the network to work, and this requires offering a proper abstraction.

- Adding human interventions slows down the process and only automated networks can implement this paradigm properly. Adopting automation and all its implications is no longer an option.

All these drivers led to establishing the network as a product (or a bunch of them, depending on the different purposes they have). And, when something becomes a product, it has to be managed as such. This is why we move next into how we evolve from just gathering operational state data into transforming it into business-level information that represents the state of the network as a product.

Transforming data into information

A key aspect of the modern network observability approach is to focus not only on collecting and visualizing operational data but also, on going a step further, transforming the data into actual information with a purpose.

The importance of using business terms

Networks (and most IT infrastructure components) have been seen, in many cases, as necessary IT actors without a strategic role in organizations' businesses. Despite being a crucial actor in sustaining almost every part of every organization, IT infrastructure departments haven't been able to properly communicate the value they provide

IT infrastructure teams have to change this to become more relevant in the business strategy. The business' success is built around all the teams, but some of them can explain more clearly how they contribute because they speak the business language.

So, the question is, how could we speak the business language? We recommend focusing on translating the network information into something that can be translated into business terms. For instance, the next table shows a few examples:

Business Goal	Network Operational Data
How much impact a network incident has on the revenue	How much network downtime there was and which business services were impacted
How much better users' experiences are, and how this translates to customer satisfaction	The level of packet loss and latency seen by the end users
Control the revenue per utilization of services	Use network stats to charge users per consumed bandwidth
How much faster the company processes can be delivered and contribute to increasing the company revenue	How many times a network service is automated versus run manually

Business Goal	Network Operational Data
How much spare network capacity is available to expand the business?	How many access ports are available, and how much capacity in the uplink links is left
When will we need to increase the capacity of the network according to the current usage trend?	Use data forecasting analysis to infer the trends of data consumption and determine the breaking point in the future.

Table 2.1 – Business goals mapped to network operation data examples

Mapping the business goals into network observability data is a key task that shouldn't be procrastinated. In this section, we will use **Key Performance Indicators** (**KPIs**) to help others understand how good (or bad) the network services are doing, and how these impact directly to the business.

Defining KPIs

KPIs help with translating data into information. In the networking context, KPIs transform network operational data into something that can be understood by other teams, technical or not.

There is a lot of literature about KPIs as they may apply to everything. In the service level terms (remember that we want people to see the network as a product/service), it's useful to introduce three basic service-related concepts:

- **Service-Level Indicator** (**SLI**): The metric that defines the quality of the service perceived by the users.

- **Service-Level Objective** (**SLO**): The target goal for our SLI.

- **Service-Level Agreement** (**SLA**): The SLO translated into a formal commitment. If it's breached, legal claims are supported.

The following figure shows the order of how the levels of service are defined:

Figure 2.2 – How to define the levels of service

> **Note**
> You can get more information about service level definitions in the Google SRE guide: `https://sre.google/sre-book/service-level-objectives/`.

Let's get now into each one's details to map them into network examples, starting from the beginning.

SLI

You can't commit to something (e.g., an SLA) without understanding what you can measure. In networking, an SLI measures some aspect of your network state in terms of the service it is providing. For example, a common SLI is the availability of your network service.

However, the *availability* SLI is an abstract term that requires different data depending on each use case. A network may be available in some terms, but every network has different expectations for the service it provides. For instance, knowing that the end-to-end reachability is working may not be enough. You may want to ensure that the communications are below a certain level of packet loss or average latency. Therefore, the *availability* SLI may be composed of a few metrics that combined create the final indicator:

- Average percentage of packets lost on a transfer
- Time spent by packets moving from one end to the other in the 95th percentile

> **Note**
> All these metrics must be related to the defined interesting traffic. This means that the packets may be related to the traditional ICMP protocol, or a synthetic application running on a specific network port (e.g., TCP/443) with some traffic pattern.

The list of metrics and other operational data could go on, adding more and more dimensionalities to define the SLI. This is a trade-off; more data would require more resources, meaning more complexity to process, and sometimes not adding relevant information. Thus, not overcomplicating but getting the right level of data is key to successfully defining the SLIs of your network.

Once you understand the power of SLIs, you may try to transform every metric into an SLI. This will overwhelm you. We recommend focusing on relevant information that adds value, and not over-monitor. The question, then, is *"How do you know what adds value?"* Well, put yourself in your users' shoes, and try to understand the key indicators that will help them understand how your service is expected to behave, so they can make decisions on top of them.

When you know which information to track (i.e., the SLIs), it's time to define what is the expected value of the SLI that you want to target, the SLO.

SLO

The SLO is the target value of your SLI – in other words, the value of the SLI that you want your network service to match. The SLO is not a hard commitment but a soft one that should communicate, internally and externally, what the network service is aiming for. Externally, it helps your users understand what they can expect from your service, and internally, it helps your team work toward building a network that delivers the expected level of service.

Continuing with the previous example about the network availability, once you have set the SLI measurement in place, you will start having a history of data to understand the actual value of your SLI over a while. For example, in the following graph, you can observe 12 data points of the availability SLI ranging from 94.5 to 100 over a period of time:

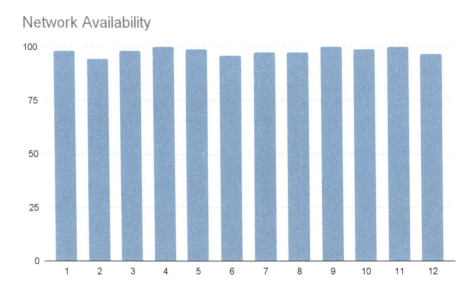

Figure 2.3 – Example of network availability graph

Having a reference is crucial to understanding the actual level of service that you may offer. You should target an SLO that is realistic (or achievable) to align expectations from the internal team and the external users.

But, what is the value you should set for the SLO? In the most conservative approach, you may look for the worst SLI value, but this could be impacted by some bad times. For example, in the previous example, the lowest number is 94.5%, and it could be a valid SLO to communicate externally. However, internally, you could work toward a higher one as the end goal is to improve your network service. In the graph, we could target 97.5% as the internal SLO that is achieved in 70% of the samples, and work toward it.

> **Tip**
> Overdelivering is always better than overcommitting.

Every SLO comes with a cost in terms of resources and attention. It's important to choose carefully the right level of service for your network, neither more nor less. This means that having an SLO of 99.99999% may come with a cost that is not necessary for your users because a 99.5% availability is enough to run their business applications (which are resilient to transient network issues).

The SLOs are useful to explain the network roadmap. Every decision you take on the network design and operation will have an impact on them (positive or negative), so relating one to the other can help you justify your proposals.

Finally, when your network service is used by others (internally or externally) and you want to commit to them (usually, external users), you have to evolve from the SLOs into the SLAs.

SLA

The SLA goes one step beyond the SLO. It's no longer a goal, it's a commitment, which, if not matched, may originate legal actions against your team. So, you should take it seriously, and require support from the legal department to write SLAs.

However, you won't always need to provide SLAs. Only when your network users require it. It's pretty common within companies to use only SLOs as a soft commitment without official SLAs.

SLAs are useful to define the right level of expectations for the customers, so they can use the network service with enough confidence. This usually takes a pyramid approach. To define the SLA for your network, you may require other services that will come with their SLAs (e.g., external network services such as WAN lines, or virtual network services in the cloud), and your final level of service is influenced by these SLAs and the way you design your network. Your work as a network engineer is to build/architect network services that match your SLAs, taking into account the others.

If you have been reading through, it shouldn't come as a surprise that the SLAs should be less ambitious than your SLOs (if you don't want to get buried under many legal issues). Why? You shouldn't commit to something that is beyond your real capabilities.

SLAs are legal contracts so they have to be concise and clear to understand by others. Also, even though SLAs are usually based on aspects that you control, they are influenced by other external elements. For example, there may be some issues that require consumer involvement to resolve them, and this must be clearly defined.

The SLAs transform your network into a product upon which others can rely. Thus, in most cases, if you want your network to be taken seriously, you have to get to them (or robust SLOs if it's an internal service). However, you should wisely define them to provide the expected service level without putting too much pressure on the team. Ideas such as an error budget can help with that.

Error budgets

You always want to provide the highest SLA to convince your network users that you are providing a great service. But adjusting the SLA or SLO too close to your actual best SLI is putting a lot of pressure on your team because you have little margin when something goes wrong.

The Google SRE guidelines (referenced in a note earlier) recommend creating an **error budget**, or in other words, adding padding between what you commit and what you can achieve. This gives you some extra downtime margin to help you identify unknown situations and improve, while you still match your users' expectations.

In addition, don't overachieve! It's common in infrastructure services such as networks to provide a consistently higher level of service than the actual SLO. However, when this situation extends for too long, your users will get used to it, and they could assume that your performance is higher than your actual SLOs, creating an over-dependency on it. It's not a bad idea to use your error budget in this case, and introduce some turbulences (without breaking your SLO) to *encourage* your users to build a more resilient service around.

A related approach is the **Chaos Monkey** idea, promoted by Netflix, which created some jaw-dropping effects in SRE conferences in the middle of the 2010s. The idea was simple and radical: instead of waiting for the relying infrastructure to fail (in their case, AWS cloud services), let's introduce random issues so the application running on top must adapt from the beginning to unexpected service level conditions.

From data to information

So far, we understand that defining our service levels in business terms helps everyone understand the value of the network as a product.

However, the operational data extracted from a network needs a curation process before turning it into information that helps others understand how the network is running.

This book is all about this. We will focus in *Chapter 3* on understanding which types of operational data are available, and in *Part 2 and Part 3* of the book, how to collect, process, store, visualize, and act on this data to properly communicate relevant information to different audiences.

The audience of a network observability solution may vary from internal network engineers to other IT departments, or even business-related teams. Everyone will have different expectations, as we will explain in the next section.

In all cases, the process of transforming data into information is putting ourselves on the receiver's side, enunciating the questions the receiver may have, and then coming up with the proper answer to that question.

> **Note**
>
> An interesting approach to consider is the **Goal Question Metric (GQM)** approach about reusing the questions and metrics for different business goals. You can read more here: `https://www.cs.umd.edu/users/mvz/handouts/gqm.pdf`.

For example, in a content delivery internet network, providing a trend of network transit link utilization correlated with the utilization of compute ports could lead to different business decisions, such as increasing the transit link capacity (if there is the availability of access ports) or to look for a new location to create a new **Point of Presence (PoP)** if we are reaching the capacity limit of compute ports of the network architecture per site.

Having introduced the context of networking in the 2020s and how we can transform the data into relevant information, the final section of this chapter will present you with the most relevant expectations for network observability in this context.

Expectations for network observability

The focus on useful information, combined with the key role of network observability within the automatization of network operations, has raised the expectations that these solutions should fulfill.

Heterogeneous and enriched data

As we have mentioned in the *Technological changes* subsection, the networks are more diverse nowadays. This means that the data obtained from the different network systems and platforms is no longer as it was in the past. We have been used to leverage mostly on **Simple Network Management Protocol (SNMP)**, and its standard **Management Information Bases (MIBs)**, which made the data gathering and processing pretty consistent.

Today, the information we are looking for comes from the combination of operational data of different types. For example, you may have a **Border Gateway Protocol (BGP)** session running between an on-premises router connected via VPN to a cloud provider. On each side, you have to use the proper collection methods (e.g., SNMP, gNMI, CLI, REST API, logs, etc.) to get the operational data, which is not usually directly comparable.

Moreover, it not only has different types of data to be compared but also to be combined to increase the understanding of the network state. For example, combining the state of the BGP peering with the logs from the interface under which the peering is established can provide more context to understand what's going wrong.

In modern network observability, the flexibility to integrate many different types of operational data, and to enrich the information by combining them, is a must.

Proactive role in network automation

Traditional network monitoring was mostly focused on providing good visualization and reporting, and notifying alerts when some conditions were met (e.g., threshold violated). That's still in the current scope, but within the scope of network automation, network observability requires a much more active role.

As we will see in *Part 2* of the book, the key role of network observability within a network automation solution is to create the feedback loop between what we expect the network to be (i.e., the intended state) and what it is (i.e., the actual state).

To achieve this, it is necessary to integrate with the network source of truth that contains the network's intended state. The source of truth is an abstract concept that encompasses all kinds of data that defines the state we want the network to be. It can be represented by simple data-structured files (e.g., YAML or JSON) or by sophisticated systems backed up by databases.

Comparing both states, intended and actual, the network observability solution should connect to other systems (e.g., a workflow orchestrator) to start the automated process of moving the network back to the desired state or breaking a continuous deployment process to release new changes if something is wrong. All these flows should ideally be automated, but they could also contain some manual judgment steps when needed.

Full visibility of network state

It's an industry-accepted rule to focus on the **golden signals** to monitor an application or a network. These golden signals are the latency, the amount of traffic, the errors, and the saturation. These signals provide key information to understand the state of your network, and they have been used in network monitoring for years.

> **Note**
>
> In this book, you will see concrete network-related examples of what you should care about when collecting metrics and generating alerts.

However, networks have some characteristics that require an extra level of visibility. Networks are complex distributed systems that run protocols between the nodes, and some of these nodes are out of your control (i.e., BGP peering with another network). Also, the network transports many different types of applications on top, and each one may have different behavior depending on the network configuration and its needs (i.e., using **Quality of Service** (**QoS**) per application, or having firewall rules blocking some traffic.

Thus, to get capable information about the network state, it's necessary to get data from many sources to provide assurance that the network is behaving as expected for all the different use cases it is expected to support (i.e., you can't control 100%, focus on what you are committed to). To achieve it, network observability would require collecting and combining different types of data, as we will explore in the next chapter.

Faster, more accurate, and at scale

So far, we have already introduced some hints about the explosion of data to be considered, and how this data needs to provide information that eventually will become knowledge. Here are some of the expectations:

- Bigger and more heterogeneous networks, with more nodes and different platforms
- Useful SLIs combining different data to provide a more complete visibility
- Active role in network operations via network automation
- More data samples due to a higher frequency of data retrieval

If we try to delegate to humans the task of processing the data to manage these expectations, you easily understand that it's not doable. For example, in a network with thousands of interfaces, how could you correlate an unexpected traffic pattern in real time and be able to conclude the root cause via understanding the dependencies?

For sure, you could try to create a logic at a small scale that works, but to apply it at a bigger scale would require using **Artificial Intelligence** (**AI**) and delegating this work to computers. This is not new; we have used AI for a long time. The simple **If This Then That** (**IFTTT**) approach is AI. It's a logic that a computer can apply. However, using simple logic for AI has some limits, as we have to provide concrete patterns to be understood by computers. The next iteration is to allow the systems to learn from historical data and define knowledge models that can help infer future patterns, not explicitly defined. This is known as **Machine Learning** (**ML**).

In network observability, we can find many use cases where AI/ML can help to provide educated suggestions (notice that learning is never 100% accurate). For instance:

- Use general-purpose LLMs to obtain log summarization and to propose a potential **Root Cause Analysis** (**RCA**)
- Leverage dynamic traffic thresholds forecasting what the threshold under some conditions would be
- Determine traffic patterns on interfaces to detect anomalies (e.g., higher or lower bandwidth for the upstream/downstream interface types)

These are just a few examples; we will cover more AI/ML use cases in *Chapter 13*.

With this we end this section, in which we have introduced a few of the most relevant new expectations that modern network observability solutions have to face.

Summary

In this chapter, we have briefly introduced the state of networking nowadays to understand how this impacts network observability and then presented the challenge to transform the network operational data into information that the business can understand to realize the role of the network in the company operations. Finally, taking all into account, we have enunciated the top expectations for network observability, which will be addressed with specific examples in this book.

Next, to conclude *Part 1* of the book, we will provide more context about what modern network observability means and the different types of operational data that may be in its scope.

3

Data's Role in
Network Observability

As we introduced in *Chapter 1*, network observability aims to provide answers to many different questions by getting visibility of the operational state of the network, from different perspectives. This visibility comes from many distinct types of operational data, each one requiring different collection methods.

In this chapter, you will learn about the following:

- The basic concepts around traditional network monitoring and the new telemetry approach, together with a classification by which network planes (i.e. management, control, and forwarding) are targeted by each monitoring protocol.

- Recap of available data types and how to collect them with programmatic hands-on examples leveraging Python libraries. Having a basic understanding of Python will help you navigate the examples better, but the explanations should make it understandable either way.

> **Note**
>
> We use the informational RFC 9232, *Network Telemetry Framework* (`https://datatracker.ietf.org/doc/rfc9232/`) as a reference in this chapter because it offers a solid framework and taxonomy of this topic. Please, check it out.

This chapter should serve as the basis for getting into *Part 2* of the book, where you will see how the collected data is being processed, stored, and consumed to realize the goal of network observability. However, due to scope limits, we won't cover all the options described here in the rest of the book.

> **Note**
>
> In this chapter, we will only focus on the protocols, not on the complete architecture of the observability solutions, which we will tackle in *Chapter 4*.

The chapter covers the following topics:

- Network monitoring and telemetry
- Collecting data, in practice

Let's start by understanding the evolution of traditional network monitoring to network telemetry.

Network monitoring and telemetry

Network monitoring, as you already know, is not something new. Network administrators have been gathering network operational data for years to understand how the networks behave. This approach, which we could define as classic network **Operations, Administration, and Management (OAM)** methods, has encountered some challenges that stand in the way of meeting current expectations. To address some of the limitations of the previous approach, a new technical approach, network telemetry, has emerged with different techniques.

It's not one or the other. To support modern network observability, we will leverage all the available options. So, let's start with a recap of the challenges of traditional monitoring.

Challenges of traditional network monitoring

When someone asks you about network monitoring, the first two protocols that will come to your mind are probably **SNMP** (for metrics) and **Syslog** (for logs). Both have been the foundation of every network monitoring system. Then, other OAM methods, such as Ping, Traceroute, and BFD, have complemented them to get complementary insights.

Even though these protocols have been serving monitoring purposes relatively well, there are a few problems that have not been fully solved by them:

- Poll-based low-frequency data collection doesn't provide enough data or precision for some monitoring applications
- Heterogeneous operational data (e.g., user flows, device configuration, and line card metrics) are only covered partially, and traditional network devices are not programmable enough to adapt to
- Some solutions (e.g., Syslog or **command-line interfaces (CLIs)**) lack a formal data model, and machines can't work without structured data (well, with ML techniques, they can at least try)
- Data push solutions have limited capabilities with some predefined management plane events (e.g., SNMP Trap) or sampled user packets (e.g., sFlow)

The network industry has been aware of these limitations for years and has worked to solve them with new protocols that we include in the network telemetry umbrella, which we will describe in the next section. However, do not disregard these solutions; they may still be the only way to get data from some network platforms, so there is still a role for them in your observability solutions.

Network telemetry

Under the wide network telemetry umbrella, we can include several technical technologies that were created to improve data collection and consumption. Just to name a few (we don't pretend to be exhaustive at all), there are:

- Netflow/IPFIX (`https://www.rfc-editor.org/rfc/rfc7011.html`), gRPC-based (e.g., JSON-RPC, gNMI)
- Packet Sampling (PSAMP) (`https://datatracker.ietf.org/doc/html/rfc5476`)
- BGP Monitoring Protocol (BMP) (`https://datatracker.ietf.org/doc/html/rfc7854`),
- YANG-Push (`https://datatracker.ietf.org/doc/rfc8641/`)

Some of them are well established and popular, and others are still in the early adoption stages. Each one has its own focus and implementation differences, but as a group, they share some characteristics when compared to the traditional monitoring technologies:

- **Push model and streaming**: Instead of polling data from network devices, telemetry allows subscription mode to stream data from the network devices to the collectors
- **Volume and velocity**: The target consumers of observability data are machines that *love* consuming huge volumes of data, and also allow real-time processing optimization
- **Normalization and unification**: There is *some* push towards unified data representation to simplify how to process data from different platforms/vendors
- **Model-based**: Data is structured and modeled, allowing consumers to know how to consume it
- **Data fusion**: The data from a single application or use case can come from multiple data sources, so sharing a common identifier improves data correlation
- **Dynamic and interactive**: Data collection should adapt continuously to the needs of network automation when used in closed-loop systems, so automated management is needed

This is a basic classification that should help you visualize the delineation between traditional monitoring protocols and telemetry-based ones. Another useful perspective is to analyze it by the "source" type of the data collected, as we will do next.

Network observability framework

At a high level, network devices implement three functional planes:

- Management
- Control
- Forwarding

This is the traditional approach to breaking down the responsibilities within network devices, and each plane has different protocols focused on collecting its related data. Aside from these planes, we can include the external data that represents the surroundings of the network device:

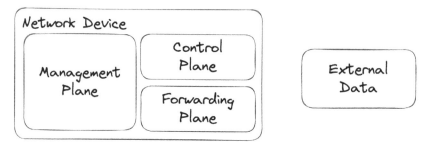

Figure 3.1 – Network planes

Let's get a quick definition of each plane and a few examples:

- **Management plane**: This is for interfacing humans or **Network Management Systems (NMS)** with the network device to change the state of the device, or to read data about the configuration, logging, or network statistics. Protocols in this group are SNMP, Syslog, CLI, gNMI, NETCONF, and RESTCONF.

- **Control plane**: This is where the distributed protocols that control the state of the network, such as layer 2 or layer 3 forwarding tables, run. A few examples of control plane protocols are OSPF, IS-IS, and BGP. These planes can be observed via techniques such as Ping or Traceroute, or telemetry protocols such as BMP.

- **Forwarding plane**: This plane is where the packets are moved (e.g., network interfaces), and it is the most demanding in terms of data volume and velocity. Naturally, when observing it, it's also crucial to not impact the primary goal of the plane, which is to forward packets. In this group, we have tools such as TcpDump, IPFIX, sFlow, Netflow, Cisco SLA, PSAMP, and eBPF.

- **External data**: This category includes everything that is not network device-specific. For instance, the circuit provider information and the contact for a given interface, coming from an external asset manager system or physical **Internet of Things (IoT)** sensors could fit into this broad field.

Because there is no better way to learn about something than to use it, in the next section, we will explore a few popular data collection techniques that we will use in the rest of the book.

Collecting data, in practice

There are many ways to collect operational data from network devices (corresponding to the planes mentioned earlier), each with different characteristics. In this section, we will give you a taste of the most popular ones to give you an overall understanding of them.

This book does not aim to cover every protocol or technique in detail. This section aims to provide one example for each type, with more detailed coverage in the lab scenarios later in the book.

But, before starting with the examples, it's worth adding yet another classification method, depending on where the collector runs.

Agent-based versus Agentless approach

The two options are collecting the data and sending it to the persistence layer from within the observed system/network device or from an external agent:

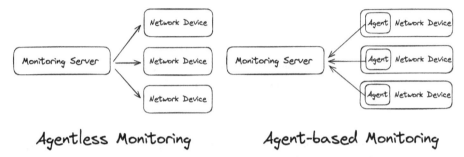

Figure 3.2 – Agent-based versus Agentless data collection

Let's look at these in detail:

- **Agent-based** monitoring collects the data from the observed system and provides several benefits because you have access to deeper and richer data than you do from outside (where only the exposed data is available), more effective bandwidth consumption (only relevant collected data is forwarded), and a better security and reliability posture. eBPF (explained later in this chapter) is an example of a mechanism that collects data running as an agent in the target device.

- **Agentless** is the most common approach in traditional network devices because of the limitations of running processes inside the boxes. The collector agent runs outside of the target system, and it retrieves the data remotely via different means. Most of the traditional options fall into this bucket. This is less intrusive than the agent-based and has lower maintenance costs, has less resource consumption in the production nodes, and is easier to maintain, so it is suitable when running collector processes within the observed system is not a viable option. There are many examples in this category, from traditional SNMP to the new streaming telemetry.

Let's get a bit more detail on both approaches, starting with agent-based monitoring.

Agent-based approach

Running the collector process from within the observed system allows faster detection, deeper control, and better monitoring capabilities because it has full access to any data that can be observed.

Traditional **network operating systems (NOS)** have not allowed access to the network administrator to run custom code in the boxes. This is something that new NOSs are addressing (at different levels), but there are still many network services, such as cloud networking ones, where you can only observe what the service provider is exposing.

As an example of a network agent-based approach, this blog by Arista Networks (`https://arista.my.site.com/AristaCommunity/s/article/streaming-eos-telemetry-states-to-prometheus`) describes how to run a Prometheus agent in the box. Then, this process sends the metrics to the external Prometheus storage. Another agent-based example is collecting stats from **Vector Packet Processing (VPP)**, as described in this other blog (`https://ipng.ch/s/articles/2023/04/09/vpp-stats.html`).

Agentless approach

As mentioned previously, the agentless approach was the only way to retrieve data programmatically from network devices for a long time. Even though modern NOSs can run processes directly, most networking monitoring solutions still promote this approach because it provides backward support and due to simplicity reasons: the common supported solution is selected.

The agentless approach moves most of the complexity outside of the network infrastructure, but it's limited by the amount of data exposed by the observed system. This said, in most cases, the data exposed should be sufficient to cover your observability needs, and sometimes, it's the only way to get what you need.

Because of its ubiquity, we will use this approach in the examples in *Part 2* and *Part 3* of the book. The agent-based approach is also supported by the proposed architecture; the only difference is how the data is collected.

Next, we will give you an overview of the different methods of collecting operational data from your network.

Network data collection methods

There is no single way to collect operational data. There are many different options (maybe more than you have in mind right now), and you will need to evaluate which ones to use in each case based on what your network infrastructure supports and the data that you need to understand the state of your network.

In this book (*Part 2*), to illustrate how to build a modern network observability stack, we will leverage only a few of them, but to give you an overview of the different options available, we present, in this section, a glance at a set of methods of collecting operational data. It doesn't mean that you have to implement all of them every time. Just pick the ones that provide the data that serves your observability purpose.

We don't want to confuse you with product-specific implementations. Because of this, we decided to provide short and simple examples of how to interact programmatically with these protocols (you don't need to master them to go through the rest of the book).

For easier comprehension, we picked a single programming language for all the examples, Python, because it comes with a great variety of libraries that will support different use cases. However, the same approach could be implemented with other programming languages (e.g., Go or Rust).

Note

This book is not a Python development book. For best practices (i.e., using `virtualenvs` to isolate dependencies), we recommend using other reference material, such as *Mastering Python Networking* by Eric Chou (`https://www.packtpub.com/product/mastering-python-networking/9781784397005`).

Setting up the lab environment

All the examples in this chapter can be reproduced easily using the lab topology used in the book. In this lab, the `ch3-completed` scenario, we will focus on gathering data from two Arista devices, `ceos-01` and `ceos-02`, by using a few Python scripts that leverage different technologies.

Before starting the lab, you need to prepare a machine following the steps outlined in the *Appendix A*. There is a manual guided process or an automated one to deploy a server in DigitalOcean cloud infrastructure. From there, you will be able to set up all the containerized environments, with the two network devices and a container with the Python code and necessary libraries already installed.

As you will see in *Part 2*, all the labs in the book will follow a similar pattern. You have to install the `netobs` command tool as explained in the *Appendix A*. With this tool, you can automate the server's setup, and within the server, deploy the scenario with `netobs lab prepare --scenario=ch3-completed`.

Then, within the server, you can list all the running containers with `docker ps` and get access to the container's shell with `docker exec -it <container-id> bash`.

Note

All the following examples are available at `https://github.com/network-observability/network-observability-lab/tree/main/chapters/ch3-completed`, and you can run them from within the server containing the lab according to the setup in the *Appendix A*.

So, let's get started with the pioneer in network monitoring. We bet that you have heard about SNMP before.

SNMP

Since the late 1980s, SNMP has been the most widely used protocol for retrieving data from network devices. From its initial version, SNMPv1 (`https://datatracker.ietf.org/doc/html/rfc1157`) in 1990, to SNMPv2 and SNMPv3 in 1998, many different enhancements have been added to overcome limitations found in its operation.

> **Note**
>
> A curious limitation of SNMPv1 was the usage of only 32-bit counters, which made it impossible to handle Gigabit interface traffic until SNMPv2 upgraded to 64-bit.

We can't provide a detailed SNMP explanation in a few pages. Indeed, a lot of literature exists about the topic. Check out this list of references (`https://snmp.com/protocol/snmpbooks.shtml`). Nevertheless, we want to give you the basic concepts before demonstrating how to use it programmatically:

- **Architecture**: SNMP defines two roles to implement a client-server model:

  - **Managers** are in charge of collecting the data from the agents. This could cause performance issues in large networks due to the continuous polling of agents.

  - **Agents** are any device or component that exposes data via SNMP.

  - SNMP uses the UDP protocol in ports `161` and `162`.

- **Management Information Base** (**MIB**): SNMP defines a hierarchical tree-like data structure containing scalar and tabular values. Every level of the hierarchy is defined as an **Object Identifier** (**OID,** `https://datatracker.ietf.org/doc/html/rfc1157#section-3.2.6.3`), in a dotted notation. Standard MIBs cover the general characteristics, but almost every vendor provides private MIB tables to expose their particular features. This makes it hard to homogenize the different data because you must know which OIDs to request data from. For example, these are two examples of OIDs, one standard and the other from a vendor:

  - **Standard**:

    - `/iso/org/dod/internet/mgmt/mib/system/sysDescr`

    - `1.3.6.1.2.1.1.1`

  - **Vendor** (from Juniper)

    - `/iso/org/dod/internet/private/enterprises/juniperMIB/jnxNsm`

    - `1.3.6.1.4.1.2636.6`

It's interesting to notice that the vendor/private MIBs are usually organized under the internet OID, and then, there is a specific identifier for each enterprise. In our example, Juniper uses the `2636` identifier.

- **Protocol commands**: SNMP provides different methods for getting and setting data from and in the MIB. It was designed as a network management protocol, so both operations are possible. However, it has mostly been used in read-only mode. These are the messages that SNMP supports:

  - `Get`: A manager requests the value of a specific OID from an agent, which responds with a response message.

  - `GetNext`: A manager can request the next sequential object in the MIB, so it can traverse the MIB without knowing its structure.

  - `GetBulk`: Similar to `GetNext`, but instead of getting one response per query, like `GetNext`, the agent replies with the minimum responses needed to group all the requested data, which is better.

  - `Set`: The manager sends the value to be updated in the MIB.

  - `Response`: This is the message that agents send back to the manager to transfer the data.

  - `Trap/Inform`: A trap is an asynchronous message sent by an agent to a manager to notify about an event without an explicit request by the manager. `Inform` is the response from the manager to acknowledge the reception of the trap.

- **Security**: SNMPv1 introduced the (still) popular concept of communities to identify the permissions of a manager to perform SNMP operations. However, this mechanism had many security issues, including authentication in plain text or non-encrypted communication, which required revisions in subsequent protocol versions:

  - Communities, used in SNMPv1 and SNMPv2, are strings used to identify the relationship with an MIB view, either read-only or read-write

  - SNMPv3 introduced a user-based security system that allowed authentication and encryption of the data

As its widespread adoption demonstrates, SNMP has been an effective method for network monitoring, but its focus on device-level information doesn't allow it to provide visibility into network traffic patterns.

To illustrate how to interact with SNMP via Python, we use the `pysnmplib` library (`https://github.com/pysnmp/pysnmp`), which is a fork from Ilya Etingof's project `etingof/pysnmp` (Ilya sadly passed away in 2022). The library is already installed within the lab preparation for this chapter, but if you need to install it, you can do so with `pip install pysnmplib`.

> **Note**
>
> Since July 2024, the original library, `pysnmp`, has been resuming releases adopting `asyncio`, the asynchronous code execution in Python. Thus, the examples in the book are not 100% compatible with the other library.

As mentioned earlier, there are different methods supported by SNMP, but we are going to focus on the ones used to retrieve data (remember, SNMP can be used to write data too, with the `Set` command). And, to follow a systematic approach, we start creating a script that uses the `Get` method to obtain data from two OIDs, the system description and the system uptime:

```python
import sys
from pysnmp.hlapi import * (1)
iterator = getCmd(
    SnmpEngine(),
    CommunityData("public"), (2)
    UdpTransportTarget(("ceos-01", 161)), (2)
    ContextData(),
    ObjectType(ObjectIdentity("SNMPv2-MIB", "sysDescr", 0)), (3)
    ObjectType(ObjectIdentity("SNMPv2-MIB", "sysUpTime", 0)), (3)
)
errorIndication, errorStatus, errorIndex, varBinds = next(iterator)
(4)
if errorIndication or errorStatus:  # SNMP engine or agent errors
    print("Something went wrong")
    sys.exit(1)
else:
    for varBind in varBinds:  # SNMP response contents
        print(" = ".join([x.prettyPrint() for x in varBind]))
```

Let's run through what this code does:

1. It imports all the Python objects (e.g., `SnmpEngine`, `CommunityData`, and so on) for simplicity, but you can import only the necessary ones.

2. The target network device is the Arista `ceos-01` device, one of the two switches spun in the lab scenario, using the *public* community.

3. We use the SNMPv2-MIB notation to refer to the OID, but we could have used the raw `1.3.6.1.2.1.1.1.0`.

4. The SNMP iterator created collects the SNMP data and stores it in `varBinds`, which we print at the end.

When you run the script, you will get the content of the MIB for these OIDs (using the notation we defined in the `print` statement):

```
root@ab7700def797:/usr/src/app# python snmp_get.py
SNMPv2-MIB::sysDescr.0 = Arista Networks EOS version
4.29.2F-30640700.4292F (engineering build) running on an Arista
cEOSLab
SNMPv2-MIB::sysUpTime.0 = 49504655
```

The SNMP GET requests work well for simple SNMP queries when only a single piece of information is required, but when you need to retrieve many data points, this operation is not efficient because of the many request and answer communications (and the related latency).

Lucky for us, the SNMP GetNext and GetBulk operations retrieve a list of data points in the answer without specifying them (using the hierarchy of the OIDs in the MIB). To illustrate this, we will use a popular MIB object, the interfaces table, or ifTable (http://www.net-snmp.org/docs/mibs/interfaces.html). This object contains the index (the identifier of the interface in the MIB), the description, the **Maximum Transmission Unit (MTU)**, the physical address, and much more (check out the ifTable reference link for more information).

In the pysnmp library, getCmd() and bulkCmd() are used for GetNext and GetBulk operations, respectively. While GetNext requires a request and response for each MIB variable, GetBulk queries multiple variables in one packet. Due to its efficiency, we use GetBulk in the next example.

> **Note**
>
> The GetBulk operation uses two parameters: non-repeaters and max-repetitions, which are explained in detail in RFC 1905 (https://datatracker.ietf.org/doc/html/rfc1905.html#section-4.2.3), to define how much data we get per interaction.

In the next example, which extends the previous code, we replace the iterator definition with bulkCmd:

```
iterator = bulkCmd(
    SnmpEngine(),
    CommunityData("public"),
    UdpTransportTarget(("ceos-01", 161)),
    ContextData(),
    0,
    50,
    ObjectType(ObjectIdentity("1.3.6.1.2.1.2.2.1.2")), (1)
    lexicographicMode=False, (2)
)
```

Let's run through the code snippet:

1. The base OID (using raw notation) defines where the starting point of the exploration is All the hierarchically dependent OIDs will be explored.

2. The `lexicographicMode` option (by default, `True`) limits the query exploration to only descendants. If you remove it (or set it to `True`), you will get the full MIB.

If you run it, you get the interface descriptions:

```
root@ab7700def797:/usr/src/app# python snmp_getbulk.py
SNMPv2-SMI::mib-2.2.2.1.2.1 = Ethernet1
SNMPv2-SMI::mib-2.2.2.1.2.2 = Ethernet2
SNMPv2-SMI::mib-2.2.2.1.2.999000 = Management0
SNMPv2-SMI::mib-2.2.2.1.2.5000000 = Loopback0
SNMPv2-SMI::mib-2.2.2.1.2.5000001 = Loopback1
```

In this output, the last numbers (i.e., 1, 2, 999000, 5000000, and 5000001) represent the index of the interface. So, the `Management0` interface is referenced in the MIB with the index 999000. If you explore other OIDs, you will see that these indexes are referenced, so when you need to get data for a specific interface, knowing its index is necessary.

Logs

A **log**, or **event log**, is a very common mechanism in IT systems for providing information about any type of event that may be of interest. For example, in network devices, we can find logon sessions, interface operational state changes, or routing protocol updates. To generate these events, every system must implement its internal logic to trigger a log and transform it into exported data.

The de facto standard for logs is the **Syslog** message logging (released in the 1980s and with the latest update in RFC 5424, `https://datatracker.ietf.org/doc/html/rfc5424`), which defines some parts of the log message that are usually common, but it leaves the *message* content very open:

*   **Facility**: The type of system that is logging the message, such as security, kern, user, auth, or local ones to allow customization.

*   **Severity level**: To express how relevant a log is. The levels range from Emergency to Debug.

*   **Message**: This contains the message of the log.

Log collection is not implemented as a client-server mechanism. The logs are sent by the log *originator* when they are generated, and it is up to the collector to get them. There are many transport options for syslog messages, but UDP is the most widely adopted one (with all its pros and cons).

To give a hands-on overview of how to collect logs, we will build a basic syslog receiver in Python to collect syslog messages from our network lab.

The following code creates a simple Syslog server using the Python standard library `socketserver`, which facilitates building TCP and UDP server applications. The example code listens at socket

`198.51.100.1:1515` (the server IP in the management network of the virtual lab mapped to the container where the examples are run). Both network devices (`ceos-01` and `ceos-02`) are already configured to send their syslog message to this socket via TCP, thus we use TCP in this case, instead of UDP:

```
import socketserver
class SyslogHandler(socketserver.BaseRequestHandler): (1)
    def handle(self):
        self.data = self.request.recv(1024).strip()
        print("{} sent:".format(self.client_address[0]))
        print(self.data)
with socketserver.TCPServer(("198.51.100.1", 1515), SyslogHandler) as
server: (2)
    print("... listening for Syslog messages ...")
    # Activate the server; interrupt the program with Ctrl-C
    server.serve_forever() (3)
```

Let's run through what we did:

1. The `SyslogHandler` manages the socket requests with the `handle` method that simply prints the data received on the screen.

2. The context manager `socketserver.TCPServer` creates a TCP server for the server IP and attaches the custom handler.

3. When you run the script, it's going to keep listening forever (until you break it).

At this point, you can start our TCP Syslog server in one terminal, and let's do some actions in the network device that generate Syslog messages:

```
root@ab7700def797:/usr/src/app# python syslog_receiver.py
... listening for Syslog messages ...
```

There are several types of actions that generate events that translate into Syslog messages in the network devices, and they are usually configurable. In our lab, an easy one is to get into the router CLI and access the configuration mode.

On another screen, connect via SSH to one of the routers using the device credentials defined in the lab (by default, `netobs/netobs123`):

```
root@ab7700def797:/usr/src/app# ssh netobs@ceos-01
(netobs@ceos-01) Password:
Last login: Sat Oct 21 14:03:35 2023 from 198.51.100.1
ceos-01>enable
ceos-01#configure
ceos-01(config)#
```

Once you activate configuration mode, on the screen where you had the Syslog server running, you can visualize the event sent:

```
root@ab7700def797:/usr/src/app# python syslog_receiver.py
... listening for Syslog messages ...
198.51.100.1 sent:
b'<61>Oct 21 14:06:11 ceos-01 ConfigAgent: %SYS-5-CONFIG_E: Enter
configuration mode from console by netobs on vty20 (198.51.100.1)'
```

You have just collated our first Syslog message! However, you should notice an important limitation: it's a raw string, not structured data.

So, you need some way to extract the data we want from the text. A well-known approach is using **Regular Expressions**, or **regex**, a sequence of characters that specify a match pattern in text. There are other options available, such as template engines, which we will cover in the *CLI parsing* subsection.

In Python, the re library, which is also in the standard library, provides all the features required to use regex, so we are going to refine our Syslog handler to get the exact information from the unstructured message.

However, creating the right regex pattern is not an easy task. You could argue that this is more an art than a science, and it requires time to master. Another handy option is to use machine learning generators to provide you with some educated suggestions to get started.

> **Note**
>
> We recommend the book *Mastering Regular Expressions, 3rd Edition* (https://learning.oreilly.com/library/view/mastering-regular-expressions/0596528124/) by Jeffery Friedl to learn more about regex, and the online https://regexr.com/ to play with it.

In this case, a simple regex to capture the IP address within parentheses is \ (([^)] *) \ ) $. Its simplicity means that it could result in mismatches in some cases, but for our message type, it simply looks for the last word within the parentheses (without checking if it's a valid IP address).

Let's modify the previous Python code to use the re library and extract the source IP address:

```
import socketserver
import re
class SyslogHandler(socketserver.BaseRequestHandler):
    def handle(self):
        self.data = self.request.recv(1024).strip()
        print(f"{self.client_address[0]} sent:")
        print(self.data)
        pattern = r'\((([^)]*)\)$'  (1)
        source_ip_addresses = re.findall(
```

```
                pattern, self.data.decode("utf-8")
            ) (2)
            print(f"The source IP was {source_ip_addresses[0]}")
```

Let's run through the extended code:

1. The regex pattern is defined in a variable.

2. The code extracts the objects that match the regex pattern.

If you start the regex-powered Syslog receiver and log out from the configuration mode in the router (using exit in the previous CLI session), you can observe how, now, the IP address of the connected client is extracted and exposed according to the print statement:

```
root@ab7700def797:/usr/src/app# python syslog_receiver_regex.py
... listening for Syslog messages ...
198.51.100.1 sent:
b'<61>Oct 21 14:35:59 ceos-01 ConfigAgent: %SYS-5-CONFIG_I: Configured
from console by netobs on vty20 (198.51.100.1)'
The source IP was 198.51.100.1
```

This example shows how using regex allows you to capture the information you want, overcoming the limitation of receiving raw text.

Next, still within the realm of text manipulation, we will explore how to get the data from **command-line interfaces (CLIs)**.

CLI parsing

CLIs have been the main interfaces for network engineers for decades, both for configuring and obtaining operational data from devices.

The main users of CLIs are humans, and the information available via this interface is not ideal to use via programmatic means. However, with the advent of network automation, several methods have appeared to overcome these traditional challenges.

One project with a lot of traction is the Python Netmiko library (https://github.com/ktbyers/netmiko), which supports almost any platform (https://ktbyers.github.io/netmiko/PLATFORMS.html) you could imagine. The library uses SSH to connect to the devices and to interact with them. It comes already installed in the lab container, but you can install it with pip install netmiko.

Now, you can create a script that connects to our network devices and collects the status of the interfaces using the show interfaces status command (the CLI commands are specific to each platform syntax):

```
from netmiko import ConnectHandler
device = ConnectHandler( (1)
```

```
            host='ceos-01',
            username='netobs',
            password='netobs123',
            device_type='arista_eos'
    )
    show_run_output = device.send_command('show interfaces status')  (2)
    print(show_run_output)
```

Let's run through the code

1. The script uses the Netmiko `ConnectionHandler` to create a custom Arista EOS connection.

2. It executes the command to retrieve the information that is outputted on the screen.

3. If you run the script, you will get the same output if you were running the command in the CLI session over SSH:

```
root@ab7700def797:/usr/src/app# python netmiko_intf.py
Port       Name    Status        Vlan      Duplex
Speed  Type              Flags Encapsulation
Et1            connected     routed   full    1G      EbraTestPhyPort
Et2            connected     routed   full    1G      EbraTestPhyPort
Ma0             connected     routed   a-full
a-1G   10/100/1000
```

With Netmiko, you have full access to all the information available on your device (that is exposed via CLI)! However, as we have seen in the logs, the data is not structured, so using it for programmatic purposes can become nightmare.

One way to overcome these limitations is by using the `ntc-templates` library (`https://github.com/networktocode/ntc-templates`), built on top of the TextFSM template-based parsing library (`https://github.com/google/textfsm`). Luckily for us, `ntc-templates` are installed by default in Netmiko and are ready to use. By setting the `use_textfsm` argument to True, you get a structured output:

```
show_run_output = device.send_command('show interfaces status', use_
textfsm=True)device.send_command(
    "show interfaces status", use_textfsm=True
)
```

So, the output now is not a `string`, but a `list`, and the structured output provides all the necessary information per interface (a total of three interfaces, as we saw before): `port`, `name`, `status`, `vlan_id`, `duplex`, `speed`, and `type`:

```
root@ab7700def797:/usr/src/app# python netmiko_intf_parsed.py
[{'port': 'Et1', 'name': '', 'status': 'connected', 'vlan_id':
'routed', 'duplex': 'full', 'speed': '1G', 'type': 'EbraTestPhyPort'},
{'port': 'Et2', 'name': '', 'status': 'connected', 'vlan_id':
```

```
'routed', 'duplex': 'full', 'speed': '1G', 'type': 'EbraTestPhyPort'},
{'port': 'Ma0', 'name': '', 'status': 'connected', 'vlan_id':
'routed', 'duplex': 'a-full', 'speed': 'a-1G', 'type': '10/100/1000'}]
```

> **Note**
>
> Notice that the same device provides different data and naming (e.g., Ethernet1 versus Et1) via SNMP and CLI.

Feel free to test other commands to see different outputs. For instance, if you try the command show ip bgp summary (instead of show interfaces status), you get all the BGP neighbors properly parsed:

```
[{'router_id': '10.17.17.1', 'local_as': '65111', 'vrf': 'default',
'description': '', 'bgp_neigh': '10.1.2.2', 'neigh_as': '65222',
'msg_rcvd': '246', 'msg_sent': '246', 'in_queue': '0', 'out_queue':
'0', 'up_down': '04:01:50', 'state': 'Estab', 'state_pfxrcd': '1',
'state_pfxacc': '1'}, {'router_id': '10.17.17.1', 'local_as': '65111',
'vrf': 'default', 'description': '', 'bgp_neigh': '10.1.7.2', 'neigh_
as': '65222', 'msg_rcvd': '0', 'msg_sent': '0', 'in_queue': '0', 'out_
queue': '0', 'up_down': '04:01:59', 'state': 'Active', 'state_pfxrcd':
'', 'state_pfxacc': ''}]
```

However, there are some drawbacks to this approach. The output can change from platform to platform, or even from version to version, which means that a constant update (and versioning) is needed to keep parsing effective. Also, there is not a full template coverage of all the CLI commands available (Netmiko will simply output raw text when not available).

> **Note**
>
> There are other related challenges, such as the high number of requests to external authentication systems (e.g., TACACS or RADIUS) because every connection requires an authenticated connection (e.g., SSH).

Notwithstanding all the limitations of this approach, sometimes this is the only way to retrieve certain data, so do not discard it. Keep this card up your sleeve, just in case.

To solve the limitation of the structured output, the industry came out with an innovative approach where everything is built around models: **model-driven telemetry**.

Model-driven telemetry

In the 2000s, as an improvement to SNMP for configuration management, a new network management interface, **NETCONF**, was defined by the IETF, RFC 6241 (https://datatracker.ietf.org/doc/html/rfc6241). And, to complement it, a new data-modeling technology, **YANG**, was created in RFC 6020 (https://datatracker.ietf.org/doc/html/rfc6020). This data-modeling language evolved independently from NETCONF, and it's used by other management interfaces, such as **RESTCONF** and **gNMI**.

With YANG, you can describe the data structures and mechanisms provided by a system to be consumed by other systems. The data can represent configuration and operational state, so it's yet another valuable source of information for our network observability systems. Different organizations (e.g., IETF and OpenConfig Consortium) have created YANG models relating to networking protocols and features that can be implemented by network devices. Each model has a unique focus. For example, the OpenConfig models focus on operational state, in contrast to IETF models, which are more generic.

> **Note**
>
> A good reference for learning more about YANG is the *Network Programmability with YANG* book, by Benoi Claise, Joe Clarke, and Jan Lindblad (`https://www.pearson.com/en-us/subject-catalog/p/network-programmability-with-yang-the-structure-of-network-automation-with-yang-netconf-restconf-and-gnmi/P200000009517`).

In parallel with this modeling effort, to overcome the poll method approach of SNMP, a new paradigm called **streaming telemetry** appeared. It implements a push model to stream data from the network devices to the collector, continuously. The goal is to provide near real-time access to the operational data from the devices.

In the following figure, you can identify two paradigms for establishing the streaming sessions: as an active collector using the dial-in approach or as a passive receiver using the dial-out approach. In both cases, there is an established connection (in UDP mode, obviously not, but you get a similar behavior). In the dial-in approach, the collector establishes the connection, and in the dial-out approach, the network device is responsible for establishing it:

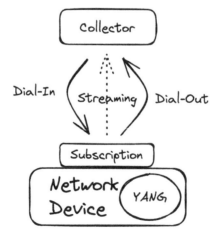

Figure 3.3 – Streaming telemetry: dial-in and dial-out

In both cases, it's possible to define the *subscription* (this term refers to establishing a streaming telemetry session) to either send data at regular intervals or to only send data when the data changes.

In *Chapter 5*, we will use the Telegraf collector to collect streaming telemetry using the gNMI dial-in plugin (https://github.com/influxdata/telegraf/blob/master/plugins/inputs/gnmi/README.md). For the dial-out approach, you can check out other examples, such as this one by Cisco for gRPC dial-out: https://github.com/influxdata/telegraf/tree/master/plugins/inputs/cisco_telemetry_mdt.

YANG models are structured in data containers with path references. They are similar to SNMP OIDs, but more human-friendly. As an example, the OpenConfig Interfaces model is defined here: https://github.com/openconfig/public/blob/master/release/models/interfaces/openconfig-interfaces.yang. With the tree visualization, you can use different tools to visualize the data model, such as the YANG Catalog.

To explore model-driven telemetry, we are going to use the gNMI interface that is enabled in our lab network devices with the Python pygnmi library (https://github.com/akarneliuk/pygnmi). It is also installed in the lab but you can install it via pip install pygnmi.

> **Note**
>
> Even though this example is in Python, the gNMI ecosystem is mostly implemented in the Go programming language.

Then, for the first example, instead of starting with streaming telemetry, we implement a GET operation to obtain the data at a specific moment (similar to an SNMP GET operation):

```python
from pygnmi.client import gNMIclient
import json
paths = ['openconfig-interfaces:interfaces']
with gNMIclient( (1)
    target=("ceos-01", 50051), username="netobs",
    password="netobs123",
    insecure=True
) as gc:
    result = gc.get(path=paths, encoding='json') (2)
    print(
        json.dumps(
            result["notification"][0]["update"][0]["val"]["
                openconfig-interfaces:interface"][0]["state"], (3)
            indent=True
        )
    )
```

These are the most relevant parts in the code:

1. The gNMIclient context manager establishes a gNMI session with the Arista device. Notice that we are using port 50051, which is the one configured for this purpose.

2. The gNMI GET operation can target several paths. In our example, it only targets the interface container.

3. The operational data (state) for the first interface (index 0), which corresponds to the Management0 interface. The result is encoded as JSON during transport, and it is converted to a native Python object by the library. We convert to a string (with json.dumps) to render it.

When running this script, ceos-01 returns all the attributes in the interfaces, which contain the operational status and the counters. Notice that it also contains other data that is part of the configuration, such as admin-status, because it's also in its operational state. If instead of state we targeted config, we wouldn't get admin-status or the counters keys:

```
root@ab7700def797:/usr/src/app# python gnmi_get_state.py
{
"admin-status": "UP",
"counters": {
            "in-discards": "219",
            "in-errors": "0",
            "in-octets": "281662",
            "in-pkts": "3991",
            "out-discards": "0",
            "out-errors": "0",
            "out-octets": "493488",
            "out-pkts": "3970",
            ... omitted output ...
        },
        "ifindex": 999999,
        "arista-intf-augments:inactive": false,
        "last-change": "1697984624461651968",
        "management": true,
        "mtu": 0,
        "name": "Management0",
        "oper-status": "UP",
        "type": "iana-if-type:ethernetCsmacd"
}
```

As an exercise, you can modify the previous script by querying different OpenConfig paths or navigating the response structure.

At this point, you may be wondering, what data models are provided by the network devices? One answer to this would be to check the vendor documentation (but this is not always a reliable option).

The most accurate way is to use the gNMI capabilities, which expose all the models and operations supported via gNMI.

> **Note**
>
> We recommend exploring the result of the script (instead of only printing the result on the screen) using Python's interactive mode with `python -i gnmi_get_state.py`. When you run in interactive mode, when the script ends, you are left in the Python interpreter with full access to the variables created.

The following script gets all this information in one go:

```python
from pygnmi.client import gNMIclient
import json
with gNMIclient(
    target=("ceos-01", 50051),
    username="netobs",
    password="netobs123",
    insecure=True
) as gc:
    result = gc.capabilities()
print(json.dumps(result, indent=True))
```

When you run the script, you will get an exceptionally long list of the supported data models. Some of them are official ones from standardization groups such as IETF or OpenConfig, and others are custom extensions from the vendor. To get detailed information, you can go directly to the model definition, for example, IETF and others (`https://github.com/YangModels/yang`) or OpenConfig (`https://github.com/openconfig/public/tree/master/release/models`). Moreover, you can also use helper tools, such as the YANG Catalog mentioned previously.

```
root@ab7700def797:/usr/src/app# python gnmi_capabilities.py
{
    "supported_models": [
        {
            "name": "openconfig-rib-bgp-types",
            "organization": "OpenConfig working group",
            "version": "0.5.0"
        },
        {
            "name": "arista-lacp-augments",
            "organization": "Arista Networks <http://arista.
com/>",
            "version": "1.0.1"
        },
```

```
... omitted output ...
    ],
    "supported_encodings": [
        "json",
        "json_ietf",
        "ascii"
    ],
    "gnmi_version": "0.7.0"
}
```

As mentioned previously, what makes model-driven telemetry more relevant is its capacity to stream data without asking for it (aside from providing the right configuration). This is done using the SUBSCRIPTION operation, which is also using the same path definition to target the data node are we interested in:

```
from pygnmi.client import gNMIclient, telemetryParser
import json
with gNMIclient(
        target=("ceos-01", 50051),
    username="netobs",
    password="netobs123",
    insecure=True
) as gc:
    telemetry_stream = gc.subscribe( (1)
        subscribe={
            "subscription": [
                {
                    "path": (
                        "interfaces/interface[name=Management0]/"
                        "state/counters/in-pkts" (2)
                    ),
                    "mode": "sample", (3)
                    "sample_interval": 10000000000, (3)
                },
            ],
            "mode": "stream",
            "encoding": "json"
        })
    for telemetry_entry in telemetry_stream:
        print(json.dumps(
    telemetryParser(...),
    indent=True
  ) (4)
```

Let's run through what we did here:

1. The gNMI SUBSCRIBE operation creates a dial-in session to retrieve data.
2. The path (interfaces/interface[name=Management0]/state/counters/in-pkts) determines which part of the data model we are interested in.
3. The sample mode means that every 10 seconds, we will get an update.
4. Finally, the telemetry stream will keep on going and yielding data as it is retrieved over the subscription. Notice that this happens within a Python context manager (https://docs.python.org/3/reference/datamodel.html#context-managers). It closes the connection automatically when the code block ends during the same session established by us (the collector).

In this example, you can see that in the path, we have used an inline filter to only focus on the data related to the Management0 interface. This notation is known as **XPath filtering**.

When you run this script, the script will be running continuously and, every 10 seconds (according to the previous configuration in the sample_interval), you will obtain the value of the incoming packets. You can stop the script with *Ctrl + C* at any time:

```
root@ab7700def797:/usr/src/app# python gnmi_subscription.py
{
        "update": {
            "update": [
                {
                        "path": "interfaces/interface[name=Management0]/
                        state/counters/in-pkts",
                        "val": 5687
                }
            ],
            "timestamp": 1697993862807187267
        }
}
...   after 10 seconds ...
{
        "update": {
            "update": [
                {
                        "path": "interfaces/interface[name=Management0]/
                        state/counters/in-pkts",
                        "val": 5709
                }
            ],
            "timestamp": 1697993872826209305
        }
}
```

Next, we will review a common programmatic approach for interacting with data: the REST APIs that have become relevant in networking environments.

REST APIs

HTTP-based **Application Programming Interfaces (APIs)** are one of the most common way to interact with external systems. Within this group, the REST APIs are the most popular subset because they use the de facto standard internet application protocol, HTTP, and provide basic methods to interact with objects behind that interface: the **CRUD** operations (**Create, Read, Update, Delete**) represented by the GET, PUT, PATCH, POST, and DELETE HTTP request types.

In today's heterogeneous networks that include controller-based or cloud provider networking services, REST APIs are the way to interact with them (i.e., changing the state or retrieving the operational state). For example, if you have to create a virtual network in a cloud provider such as AWS or Azure, the cloud provider REST API (or wrappers around it, such as Terraform, Pulumi, or CloudFormation) is the only way to make it happen.

This trend has also reached network management, which (besides the traditional interfaces, such as SSH CLI, SNMP, or new ones, such as NETCONF or gNMI), can also provide REST APIs. Aside from the vendor-specific ones, there is the standard RESTCONF protocol, which has some similarities with NETCONF and gNMI.

To test the RESTCONF API in the lab environment, we are going to enable it by following this guide: https://aristanetworks.github.io/openmgmt/configuration/restconf/. It generates a self-signed certificate for HTTPS, and enables the management interface over the VRF MGMT:

```
root@ab7700def797:/usr/src/app# ssh netobs@ceos-01
(netobs@ceos-01) Password:
ceos-01>enable
ceos-01# security pki certificate generate self-signed restconf.crt
key restconf.key generate rsa 2048 parameters common-name restconf
certificate:restconf.crt generated
ceos-01#conf t
ceos-01(config)#management security
ceos-01(config-mgmt-security)#ssl profile restconf
ceos-01(config-mgmt-sec-ssl-profile-restconf)#certificate restconf.crt
key restconf.key
ceos-01(config-mgmt-sec-ssl-profile-restconf)#exit
ceos-01(config-mgmt-security)#management api restconf
ceos-01(config-mgmt-api-restconf)#transport https test
ceos-01(config-restconf-transport-test)#ssl profile restconf
ceos-01(config-restconf-transport-test)#port 5900
ceos-01(config-restconf-transport-test)#vrf MGMT
ceos-01(config-restconf-transport-test)#exit
```

After this initial setup, you can validate that the RESTCONF API is up and ready (the port was already exposed by the `containerlab` setup) by checking out the RESTCONF state in the `ceos-01` switch:

```
ceos-01#show management api restconf
Enabled: yes
Server: running on port 5900, in default VRF
SSL profile: restconf
QoS DSCP: none
```

Once the RESTCONF API is ready, you can create a Python script using the `requests` library to interact via HTTP (library already installed in the lab).

Then, you can create a script to interact with the RESTCONF API:

```
import requests
from requests.auth import HTTPBasicAuth
import json

router = "ceos-01"
restconf_port = 5900
result = requests.get(  (1)
    (
        f"https://{router}:{restconf_port}/restconf/data/",
        "openconfig-interfaces:interfaces/interface=Management0/state"
        (2)
    ),
    auth=HTTPBasicAuth("netobs", "netobs123"),
    headers={
        "Content-Type": "application/yang-data+json",  (3)
        "Accept": "application/yang-data+json"  (3)
    },
    verify=False
)
print(json.dumps(
    result.json()['openconfig-interfaces:counters'], indent=True
))
```

Let's look at what we just did:

1. The `requests` library provides a simple and elegant way to run HTTP methods, such as GET. The RESTCONF API is just another REST API that uses the YANG models as the URL path.

2. We are crafting a URL query that targets the RESTCONF API path, and it's looking exactly for the same OpenConfig data path that you saw in the model-driven telemetry example, when you ran a GET gNMI operation.

3. The HTTP headers are used to specify the content type as YANG data serialized as JSON.

If you run the script, you will get a similar response to the counters output for the gNMI GET example. The data structure is the same (the numbers are probably different), but it is consumed over a different method:

```
root@ab7700def797:/usr/src/app# python restconf_get.py
{
"in-broadcast-pkts": "0",
"in-discards": "357",
"in-errors": "0",
     "in-octets": "535881",
     "in-pkts": "7274",
     "out-discards": "0",
     "out-errors": "0",
     "out-octets": "849743",
     "out-pkts": "6988",
     ... omitted output ...
}
```

One important consideration here is that RESTCONF doesn't support dial-in streaming telemetry because keeping an established session goes against the REST principles of not keeping a state. However, it could be used to configure dial-out streaming telemetry.

The next data type is very network-specific: **network flows**.

Flows

Flow-based monitoring emerged as a complement to SNMP monitoring to get an understanding of network traffic patterns. In the late 1990s, Cisco Systems created the first proprietary implementation of **Netflow**, and it evolved to the **Internet Protocol Flow Information eXport** (**IPFIX**) standard by IETF, which is almost equivalent to NetFlow v9.

Network flows are high-dimensional structured data used to provide information about a network flow from a source socket to a destination socket (a socket is the combination of an IP address and a TCP/UDP port). It also contains the timestamp or the bytes transferred, and other custom information depending on the protocol and implementation.

> **Note**
>
> A book about this topic is *Network Security with NetFlow and IPFIX: Big Data Analytics for Information Security* by Omar Santos.

The most common way to collect this flow information is to configure an exporter in the network device that samples the packets to collect flow information and send it using Netflow or IPFIX to a collector. Compare the pattern to SNMP: the data is sent by the network devices to the collector.

There are a few commercial and open source collectors, such as PMACCT (http://www.pmacct.net/), akvorado (https://github.com/akvorado/akvorado), and goflow2 (https://github.com/netsampler/goflow2), but to illustrate the protocol usage, we will use a simpler tool, the Python Scapy library (https://scapy.net/), which allows packet-level management.

Flows are useful for many different purposes in the realm of network observability because they offer deeper visibility. For example, flows could be used by network providers to calculate the billing based on the usage of the related flows, to understand how the network traffic is routed, or to do security analysis. However, exporting flows has a high computational cost, so you must select which interfaces it's necessary to enable.

As we mentioned, we will use the Scapy library to analyze an existing packet (an IPFIX packet with flow information) instead of receiving flows from our network lab. This library is widely used for many different use cases, such as pen testing, and we will use it later to create custom packets for synthetic monitoring purposes.

Scapy comes pre-installed in the lab, but you can install it with pip install scapy.

Instead of using a Python script, let's use an interactive session and load an existing PCAP containing IPFIX flows (the PCAP file and the related script, ipfix_receiver.py, are in the examples repository):

```
root@ab7700def797:/usr/src/app# python
Python 3.11.4 (main, Jul  5 2023, 13:45:01) [GCC 11.2.0] on linux
Type "help", "copyright", "credits" or "license" for more information.
>>> from scapy.all import *
>>> capture = rdpcap('examples/ch03/ipfix.pcap')
```

Here, rdpcap reads all the packets of the PCAP file.

Then, you can use the netflowv9_defragment method to parse (i.e., interpret) the capture and review of the content of the first packet (notice the index 0):

```
>>> pkt = netflowv9_defragment(capture)[0]
>>> pkt.summary()
'Ether / IP / UDP 204.42.110.30:33092 > 204.42.111.229:9995 /
NetflowHeader / NetflowHeaderV10 / NetflowFlowsetV9'
>>> pkt.show()
###[ Ethernet ]###
  dst        = 00:50:56:84:63:c9
  src        = 84:78:ac:00:0b:0a
  type       = IPv4
###[ IP ]###
     ... output omitted for brevity ...
     proto     = udp
     chksum    = 0x4704
     src       = 204.42.110.30
```

```
       dst        = 204.42.111.229
       \options   \
###[ UDP ]###
         sport     = 33092
         dport     = 9995
         len       = 124
         chksum    = 0x0
###[ Netflow Header ]###
           version   = 10
###[ IPFix (Netflow V10) Header ]###
             length    = 116
             ExportTime= Tue, 29 Nov 2016 20:08:55 +0000 (1480450135)
             flowSequence= 3791
             ObservationDomainID= 0
```

The packet summary (pkt.summary()) provides readable information about the packet, but you can access each layer of the packet programmatically. In the next example, notice the difference between pkt.dst (Ethernet address) and pkt[IP].dst (IP address), and also the capacity to access the different layers of the packet. Finally, we go into the UDP layer to check the source port:

```
>>> pkt.dst
'00:50:56:84:63:c9'
>>> pkt[IP].dst
'204.42.111.229'
>>> pkt.layers()
[<class 'scapy.layers.l2.Ether'>, <class 'scapy.layers.inet.IP'>,
<class 'scapy.layers.inet.UDP'>, <class 'scapy.layers.netflow.
NetflowHeader'>, <class 'scapy.layers.netflow.NetflowHeaderV10'>,
<class 'scapy.layers.netflow.NetflowFlowsetV9'>]
>>> pkt["UDP"].sport
33092
```

As expected, you can also get into the actual Netflow/IPFIX headers, and extract, directly, the information parsed, such as the flowSequence.

```
>>> pkt["NetflowHeaderV10"].flowSequence
3791
```

Then, we continue with the packet analysis, generalizing it to any kind of data.

Packet capture and deep packet inspection (DPI)

There is nothing more accurate than analyzing the packets forwarded in the network. You can examine the headers, payload, and application data to fully understand what your network is doing. It is a heavy computational task, though, so it's not something you will do as a general approach. However, on some special occasions (such as security incidents), having access to this level of information, such as understanding how your QoS policy impacts your applications or detecting security threats, could save the day.

> **Note**
>
> The book *Practical Packet Analysis, 3rd Edition: Using Wireshark to Solve Real-World Network Problems* by Chris Sanders is a good reference for understanding how to analyze packet captures.

The Scapy library also serves this purpose, so we will continue with the previous example. There, you loaded an existing capture, but you can also use Scapy sniff to get traffic from the network. In the next example, we will sniff packets based on a capture filter (so we don't capture everything) and limit it to only two packets:

```
>>> pkts = sniff(filter="icmp and host 1.1.1.1", count=2)
```

This command will leave the Python interpreter waiting for the two packets that match the capture filter. Thus, it's time to move to another terminal and run a ping that matches that filter. The Python interpreter will unblock and sniff the two packets:

```
root@ab7700def797:/usr/src/app# ping 1.1.1.1
PING 1.1.1.1 (1.1.1.1) 56(84) bytes of data.
64 bytes from 1.1.1.1: icmp_seq=1 ttl=59 time=1.96 ms
```

In the Python interpreter, you can explore the two captured packets (in the pkts variable), as we did in the *Flows* subsection:

```
>>> pkts
<Sniffed: TCP:0 UDP:0 ICMP:2 Other:0>
>>> pkts.summary()
Ether / IP / ICMP 167.172.190.221 > 1.1.1.1 echo-request 0 / Raw
Ether / IP / ICMP 1.1.1.1 > 167.172.190.221 echo-reply 0 / Raw
```

This was a very simple example of how to capture and analyze packets as they are received. You could elaborate on it and try to decode the information to get all the information you need.

Next, still using Scapy, we will figure out how to craft network traffic and check its behavior to observe how your network behaves under specific circumstances.

Synthetic monitoring

Sometimes, you don't want to wait and see what's going on in your network. You need to test proactively how some network traffic behaves to understand, in detail, the impact on your key business applications. For example, if your network's main purpose is to serve DNS service, then it makes sense to validate your network from the perspective of DNS traffic.

One obvious (and valid) approach is to observe the actual DNS traffic. But, in this case, there may be external factors influencing its behavior that combine different variables.

Another option is to use a DNS client to emulate real DNS requests. This approach is called **synthetic monitoring** because we are using production-like traffic to understand how the network behaves.

Within the synthetic monitoring approach, you can craft any kind of network packets that you are interested in to emulate existing network applications. This gives you full control over what you are testing your network for.

The Scapy library, which we used in the previous sections, enables us to build any network packet (this is the reason that it's part of any basic hacker's toolkit).

First, let's try to simply send one ICMP packet with an IP packet of type ICMP with the send function:

```
>>> from scapy.all import *
>>> host = "1.1.1.1"
>>> send(IP(dst=host)/ICMP())
.
Sent 1 packets.
```

As you can see, there was no response because we only *sent* a packet. Scapy comes with the *send'n'receive* functions family, which allows sending and receiving packets that have been defined. The sr() and sr1() (the sr variant that only returns one packet) functions allow us to not only send packets but also receive responses.

Thus, with sr1(), you send and receive the ICMP packet, and can analyze the response:

```
>>> response = sr1(IP(dst=host)/ICMP())
Begin emission:
Finished sending 1 packets.
*
Received 1 packets, got 1 answers, remaining 0 packets
>>> response
<IP  version=4 ihl=5 tos=0x0 len=28 id=31789 flags= frag=0
ttl=59 proto=icmp chksum=0x9b28 src=1.1.1.1 dst=167.172.190.221
|<ICMP  type=echo-reply code=0 chksum=0x0 id=0x0 seq=0x0 |>>
>>> response[ICMP]
<ICMP  type=echo-reply code=0 chksum=0x0 id=0x0 seq=0x0 |>
>>> response[ICMP].code
0
```

We can create more complex packets. Going back to the previous DNS example, let's craft a DNS request for the www.example.org domain (using the 1.1.1.1 DNS server by Cloudflare). In the response, you can observe how the DNS data (e.g., the type of the DNS entry and the value) have been returned, and how you can access it:

```
>>> response = sr1(IP(dst=host)/UDP()/DNS(rd=1,qd=DNSQR(qname="www.
example.org")))
Begin emission:
Finished sending 1 packets.
*
Received 1 packets, got 1 answers, remaining 0 packets
>>> response.show()
###[ IP ]###
  version   = 4
  ... omitted some output ...
  proto     = udp
  chksum    = 0x793f
  src       = 1.1.1.1
  dst       = 167.172.190.221
  \options   \
###[ UDP ]###
     sport   = domain
     dport   = domain
     len     = 57
     chksum  = 0x1166
###[ DNS ]###
        opcode    = QUERY
        ... omitted output ...
        \qd          \
        |###[ DNS Question Record ]###
        |  qname      = 'www.example.org.'
        |  qtype      = A
        |  qclass     = IN
        \an          \
        |###[ DNS Resource Record ]###
        |  rrname     = 'www.example.org.'
        |  type       = A
        |  rclass     = IN
        |  ttl        = 83656
        |  rdlen      = 4
        |  rdata      = 93.184.216.34
     ns       = None
     ar       = None
>>> response[DNSRR].rdata
'93.184.216.34'
```

These were only two basic examples of the many sophisticated tests you can create with Scapy to validate your network behavior for some specific network traffic.

One well-known example of a synthetic monitoring tool is the IP SLA protocol (created by Cisco), standardized as RFC 6812 (`https://datatracker.ietf.org/doc/html/rfc6812`). The Cisco SLA protocol allows running a set of measurements such as test connectivity (ping-like) or UDP probes, with latency and packet loss measurements.

Now, let's move on to a broad group of network control protocols and their monitoring variants.

Network control monitoring protocols

Yet another network-specific field for monitoring is observing how the network control protocols are behaving. In the network, we have many control protocols that have a crucial impact on the forwarding plane behavior, and being able to understand how they work can be very important in some cases.

The most common way to get the data has been via checking its state via a CLI or participating in the routing protocol interaction. However, often this provides only partial information.

An example of a monitoring protocol around a network control protocol is the **BGP Monitoring Protocol** (BMP, RFC 7854: `https://www.rfc-editor.org/rfc/rfc7854.html`). BMP provides information (without intervening on the routing protocol) about data used by the protocol that was not available outside of the devices (e.g. the adj-rib-in, loc-rib, and adj-rib-out BGP tables). This allows a very granular understanding of the BGP function in network devices.

There are specialized tools for analyzing this protocol (e.g., PMACCT), but you can use lower-level tools such as Scapy, or other tools able to decode/understand the payload. In this case, we will use pyshark (`https://github.com/KimiNewt/pyshark`), a wrapper around the Wireshark CLI interface, Tshark (`https://tshark.dev/setup/install/`). There are similarities between pyshark and Scapy, but pyshark shines when it uses Wireshark's decode capabilities.

First, let's install PyShark and Tshark. Check out the documentation for your OS (`https://tshark.dev/setup/install/#install-wireshark-with-a-package-manager`). The container image for the lab already includes both, so no action is required.

Now, you can import a BMP capture to start exploring it (it's included in the lab). Notice that we are using FileCapture with a specific decode_as option for BMP decoding, and the TCP port:

```
>>> import pyshark
>>> capture = pyshark.FileCapture("bmp.pcap", decode_as={"tcp.
port==6666":"bmp"})
```

The capture includes several packets. Without using the load_packets function, the packets are loaded on demand when accessed, which is very convenient for large captures:

```
>>> capture.load_packets()
>>> len(capture)
```

```
50
>>> for packet in capture:
...        print(packet.layers)
...
[<ETH Layer>, <IP Layer>, <TCP Layer>]
[<ETH Layer>, <IP Layer>, <TCP Layer>]
[<ETH Layer>, <IP Layer>, <TCP Layer>]
[<ETH Layer>, <IP Layer>, <TCP Layer>, <BMP Layer>]
... omitted for brevity ...
```

In the previous output, you can observe that some packets contain BMP information, and others do not. If we analyze these packets, there is a BMP initialization dialog (packets 4 and 6), so we focus on the first one that contains some relevant BMP update info (for instance, the eighth one), after the protocol initialization:

```
>>> packet = capture[7]
>>> packet.show()
Layer ETH
... omitted for brevity ...
Layer IP
... omitted for brevity ...
Layer TCP
... omitted for brevity ...
Layer BMP
:        Version: 3
         ... omitted output for brevity ...
         Type: Route Monitoring (0)
         Address: 10.255.0.101
         ASN: 32934
         BGP ID: 10.10.10.1
         Type: UPDATE Message (2)
         Next hop: 10.255.0.101
         Network Layer Reachability Information (NLRI)
         10.10.10.2/32
```

In the output, you can see the complete payload of a BMP update message about the 10.10.10.2/32 prefix, with all the relevant BGP details. This could be crucial for properly understanding the BGP protocol's behavior.

Next, to expand the scope of networking observability further, we want to give you a quick introduction to traces and the OpenTelemetry project.

Traces and OpenTelemetry

Even though they haven't been seen as network-related data, we want to include traces as a key component of a modern network observability system. Today, most networks have adopted some level of automation to improve their management, or the network services are implemented by modern software with many steps for each processing request. Thus, these software systems support a way to correlate all the events related to the same request.

Since the 2010s, the software community has understood that logs, which represent discrete events, are not sufficient to understand complex applications, especially distributed ones. This led to several initiatives relating to distributed tracing, such as Google's **Dapper**, and other projects, such as **Zipkin** (developed by X, formerly Twitter) and **Jaeger** (developed by Uber), which provide all the tooling required to send and analyze traces.

Distributed traces relate to the data fusion requirement mentioned in the *Network telemetry* section, where the data for an application comes from different sources but they are correlated via a common identifier. This is also known as data correlation for observability signals.

Traces are an evolution of logs to provide a continuous view of an application request through the whole flow of execution. This way, once a request enters the system, it is identified, and you can track which functions were executed, how much time they took, and even the parameters passed.

Luckily, after some consolidation efforts, since 2019, the **OpenTelemetry** (`https://opentelemetry.io/`) initiative took over most of the responsibility for implementing a standard tracing implementation and started covering metrics and logs events. Later, it became a project of the **Cloud Native Computing Foundation** (**CNCF**, `http://cncf.io/`).

The first goal of OpenTelemetry was to establish a common structure and tooling to create and manage traces. This ended with a robust implementation in many different programming languages. Similar to the previous examples, the libraries are already installed, but you can do it using `pip install opentelemetry-api opentelemetry-sdk`. Then, you can use the libraries to create traces for any request.

In this example, we are going to create a simple web app using the Python Flask application (`https://flask.palletsprojects.com/`) (already installed), which will use context managers to create a trace for every request received for the execution and for the actual business code:

```
from flask import Flask
from time import sleep
from opentelemetry import trace
from opentelemetry.sdk.trace import TracerProvider
from opentelemetry.sdk.trace.export import BatchSpanProcessor,
ConsoleSpanExporter
app = Flask(__name__)
provider = TracerProvider()
processor = BatchSpanProcessor(ConsoleSpanExporter())
```

```
provider.add_span_processor(processor)
trace.set_tracer_provider(provider)
tracer = trace.get_tracer(__name__)
@app.route("/") (1)
def entry():
    with tracer.start_as_current_span("my_opentelemetry_traces"):
        return do_something()
def do_something(): (2)
    with tracer.start_as_current_span("my_opentelemetry_traces"):
        sleep(3)
        return "job done"
```

Let's look at this in detail:

1. We create a root entry path into the web app that simply calls a business logic function within the trace context manager.

2. The example business logic function only sleeps for three seconds (to impact the timestamps) and returns a string.

Now, you can start the Flask web server, defining the file to run with the environmental variable FLASK_APP. By default, Flask runs in port TCP 5000:

```
root@ab7700def797:/usr/src/app# export FLASK_APP=opentelemetry_traces
root@ab7700def797:/usr/src/app# flask run
 * Serving Flask app 'opentelemetry_traces'
 * Debug mode: off
WARNING: This is a development server. Do not use it in a production
deployment. Use a production WSGI server instead.
 * Running on http://127.0.0.1:5000
Press CTRL+C to quit
```

Then, in another terminal, you can run an HTTP query to the server to obtain the expected string after three seconds:

```
root@ab7700def797:/usr/src/app# curl http://localhost:5000
job done
```

If you move back your attention to the Flask server terminal, you should have gotten the two expected traces, one for the root entry path and the other for the business logic:

```
127.0.0.1 - - [30/Oct/2023 05:26:13] "GET / HTTP/1.1" 200 -
{
    "name": "my_opentelemetry_traces",
    "context": {
        "trace_id": "0x3bc29be7d6b4457cd68695b8277b10c1",
        "span_id": "0x6a040974bd9f6df2",
```

```
            "trace_state": "[]"
        },
        "kind": "SpanKind.INTERNAL",
        "parent_id": "0x7337c822eeb3d199",
        "start_time": "2023-10-30T05:26:10.244114Z",
        "end_time": "2023-10-30T05:26:13.244339Z",
        ... omitted for brevity ...}
    {
        "name": "my_opentelemetry_traces",
        "context": {
            "trace_id": "0x3bc29be7d6b4457cd68695b8277b10c1",
            "span_id": "0x7337c822eeb3d199",
            "trace_state": "[]"
        },
        "start_time": "2023-10-30T05:26:10.244036Z",
        "end_time": "2023-10-30T05:26:13.244426Z",
        ... omitted for brevity ...
        }
    }
```

In the previous output, you can see all the goodies you get by default with OpenTelemetry:

- There is a well-structured envelope with key information that can be extended to included other relevant metadata.

- The trace_id is constant for all the processing of the request. This would be valid even if many backends were traversed. Within a trace, with the span_id and parent_id, you can identify the dependencies.

- The end_time represents the three seconds spent, for both traces, because the second trace was contained by the first trace. So, you can determine in which step of the process most of the time was spent.

The OpenTelemetry group aims to extend the same standardization for metrics and logs. As a reference, the next example exposes metrics via OpenTelemetry.

Building on the previous example of traces, this new code example removes the traces (for simplicity, you could use the application traces from the previous example and the metrics from this one) and uses the metrics objects to increment a metric for every request, and then another random increment when executing the code:

```
from flask import Flask
from random import randint
from opentelemetry import metrics
from opentelemetry.sdk.metrics import MeterProvider
```

```python
from opentelemetry.sdk.metrics.export import (
    ConsoleMetricExporter,
    PeriodicExportingMetricReader,
)
app = Flask(__name__)
metric_reader = PeriodicExportingMetricReader(ConsoleMetricExporter())
provider = MeterProvider(metric_readers=[metric_reader])
metrics.set_meter_provider(provider)
meter = metrics.get_meter("my_opentelemetry_metrics")
@app.route("/")
def entry():
    counter = meter.create_counter("counter")
    counter.add(1)
    return do_something()
def do_something():
    counter = meter.create_counter("counter")
    counter.add(randint(1, 10))
    return "job done"
```

Again, we start the Flask web application, setting the ENV to the new file:

```
root@ab7700def797:/usr/src/app# export FLASK_APP=opentelemetry_metrics
root@ab7700def797:/usr/src/app# flask run
 * Serving Flask app 'opentelemetry_metrics'
 * Debug mode: off
WARNING: This is a development server. Do not use it in a production
deployment. Use a production WSGI server instead.
 * Running on http://127.0.0.1:5000
Press CTRL+C to quit
```

And, in another terminal, repeat the HTTP query, which returns the same string:

```
root@ab7700def797:/usr/src/app# curl http://localhost:5000
job done
```

Now, in the Flask server terminal, the result is a metric value with the counter's value after the whole execution:

```
127.0.0.1 - - [30/Oct/2023 05:30:13] "GET / HTTP/1.1" 200 -
{
    "resource_metrics": [
        {
            "scope_metrics": [
                {
                    "scope": {
```

```
                          "name": "my_opentelemetry_metrics",
                      },
                      "metrics": [
                          {
                              "name": "counter",
                              "data": {
                                  "data_points": [
                                      {
                                          "value": 3
                                      }
... omitted output for brevity ...
```

> **Note**
>
> There are other implementations similar to traces for specific protocols, such as DNSTAP (`https://dnstap.info/`) for DNS, which enables getting more application-specific information from a request. All the options should be considered for observing your network.

Finally, before wrapping up the chapter, we will cover what can be achieved by using Linux's eBPF.

eBPF

As already mentioned, running an agent inside the target device makes it possible to observe that system with complete detail. **Extended Berkeley Packet Filter** (**eBPF**) is a Linux technology that provides access to many functions and events in the kernel with low-level observability possibilities. eBPF has evolved from BPF, which has been in use for a long time by applications such as `tcpdump`.

eBPF, at a fundamental level, allows the execution of user-space code injected directly in the kernel without the traditional approach of building and loading a kernel module. This provides the flexibility to run custom code and attach it dynamically to different kernel events (e.g., `kprobes` or `tracepoints`).

eBPF has a lot of applications, with a special mention for cloud-native environments:

- **Security**: It can influence firewall policies, isolation, or any security measure directly in the host without having to use the **sidecar pattern** (`https://learning.oreilly.com/library/view/designing-distributed-systems/9781491983638/ch02.html#idm139824744418064`).

- **Networking**: Having access to the network packets at different levels of the stack allows custom and highly performant implementation (because we can skip many layers of the kernel) of many network functionalities, such as load balancing and DDoS mitigation. As container networking happens on a host, eBPF provides control of how packets flow in these environments.

- **Observability**: Due to direct access to the kernel in an isolated and nonintrusive way, eBPF provides more visibility and better accuracy. It is a goldmine for system observability (bpftrace (`https://github.com/iovisor/bpftrace`) is a popular example in this field). For instance, we can track syscalls attaching custom logic to the `execve` kprobe or observe network packets at different points of the kernel networking path. This level of observability is crucial for container-based environments (e.g., Kubernetes) because traditional network monitoring is not effective enough.

In all these cases, eBPF provides helper functionalities to enable access to kernel tables and information, which is a good compromise for developing advanced functionalities without having to reinvent everything.

In the following figure, you can see that from the user space (using many different high-level programming languages), you can load C or Rust code into the kernel, which gets compiled with a **just-in-time** (**JIT**) compiler and verified before becoming an eBPF function, ready to be used in the BPF virtual machine. Then, this code is attached to a kernel event from the list of predefined hooks. When attached, the loaded eBPF code takes action immediately and can communicate with the user space via the BPF maps (a shared memory space):

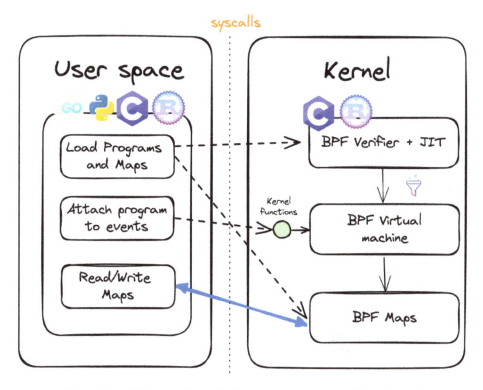

Figure 3.4 – eBPF basics (Created using the icons from https://icons8.com)

> **Note**
>
> To go deeper into this topic, we recommend the *Learning eBPF* book by Liz Rice (`https://isovalent.com/books/learning-ebpf/`), with many hands-on examples to get started.

We want to provide you with a high-level example of how eBPF works so you can understand what the applications that are based on this technology are doing. To do so, we use the BCC library (`https://github.com/iovisor/bcc`) (BCC stands for BPF Complier Collection), which has a frontend in Python.

In contrast with the other examples, we will run this directly in the server hosting the lab (i.e., not within the container).

First, we have to install the BCC library, which will take care of all the low-level tasks related to eBPF management, including installing the necessary Linux dependencies:

```
root@netobs-droplet:~# sudo apt-get install bpfcc-tools linux-headers-
$(uname -r)
```

Then, you create a Python script that will load a simple C program that prints on the screen that a packet has been received, along with its length. The first contact of a packet with the networking stack in the kernel is the **eXpress Data Path** (**XDP**) hook, right after the packet is ingresses in the interface.

The Python code is a wrapper to manage the actual C code that gets installed in the kernel as an eBPF function, and can interact with its outcomes via BPF maps (memory shared between the kernel and the user space):

```python
#!/usr/bin/python3
from bcc import BPF
program = r"""
#include <bcc/proto.h>
int hello(struct xdp_md *ctx) {
  void *data = (void *)(long)ctx->data;
  void *data_end = (void *)(long)ctx->data_end;
  u32 len = ctx->data_end - ctx->data;
  bpf_trace_printk("Got a packet %d", len);
  return XDP_PASS;
}
"""
b = BPF(text=program) (1)
fx = b.load_func("hello", BPF.XDP) (2)
BPF.attach_xdp("lo", fx, 0) (3)
b.trace_print() (4)
```

Let's look at this code in detail:

1. The C code is loaded from a string that contains the C code. It could have been loaded from a file.

2. The "hello" function is loaded as an XDP function.

3. The loaded function is attached to the loopback interface, for ingress packets.

4. The trace is outputted on the screen.

When you run the previous script, the eBPF hello function will start taking action and outputting all the received packets (in the loopback interface) and the size of each packet on the screen. You can see this by sending packets to the loopback interface, 127.0.0.1:

```
root@netobs-droplet:~/network-observability-lab# sudo python3
chapters/ch3-completed/scripts/ebpf_hello.py
... some warnings ...
b'             node-26385    [001] d.s11 846912.414330: bpf_trace_
printk: Got a packet 102'
b'             node-26036    [000] d.s11 846912.416492: bpf_trace_
printk: Got a packet 298'
b'             sshd-26281    [002] d.s11 846912.419606: bpf_trace_
printk: Got a packet 120'
... it continues until you stop it ...
```

This is a very simple example, but it should show what eBPF can offer for (network) observability.

With eBPF, we conclude the overview of the various network data collection methods, and we are ready to start using them in a complete network observability stack.

Summary

This chapter provided a high-level overview of all the observability data types with which you can build robust network observability solutions and the different methods of collecting the data (i.e., traditional and modern ones). We won't use all of them in this book, but knowing all the available approaches provides flexibility.

This chapter closes *Part 1* of the book. This initial part of the book aims to introduce you to the basic definitions and fundamentals around observability to use them in the rest of the book.

Part 2 of the book proposes an architecture for building modern network observability solutions (introduced in the next chapter, *Chapter 4*), and then, goes through component after component of this architecture and explains how to implement them, focusing on some of the most popular open source projects in each area.

Part 2: Building an Effective Observability Stack

Part 2 of the book will go into further detail about building out the stack, using primarily open source tooling. This will help to get you on the way to having a modern telemetry stack to work from. After reviewing a few of the tools available for the stack, we will dive into some of the key decisions to be made on building out of a stack to your environment.

This part contains the following chapters:

- *Chapter 4, Observability Stack Architecture*

- *Chapter 5, Data Collectors*

- *Chapter 6, Data Distribution and Processing*

- *Chapter 7, Data Storage Solutions for Network Observability*

- *Chapter 8, Visualization – Bringing Network Observability to Life*

- *Chapter 9, Alerting – Network Monitoring and Incident Management*

- *Chapter 10, Real-World Observability Architectures*

4

Observability
Stack Architecture

In this chapter, we delve into the significance of a meticulously crafted **observability stack** for modern IT setups. By adopting a standard observability architecture, we not only streamline complexities but also foster efficiency and uniformity across teams, adopting a taxonomy that everyone can understand and contribute to. From understanding the ins and outs of data pipelines and the pivotal **Extract Transform Load** (ETL) process to the challenges and best practices in setting up such an architecture, we've got you covered. As we conclude, gear up for a hands-on lab setup to truly grasp and apply the intricacies of observability, a foundation that we'll reference throughout this book.

The chapter covers the following topics:

- The components of an observability platform
- The importance of a well-designed observability stack
- Understanding data pipelines for observability
- Challenges and best practices
- Setting up a lab environment

The components of an observability platform

Throughout the history of networking, vendors have introduced monitoring solutions built either in-house or by third parties, exemplifying a monolithic approach. Similarly, traditional open source **round robin database** (RRD) tools, for example, **Cacti** (`http://www.cacti.net/`) and **MRTG** (`https://oss.oetiker.ch/mrtg/`), were crafted to address these issues, with an emphasis on being vendor-neutral and applicable across various platforms. However, these solutions often encounter difficulties in compatibility with other systems. The ability to integrate or extend their functionalities—for instance, to add new types of metrics or create new types of visualizations—has typically been constrained, necessitating custom development.

These monolithic solutions certainly have their merits and can add significant value to your organization. However, since you're reading this book, you likely have an interest in exploring a more integrated and comprehensive strategy for monitoring your network. New tools have emerged to focus on certain monitoring functionalities and with a more specialized approach. For example, some tools focus on collecting metrics and logs, others on collecting traces, and some just on data visualization and alerting.

This specialized focus on certain parts of the application has given birth to what is known as the observability stack, an architecture composed of different specialized components of the observability realm working together.

The following diagram provides a clear representation of all its components:

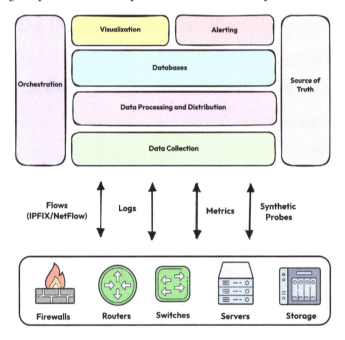

Figure 4.1 – Observability platform components (Created using the icons from https://icons8.com)

Here's a breakdown of what each element signifies:

- **Data collection**: These are the tools or processes that collect metrics, logs, and traces or perform synthetic checks across your system and network. Data collectors can be anything from agents installed on servers or network devices to systems that have been configured to scrape, pull, or listen to metrics forwarded to them. For more on data types, formats, and data collection technologies, see *Chapter 3*, and check out *Chapter 5* for a detailed look at data collectors, including popular open source tools.

- **Data processing and distribution**: After collecting the data, we often process and distribute it to the appropriate destinations. Processing can include tasks such as parsing logs, aggregating metrics, or enriching data with additional context. A common task is the normalization of the data to adhere to a specific schema. This is particularly important when merging similar metrics from other vendors to provide agnostic queries to the data. Distribution involves sending the processed data to where it will be stored, analyzed, or visualized—more on this in *Chapter 6*.

- **Databases**: This is the repository for all gathered and refined data. While it is important that it stores the collected data efficiently, it's not merely about storage. The power for developers and users largely comes from the query language used. A robust query language facilitates seamless data consumption, aggregation of multiple time series, and even differential analyses between them—all essential for visualization, report generation, and alert creation. While the needs of your observability stack might dictate using a straightforward time-series database, or perhaps a blend of various storage types, the query language is a key capability to truly unlock the potential of your observability data. For a more detailed discussion databases for the observability stack, see *Chapter 7*.

- **Visualization**: Visualization tools are used to display the data in an easy-to-understand format, often as charts, graphs, or tables. For example, these tools help network operations engineers quickly identify trends, spot anomalies, and understand the state of the network at a glance. For management, it might help them see the impact of a network outage on their revenue. It helps tell a story from the data collected. We take a deeper dive into data visualization in *Chapter 8*.

- **Alerting**: Alerting mechanisms are crucial to act upon data. It could mean notifying engineers about critical issues that need their immediate attention or triggering an automated task to handle the incident. Alerting rules define what conditions should trigger an alert, and these rules should be fine-tuned to minimize noise and ensure that only the actionable issues are brought to an engineer's attention. We explore alerting in more detail in *Chapter 9*.

- **Source of truth**: Enriching the data or comparing against thresholds requires a single reliable source of truth that provides accurate, up-to-date information about the expected state of the system. This could be a configuration database, a network management system, or another reliable data store.

- **Orchestration**: Orchestration tools help manage the deployment, scaling, and operation of the various components in your observability stack. They ensure that all these components work together smoothly and can handle the demands of your system or network.

In summary, each of these components plays its role in an observability architecture. By working together, they provide a comprehensive view of the system's or network's operation, helping engineers monitor performance, troubleshoot issues, and plan for future needs. Or, in other words, it offers the network visibility that answers the questions you have about your network.

The composable nature helps with the integration and interoperability of the tools in this space, helping different industries adapt based on their specific needs. The following sections describe each of these components in more detail with some practical examples to follow along and help you with the learning experience. But first, let's discuss the importance of a well-designed observability stack.

The importance of a well-designed observability stack

Why bother with a well-designed observability platform, and what does *well-designed* even mean? Let's break it down. It's not just about making things look nice and tidy; it's about building something that works for you. A robust platform is like a trusty toolkit—resilient, scalable, efficient, and ready to tackle your organization's unique challenges. A hastily put-together setup might leave you with a mess of data, while a thoughtfully planned one guides you to valuable insights. So, let's dive into what makes an observability platform well designed and why you should care.

> **Note**
>
> We use the terms observability *stack* and *platform* interchangeably throughout this book.

Why does an observability stack need to be well designed?

An optimally structured and finely tuned observability stack is important for a multitude of reasons:

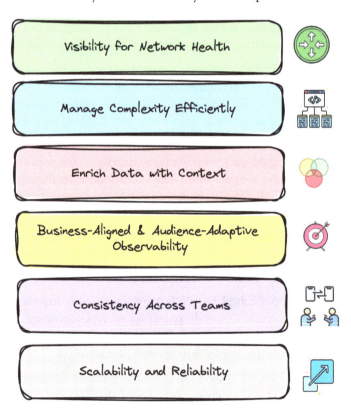

Figure 4.2 – Foundations of an effective observability platform

(Created using the icons from https://icons8.com)

Firstly, and maybe the most obvious aspect, it provides valuable visibility into the network's performance, behavior, and potential issues. In today's cloud-native and hybrid (on-premises) applications, you need real-time insights and data-driven metrics. This typically involves multiple elements of your infrastructure and applications. A robust observability stack can provide these insights and metrics, which can be used to monitor performance, troubleshoot issues, and plan for future needs.

Secondly, managing network operations in complex, large-scale IT environments is increasingly challenging. Each network vendor might offer multiple interfaces for data collection, such as SNMP, gNMI, and RESTCONF. At times, you might find it necessary to employ multiple methods concurrently to ensure comprehensive data extraction. When you add compute services or applications that also possess other data collection procedures (such as **Prometheus** and **OpenTelemetry**), the complexity can be overwhelming. An ideal observability stack must demonstrate flexibility and extensibility in data collection methods while maintaining a manageable level of complexity.

Thirdly, you need to have meaningful insights into what is currently occurring in your network. Collected data may need to be enriched with additional context to provide it. For example, an **Internet Service Provider** (**ISP**) circuit connection on a network device may be experiencing high utilization, but the utilization value alone may not prove useful unless you know the contracted bandwidth. It's important that the observability platform can connect contextual information from other data sources that we will encompass within the source-of-truth concept.

Fourthly, design with business alignment and diverse audiences in mind. While the primary users of an observability platform are often engineers, there's an emerging trend where leadership and management seek direct insights into the system's overarching KPIs. Whether it's a comprehensive report detailing an incident or a high-level view of performance metrics, the observability stack should be versatile enough to cater to both technical and non-technical stakeholders. This dual approach ensures that while engineers receive granular data vital for troubleshooting and optimization, leadership gets a distilled version emphasizing business impacts and strategic insights. Furthermore, it's imperative to fine-tune the platform's alerting mechanisms. Engineers should be shielded from redundant or irrelevant alerts, ensuring they're only notified when potential issues are detected and analyzed. This reduces alert fatigue and ensures swift response times, aligning operational efficiency with business continuity.

Furthermore, a standardized observability architecture can ensure consistency across different teams and projects within the organization. This is particularly beneficial in larger enterprises, where diverse teams may be working on various aspects of the network. Typically, a network team is responsible for managing the network infrastructure (routers, switches, cloud network instances, etc.), and a DevOps team is in charge of the compute infrastructure running applications on top of the network. The observability stack should provide a common framework in which both teams can work, ensuring consistency and, more importantly, creating a path for correlated events and more informed insights.

> **Note about data and events correlation**
>
> One of the key aspects of having an observability stack is the potential correlation of data and events happening across your network, systems, and applications. This is very important to quickly identify issues, and even predict when an issue might happen.
>
> For example, if you have an application that is experiencing high latency, you might want to know whether there are any network issues that could be causing this. Having metrics, logs, and general events correlated across your network, systems and applications can help you (and your AIOps assistant) to quickly identify the root cause of the issue.

Last but not least, an observability platform is a system in its own right that needs to be both scalable and reliable to be effective. As networks, systems, and applications expand, more monitoring targets will emerge. Particularly for applications tied to **Service-Level Agreements** (**SLAs**), which were discussed in *Chapter 2*, the reliability of the observability platform is vital to meeting commitments and maintaining operational efficiency.

In summary, a well-designed observability stack needs to provide the following:

- **Versatile data collection**: The ability to collect data from varied sources, such as network devices with unique data interfaces, applications, and other systems including compute and storage components.

- **Real-time data processing**: The ability to process and parse logs, aggregate metrics, and enrich data with additional context in real time.

- **Robust APIs and standard procedures for injecting and accessing data**: An optimally structured observability stack should provide powerful APIs that can be used to integrate with other systems and automated workflows as well as predefined methods to add new data sources into the system—overall, a framework to interact with the data.

- **Adaptive visualization**: Data should be not only accessible but also comprehensible. The platform should offer versatile visualization tools, catering to a range of formats, such as charts, graphs, and tables. More importantly, it should be adept at tailoring dashboards to diverse stakeholders—from granular details such as a specific network device's CPU usage to broader KPIs such as bandwidth utilization across a region.

- **Intelligent alerting**: Beyond just alerting users, the stack should prioritize relevance and timeliness. It should facilitate the intuitive creation and management of alerts, ensuring rapid responses to anomalies in applications, systems, or networks. This includes transcending basic trigger-based alerts to encompass advanced ones that, for instance, don't just rely on fixed thresholds but adjust based on historical data or recent trends.

Now, let's explore what it means to have a well-designed platform and how you can build one that aligns with your unique challenges and objectives.

What does it mean to be a well-designed platform?

So, what constitutes an effective observability platform? It's a matter of identifying the key components and cohesively integrating them. Here's an overview:

- **Modularity**: Think of the observability platform as a set of building blocks. The more modular it is, the easier it is to switch things around and add new pieces. This gives you the flexibility, extensibility, and scalability you need.

- **Source of truth**: In the realm of observability, clarity and consistency are essential. Your platform requires a definitive, trustworthy, and consistent source that serves as its backbone. This central repository not only identifies the critical targets for your observability but also enriches metrics and logs with contextual data, ensuring you get the full picture every time.

- **Distributed nature**: A robust observability platform should be distributed to guarantee scalability and reliability. This includes the following:

  - **Scalability**: The ability to scale up or down as needed, depending on the demands of the systems and the network.

  - **Reliability and fault tolerance**: Distributed systems are inherently redundant, enhancing reliability. If one node fails, others can take the tasks.

  - **Geographical distribution**: Deploying nodes or systems components across different physical locations improves resiliency (for instance, during a regional or site outage), and also makes possible closer data collection to remote target devices.

- **Data pipelines**: An integral yet often overlooked component, data pipelines serve as the circulatory system for your observability stack. They manage the **Extract, Transform, Load (ETL)** processes, providing the flexibility required for real-time data processing and versatile data collection. You will learn more in the *Understanding data pipelines for observability* section.

- **Orchestration and automation**: With all these moving parts, you'll need orchestration and automation to build, maintain, and operate the platform.

- **Security**: Keeping your observability platform secure is important. This means making sure that everything from how data is collected from devices to how it's shown on your dashboards is locked down and safe. It's all about making sure that both your data and the tools you use to view and manage that data are dependable and secure.

- **Cost considerations**: Keeping costs in mind is key when setting up your observability platform. There is no one-size-fits-all answer—it all depends on what your specific needs and goals are. We explore how to manage costs effectively while meeting your needs in the *Cost management* section later in this chapter.

In essence, a purpose-built observability platform functions like a cohesive unit, where each component plays its precise role. It must be modular, dependent on a reliable source of truth, efficiently distribute the workload, and coordinate all parts seamlessly. The outcome is a platform that's robust, scalable, adaptable, and prepared for the challenges ahead. However, achieving this end state is easier said than done, often requiring intricate data pipelines that manage your information flow effectively. Before we go into the challenges and best practices that come from real-world experience in building and operating observability systems, let's take a detour to discuss the critical role of data pipelines in the observability ecosystem.

Understanding data pipelines for observability

Having talked about the importance of a well-designed observability platform in the previous section, let's get into the architecture that powers it—the **data pipeline**. This framework not only gathers data but also refines and directs it where it's most needed, often in real time:

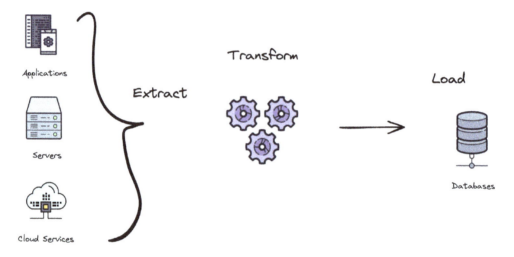

Figure 4.3 – Example of an ETL data pipeline (Created using the icons from https://icons8.com)

In the simplest terms, a data pipeline is a set of data-processing elements connected in series, where the output of one element is the input of the next. Within the realm of observability, data pipelines are the circulatory system that moves, filters, normalizes, and enriches data, allowing you to observe and make sense of your system's behavior and performance.

The versatility of data pipelines

Data pipelines can range from simple to complex. They could be as straightforward as a software agent running on a single machine or as intricate as a distributed system involving multiple specialized components.

Data pipelines are used in many areas, not just for observability. By looking at how other industries use this framework, we can learn more about them and improve how we use them for monitoring and observability:

- **Data warehousing**: Where data is extracted from multiple sources (e.g., CRM systems, financial databases, or marketing platforms) and formatted and loaded into a data warehouse for analysis and reporting.

- **Retail and e-commerce**: Data pipelines are used to aggregate data from various points in the retail chain, including inventory systems, online sales platforms, and customer databases. This data is then loaded and transformed into an analytical platform that helps with stock prediction, sales trends, and so on.

- **Finance and banking**: Data is pulled from trading platforms, customer databases, and historical financial reports to then be transformed and analyzed to detect fraudulent activities or market trends.

These components often specialize in either **Extract, Load, Transform (ELT)** or ETL functionalities. The difference between these two approaches lies, as you may have guessed, in the sequence of their operations. ELT is commonly employed with cloud-based data warehouses such as Google BigQuery, AWS Redshift, or Snowflake, where the data warehouse itself is sufficiently robust to handle data transformations. In contrast, when it comes to observability platforms and the ingestion of raw metrics and logs, ETL data pipelines are often the method of choice. We will dive deeper into ETL in the following section.

Unpacking ETL in data pipelines

ETL includes the three core stages of a data pipeline focused on collecting data from multiple sources, transforming it into a standardized format, and then loading it to a target database or data warehouse for analysis. In the realm of observability platforms, ETL pipelines are particularly used when ingesting metrics and logs data from your infrastructure and applications. Let's dive into each stage in more detail:

- **Extract**: This is the data collection stage, where raw data is gathered from various sources. For an observability data pipeline, this extraction can occur in multiple ways. Network devices may submit logs, or metrics might be pulled using SNMP and gNMI, or even fetched from external APIs.

- **Transform**: This is the stage where the raw, collected data undergoes a series of manipulations. These can include filtering out irrelevant information, normalizing the data, enriching it with contextual details, or possibly aggregating multiple data points for easier analysis. In the context of observability, this step often involves actions such as normalizing the format of interface counter metrics obtained through SNMP or gNMI and enriching them with additional context such as the interface's role or its circuit ID.

- **Load**: Finally, the cleaned and processed data is loaded into databases where it can be utilized for monitoring, analytics, or further processing. In observability platforms, these are typically time-series databases.

Here is a diagram illustrating the ETL process:

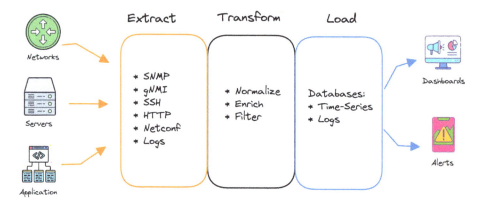

Figure 4.4 – Diagram of observability data pipeline illustrating the ETL process (Created using the icons from https://icons8.com)

Now that we've seen how a solid data pipeline is an essential piece in making your observability stack work well, it's clear that getting it right presents its challenges. These can come up in all sorts of areas, from gathering and handling data to setting up and growing your observability system. So, what's next? We're going to dig into these challenges and look at some best practices for overcoming them, based on real-world experience in building and operating these systems.

Challenges and best practices

Building a robust observability stack provides significant benefits. Yet, while implementing this system, organizations often grapple with challenges including scalability, reliability, flexibility, and cost management. In this section, we will explore these challenges, their implications, and some best practices to address them. We will conclude this section with a set of additional well-known tips to fortify your observability approach.

Scalability

Scalability isn't just about size; it's about how your observability platform evolves to accommodate growing needs. In smaller, more defined network environments, scalability might not be a pressing concern. But as networks expand and complexity escalates, the volume of data and the number of data collectors multiply.

Addressing scalability requires strategy and foresight. Here are some pointers to navigate this challenge:

- **Prioritize data collection**: Recognize what data is essential. While interface traffic might require high-resolution data, metrics such as CPU or memory usage might be best collected at broader intervals. Making these decisions eases the data management process.

- **Embrace automation**: With an expanding network or system, your observability platform's target base also broadens, making data collection more intricate. Here, automation isn't just useful; it's indispensable.

- **Opt for horizontal scaling**: There's a limit to how much you can boost performance by merely adding more CPU or merely adding more CPU or memory resources, that is, through vertical scaling (`https://en.wikipedia.org/wiki/Scalability#Vertical_or_scale_up`). Consider situations such as backend databases for your observability platform, which often fare better with horizontal scaling (`https://en.wikipedia.org/wiki/Scalability#Horizontal_or_scale_out`)—distributing data across multiple resources.

Reliability

A reliable observability platform isn't just a benefit—it's a necessity. The trust and satisfaction of your users hinge on it. So, how can you boost the reliability of your platform? Here are several approaches:

- **Adopt active/standby methodology for some observability components**: Some systems operate with a single active component at any given moment. Maintaining a standby system alongside the active one offers resilience. Should any issues arise with the primary system, the backup can seamlessly take over. Imagine you're using central data collectors, such as **Telegraf** (`https://www.influxdata.com/time-series-platform/telegraf/`), to monitor a vast array of network devices. Ideally, you'd have one collector active at any moment with a backup on standby, ready to step in as required. After all, certain network devices might not respond favorably to multiple collectors simultaneously requesting data.

- **For other components, adopt an active/active and load-balancing methodology**: Let multiple systems handle tasks simultaneously, distributing the load evenly. As an example, think about log collection agents, for example, **Logstash** (`https://www.elastic.co/logstash`). Instead of having only one system active at a time, you can have multiple sitting behind a load balancer for improved reliability.

- **Explore cluster-based systems**: Some applications embrace a cluster design, where each component contributes to the system's reliability. Many observability platform databases use this model, a combination of horizontal and vertical scaling, ensuring data is always available and operations remain uninterrupted.

- **Leverage infrastructure orchestration**: Tools such as **Kubernetes** (`https://kubernetes.io/`) are invaluable. They detect system issues and promptly deploy solutions, such as using replicas to guarantee continued operation.

These strategies can be tailored to fit different aspects of your observability stack. We'll dive deeper into these components in the next section.

Now remember, chasing 100% reliability from day one isn't necessary. If you're just launching a proof of concept or a **Minimal Viable Product** (**MVP**) for an open source telemetry and observability stack, focus on the essential components. Avoid the trap of over-architecting. Start by understanding the value and capabilities of the platform. As your needs evolve, let the reliability of your platform scale accordingly. After all, most open source tools in this space are crafted with scalability in mind.

Flexibility, extensibility, and customization

Managing diverse data collection techniques—be it SNMP or gNMI for network devices, scraping metrics from servers, or conducting HTTP health checks—is no small feat. However, the need for flexibility and extensibility extends beyond just data collection. It's equally fundamental in data processing, visibility, and alerting.

Data collection and retrieval

Having a flexible observability platform is paramount. It's not just about collecting data but collecting and processing it in a way that makes the most sense for your network, systems, and business needs. This flexibility can manifest in a few distinct ways:

- **Versatile data collection**: Whether you're dealing with SNMP, gNMI, or alternative collection methods, an adaptable platform accommodates them all. This doesn't entail relying on a singular, generic data collector agent for all your device types. At times, specialized tools, such as **gNMIc** (`https://gnmic.openconfig.net/`) for gNMI data, might be more apt. The key is identifying and utilizing the most effective tool for a given task, all while leveraging infrastructure as code tools and automation for ease of management. Reference *Chapters 3* and *5* for more on data collection.

- **Ease of data access and interaction**: Flexibility goes beyond just data collection. It also concerns how you access and engage with the gathered data. Whether you're constructing detailed dashboards or setting up alerts, the platform should allow for effortless interaction with data. Especially when integrating alerts with messaging systems such as **Microsoft Teams**, **Slack**, or **Mattermost** and **IT Service Management** (**ITSM**) platforms such as **ServiceNow**. This kind of flexibility really comes in handy when your organization goes through changes in the communications tools you are using. We will go into more detail about how to integrate observability platforms with external systems in *Part 3* of this book.

Extensibility and customization

An extensible observability platform grows with your business. It not only adapts to your present needs but also anticipates future ones. Here's how extensibility can play a critical role:

- **New alerting and incident systems**: As your operational needs evolve, you might find the need to integrate a new alert and incident system. A well-structured platform can smoothly accommodate such changes, either through inherent integrations or by scripting outputs to external channels, for example, using webhooks and templates. Stay tuned for *Chapter 9* for an in-depth look at alerting systems.

- **Advanced data processing**: Data often presents itself in varied forms. For instance, when you're collating information such as used and available storage for a network device, you may need to apply specific multipliers or thresholds and perform some math operations. It's essential that your data collection mechanism either has the capability for these advanced manipulations or that there's a layer dedicated to such processing. In *Chapter 6*, we dive into data normalization and enrichment layers describing this process, and in *Chapter 7*, we explore data aggregation and operations on top of observability data.

- **Integration with AI/ML**: With the rapid advancements in AI and **Machine Learning** (**ML**), any robust observability platform should be able to integrate with these technologies. This is where the combination of flexibility and extensibility shines. Leveraging robust APIs, platforms can seamlessly integrate AI and ML functionalities, supercharging data analysis and insights. *Chapter 13* will dive into the intricacies of incorporating AI and ML into your observability stack.

A platform's ability to adapt and extend its features is not a luxury—it's a necessity for future growth.

Cost management

Let's not forget cost management. Implementing an observability stack isn't just a technical challenge—it's a financial one, too. The more data you collect, the more you'll spend storing and analyzing it. But with a strategic approach and by applying some of the best practices we are covering in this section, you can keep your budget in check. Remember, smarter data collection and usage doesn't just enhance performance—it's also a prudent strategy for mitigating costs.

The true worth of an observability system boils down to the value it brings to your company. Sure, most companies want to keep tabs on network reliability, but let's consider some specifics. If you're a high-frequency trading firm, network latency becomes your spotlight concern. On the other hand, a content delivery network or service provider may need a real-time snapshot of the site or regional status before tweaking the network, or perhaps they even want to keep an eye on the company's stock price.

So, while there's definitely a shared core of observability needs across businesses, the specifics can vary. Different enterprises might need to collect and store different types of data (for example, metrics versus flow data), based on what's most essential to their operations. Keep this in mind when designing your observability stack—it's all about aligning with your unique business priorities:

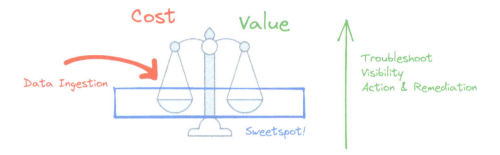

Figure 4.5 – Cost management balancing scale (Created using the icons from https://icons8.com)

We will explore this topic further in *Chapter 10*, where we discuss real-world observability platforms. Specifically, we will examine the **build versus buy debate** and the factors to consider if you decide to build your own system. Next, let's move on to some additional tips and best practices.

Other tips and best practices

Beyond tackling these typical hurdles, there are a number of best practices that can amp up the efficiency and effectiveness of your observability stack. Here are some key strategies to consider:

- **Knowing what to alert on**: Monitoring everything can cause unwanted noise for your engineers. Instead, focus on the essential data points, and review them regularly (even working on a spreadsheet and keeping track of the metrics you collect, or plan to, can prove beneficial). This way, you'll have information that matters, saving time and storage and keeping your team focused.

- **SNMP is still very much alive**: It's important to realize that SNMP is widely used in the industry and tracks a lot of different metrics from various devices, networks, and services. While streaming telemetry can collect more data from devices, it's important to note that not all vendors or their devices are set up to use streaming telemetry yet.

- **Start small with your platform**: You don't have to launch a full-fledged **high availability/ disaster recovery (HA/DR)** observability platform on day one. Use what is at your disposal and grow from there. One of the benefits of having a modularized approach for the observability platform is that you can grow horizontally or vertically depending on your environment. This strategy not only offers flexibility but also provides a practical pathway for evaluating both open source solutions and proprietary observability providers. It's a smart way to test, learn, and expand in alignment with your specific requirements and goals.

- **Infrastructure orchestration – consider automation a priority**: This means building and running each component of the observability platform manually. You could do that, but it might lead you into a complex and unmanageable situation. Instead, why not leverage some well-known tools that might already be in use at your organization? If you're a VMware (`https://www.vmware.com/`) shop, for example, Ansible might be your best path for construction and maintenance to build and maintain software in **Virtual Machines** (**VMs**). If containers are more your style, take a look at tools such as the **HashiCorp** (`https://www.hashicorp.com/`) stack, **Kubernetes**, or **OpenShift** (`https://docs.openshift.com/`). The key is to find what fits your organization's needs and can help you grow, all while maintaining a manageable and efficient system.

- **Consider the northbound interface – your observability platform API**: Developers and general users must be able to tap into your observability platform with APIs that are not only robust but human-friendly. This isn't just about accessing data; it's about querying, aggregating, and performing operations such as sums and deltas on this aggregated data. Imagine needing to compare current data with figures from a week ago or being able to check the overall throughput of a specific service provider uplinks. A well-designed API can turn these operations into routine tasks. Also among these benefits, we have the following:

  - Ease of integration with external tools from your organization (e.g., ITSM or ChatOps).

  - Enables faster and more efficient development for your network automation needs and starts tapping into **closed-loop automation**.

- **Embrace Agile development methodologies**: Utilizing an Agile methodology in the development or operation of an observability platform aligns well with the dynamic and complex environments of today. This approach cultivates a culture that values continuous improvement and adaptability—cornerstones for successful observability initiatives. By touching base with customers or stakeholders regularly, we can better understand what data to keep or let go of, optimizing the platform.

These best practices are gleaned from the experience of building and operating on open source observability platforms. It's essential to recognize the value of your operational data and thoughtfully weigh the build versus buy approach in this domain.

But remember, an observability platform is more than a mere assembly of infrastructure and technology; it's about creating a system that keeps your users engaged and satisfied. Close collaboration, active feedback loops, and a deep understanding of user workflows are critical to delivering real value through the platform you're constructing or providing.

In the end, the success of an observability platform doesn't solely depend on the technology itself, but on how well it integrates with the people and processes it serves. Keep the human aspect at the core of your strategy, and you'll be better positioned to create a platform that genuinely supports and enhances your organization's operations. Next, we'll explore the specific components of an observability stack and examine how they collectively contribute to this overarching goal.

Let's take the best practice of *starting small with your platform* that we highlighted earlier and dive into building a proof of concept. By doing this, we will not only shed light on the ideas in the upcoming sections but also transition from theoretical insights to hands-on, real-world applications. Ready to turn our discussions into tangible results? Let's get started!

Setting up a lab environment

The rest of the book is going to walk you through the stack's components, and how to build proof-of-concept observability solutions on top. Aside from the theoretical content, you will get concrete examples that can be reproduced in a lab environment to let you get hands-on. This tangible and practical environment allows you to experiment, understand, and solidify the concepts as they are introduced.

The following diagram illustrates the overall elements involved in the lab setup:

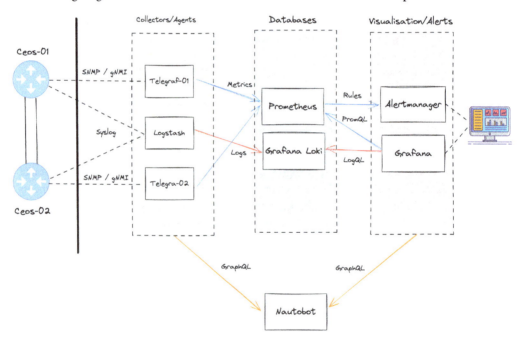

Figure 4.6 – Lab environment (Created using the icons from https://icons8.com)

The lab environment consists of the following components:

- **Network devices**: These are the devices that will be monitored and managed by the observability stack. In this case, we will be using Arista **containerized EOS** (**cEOS**) devices, but the same principles apply to other vendors as well.

- **Data collectors or agents**: These are the tools that will collect metrics, logs, and synthetic checks on the network devices. In this case, we will be using Telegraf, a popular open source data collector for fetching metrics via SNMP, gNMI, and even via the RESTCONF API of the network devices. There is also Logstash, another popular open source data collector, which will listen and parse the syslog messages sent from the network devices. Both of these topics get more attention in *Chapters 5* and *6*, where we will go into their configuration and capabilities.

- **Databases**: We will be using **Prometheus** (`https://prometheus.io/`) as the time-series database, for storing the metrics collected and processed by the Telegraf components. **Grafana Loki** (`https://grafana.com/oss/loki/`) will be used to store the parsed messages from the Logstash components. These databases not only house your data but also offer APIs and query languages for further analysis. We will go under the hood with these databases in *Chapter 7*, diving into their configurations, APIs, and best practices.

- **Visualization**: Visualization tools are used to display the data in an easy-to-understand format, often as charts, graphs, or tables. In this case, we will be using **Grafana** (`https://grafana.com/grafana/`), a popular open source visualization tool that provides means to connect to databases such as Prometheus and Loki. In *Chapter 8*, we'll go into the intricacies of setting up effective visualizations.

- **Alerting**: Alerting mechanisms are vital for notifying engineers about critical issues that need their immediate attention. In this case, we will be using **Alertmanager** components from Prometheus. *Chapter 9* will provide a detailed look at how to set up and fine-tune your alerting systems to fit your specific needs.

- **Source of truth**: In any system or network, it's important to have a single, reliable, consistent, and capable source of truth that provides accurate, up-to-date information about the intended state of the network. In this case, we will be using **Nautobot** (`https://networktocode.com/nautobot/`), a popular **Network Source of Truth** (**NSOT**) system.

- **Orchestration**: Orchestration tools help manage the deployment, scaling, and operation of the various components in your observability stack. In this case, we will be using Docker (`https://www.docker.com/`) and `containerlab` (`https://containerlab.dev/`) wrapped under a Python tool called `netobs` to ease the overall lab management process.

For an in-depth explanation of how to set up your lab environment, please refer to the *Appendix A* section of this book.

Lab scenarios

The lab scenarios detailed in the *Appendix A* of this book present various stages of the observability stack. These stages are designed to provide you with the flexibility to explore, test, and learn about the different components that make up the stack. Whether you're delving into a specific area or seeking a broad understanding, these scenarios are tailored to guide your exploration:

Scenario	Description
Batteries-included lab	Delivers a comprehensive lab setup, inclusive of configs, alerts, dashboards, and a source of truth.
Chapter lab	Customized lab environment equipped with the necessary components and configurations for executing examples from a specific chapter.

Table 4.1 – Lab scenarios

Feel free to select any chapter and run the lab environment focused on the specific component discussed in that chapter. Alternatively, if you are interested in experiencing a fully operational lab environment that includes data collection, normalization, dashboards, and alerts on a lab network, you are encouraged to try the comprehensive **batteries-included** lab.

Summary

In this chapter, we have explored the essential role that a well-designed observability platform plays in today's complex IT environments, particularly in large-scale, distributed networks. A central aspect of this platform is the ETL data pipeline, which acts as the backbone for ingesting, processing, and storing your metrics and logs.

Venturing into the realm of an observability platform is not without its challenges. From scalability and reliability to data integrity and cost, building and operating such a system is a multifaceted effort. We have outlined several best practices that can guide you through these hurdles, emphasizing the value of a reference architecture to provide a blueprint for all your observability needs.

To help you tie all these concepts together, we have included guidelines for setting up your lab environment. This lab will serve as a hands-on playground, allowing you to apply the principles and technologies discussed in this chapter, thereby setting the stage for more in-depth explorations in the rest of the book.

As we transition to *Chapter 5*, we will zoom in on the practicalities of data collection for metrics and logs. We will introduce you to tools such as Telegraf and Logstash that will play a pivotal role in your data collection efforts. Prepare to dive deep into the nuts and bolts of data collection!

5
Data Collectors

In this chapter, we'll pivot from the foundational technologies we explored in *Chapter 3* – such as SNMP, syslog, gNMI, synthetic monitoring, and others – toward integrating these technologies into telemetry and observability data collection tools. The objective is to collect and channel the data that's harvested through these technologies into your observability stack, enriching your monitoring and analytics capabilities.

To illustrate this integration, we'll delve into two prominent data collection tools: **Telegraf** and **Logstash**. Telegraf, coming from the InfluxData family, is well-known for handling network metric collection through SNMP, gNMI, and even data extraction over SSH/CLI. It is also well-adapted for synthetic monitoring. On the flip side, Logstash, a vital part of the Elastic Stack, excels in ingesting and processing log data, leveraging the syslog protocol to do so.

As we dive deeper into Telegraf and Logstash, we'll dissect their architectures and configuration specifics, providing hands-on examples so that you have a clear, practical understanding. The choice of these tools will be further justified in the subsequent sections, where their unique strengths and roles in boosting your network's observability framework will be outlined. Note that the same principles that are used for these tools could be generalized for other tools and their characteristics.

This chapter covers the following topics:

- A deep dive into data collectors
- A look into Telegraf
- A look into Logstash

A deep dive into data collectors

The first component in the observability architecture we're exploring is the Data Collection layer. Here, data collectors interface with the target infrastructure – routers, switches, load balancers, firewalls, servers, and storage appliances – using methods such as SNMP, gNMI, and syslog (as discussed in *Chapter 3*) to gather data on their health and operations:

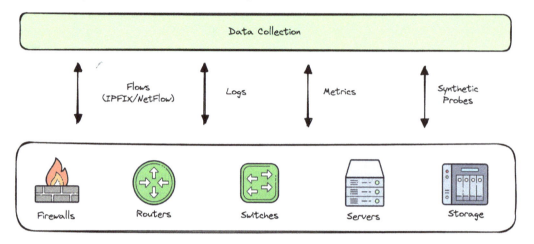

Figure 5.1 – The Data Collection layer (Created using the icons from https://icons8.com)

Let's review what we learned about data collectors in *Chapter 4*:

> *"Data collectors are the tools or processes that collect metrics, logs, and traces or perform synthetic checks across your system and network. Data collectors can be anything from agents installed on servers or network devices to systems that have been configured to scrape, pull, or listen to metrics forwarded to them."*

Note

Throughout this book, the terms *agent* and *data collector* are used interchangeably to refer to the tools facilitating our data collection efforts.

In this chapter, we will consider these technologies in practice. In particular, we will look at the following:

- Metric collection (using SNMP, gNMI, and CLI data parsing over SSH) with Telegraf

- Synthetic monitoring (using Ping) with Telegraf

- Log collection (using Syslog) with Logstash

Note

While this chapter focuses on implementing select methodologies from *Chapter 3* using Telegraf and Logstash, the approach outlined here can be applied to other data collection protocols not covered in this chapter.

Telegraf is a popular tool for data collection that's widely recognized for its diverse range of plugins that handle data fetching, processing, and exporting. In the world of networking, Telegraf stands out for its use of plugins such as SNMP and gNMI, which are essential for monitoring network devices. One of its most versatile features is its ability to customize the data collection process. Later in this chapter, we'll demonstrate this flexibility by showing you how to collect BGP data with a plugin that allows you to script interactions with network devices using SSH and their **command-line interfaces** (**CLIs**). This capability to adapt to various data collection scenarios is one of the main reasons we're highlighting Telegraf in this book.

Logstash is another data collection tool that is also well-known in the data processing arena, especially when it comes to wrangling syslog data. It's got a robust set of plugins for data collection, processing, and exporting, as well as what Logstash calls **pipelines**. One of its key features is persistent queues, which enhances reliability and resilience by storing in-flight messages on disk to prevent data loss in the event of an unexpected shutdown. This aspect is important considering the volatility of syslog data; if the backend (for example, a time series database) presents hiccups, Logstash keeps the data secure in its queue, preventing data loss for most input plugins.

We chose to focus on Telegraf and Logstash in this book because they are highly useful tools for network engineers:

- **Rich plugin selection**: Both tools offer a wide range of plugins that can gather data from various parts of your network infrastructure. For example, there are SNMP and gNMI plugins available to gather data using these protocols.

- **ETL capabilities**: They are capable of performing **extract, transform, and load** (ETL) operations, as detailed in *Chapter 4*. This involves collecting data, formatting it appropriately, and then storing it in a way that makes it accessible and useful for later use.

- **Proven use cases**: These tools are well-known in the infrastructure community for their specific strengths – Telegraf is great for collecting metrics, whereas Logstash stands out for its ability to gather and process log data.

The tools available at the Data Collection layer usually have a similar architecture to an ETL pipeline and present some key characteristics that make them an ideal agent for data collection.

Key characteristics

There are many tools available that are used for data collection, normalization, and enrichment in an observability stack. Many of these tools are designed to function as a *sidecar type of application*. This means they run alongside your main system or application, much like a motorcycle sidecar runs alongside a motorcycle. They are dedicated to collecting metrics and/or logs, working independently but closely with the main application to enhance its capabilities without affecting its primary functions.

Some of the common aspects or characteristics they share can be summarized as follows:

- **Lightweight**: it should guarantee a seamless, minimal load on the target system to ensure normal operation without interference. Here's why this is crucial:

  - **Minimal performance impact**: The agent should consume minimal system resources, such as CPU, memory, and network bandwidth and network bandwidth in order to not affect that target's primary functions.

  - **Efficient data handling**: While the agent collects valuable data, it should do so prudently. Storing an excessive amount of data in memory affects the memory consumption of the system, which is counterproductive since its role is to forward the data for further analysis and storage.

  - **Seamless operation**: Ideally, the agent's presence should go unnoticed. The users and processes of the target system should not feel the agent's impact.

- **Concurrency**: The data collection processes for various targets must operate simultaneously. In practical terms, this means you wouldn't want the data collection task for `router-02` to be queued up, waiting for the task of `router-01` to be completed if both are configured in the same agent. Instead, these tasks should run in parallel, saving time and making the data available more promptly. In situations where the agents lack inherent concurrency features, creative solutions at the infrastructure orchestration level can compensate for this. For instance, you could assign one agent to `router-01` and another to `router-02`, ensuring independent and concurrent data collection.

- **Polling with intervals**: In the realm of data collectors, especially for those not configured for streaming telemetry, the ability to specify the intervals at which data is pulled from target devices is important. This remains true even in scenarios when you are using gNMI with a subscription mode set to *sample*; there is still a need to specify the interval at which data will be sent from the target device. Therefore, having a collector that is always active and retrieves data at certain predefined intervals is a fundamental feature.

- **Modularity and extensibility**: This is the ability to extend the functionality with plugins or modules. Their modular nature is what makes them versatile and adaptable to many different environments and data sources.

Among the tools in this category, you will find gNMIc, Telegraf, Logstash, Beats (from the Elastic Stack), and Fluentd. Next, we will take a deeper look into Telegraf for network metrics collection.

A look into Telegraf

In this section, we'll delve into Telegraf, an essential tool in our data collection toolkit. This open source agent, brought to us by InfluxData, stands out for its modular design, providing a comprehensive platform for not just data collection but also for its processing and forwarding capabilities.

Built using **Golang**, Telegraf's plugin architecture allows it to have an ETL data pipeline. You can extend its capabilities through a range of existing plugins, many of which are also written in Go. However, if Go isn't your preferred language, you have the option to use `exec`/`execd` plugins that can run scripts in other languages (that is, Python, Perl, and Bash), offering you some flexibility in terms of development.

By the end of this section, you'll understand why Telegraf is such a powerful asset in the realm of observability that's capable of meeting diverse data collection and processing needs.

Telegraf architecture

First, let's talk about its architecture:

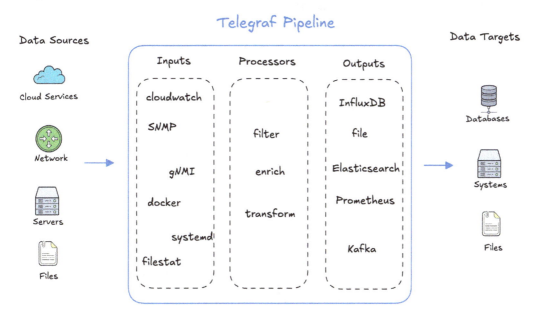

Figure 5.2 – Telegraf ETL diagram (Created using the icons from https://icons8.com)

Telegraf operates on an ETL architecture that can fit into various parts of your data pipeline. Each component in this architecture is implemented via plugins, making the tool flexible and adaptable to different environments:

- **Input plugins (extract phase)**: The extract phase is facilitated by the input plugins. These plugins collect data from various sources – from databases and hardware metrics to diverse APIs. For a more comprehensive understanding of data collection, *Chapter 3* is your go-to resource. In this chapter, however, the spotlight is on the mechanisms of extracting data from network devices via protocols such as SNMP and gNMI, as well as through their CLI, not to mention conducting synthetic monitoring with the aid of ping probes.

- **Processor and aggregator plugins (transform phase)**: Data transformation is executed through processor and aggregator plugins. Processor plugins can manipulate the data inline, while aggregator plugins can emit new aggregate metrics based on the metrics that have been collected by the input plugins. For example, they can emit the count or the mean of a set of values for a certain period. We'll dive into processing to normalize and enrich metrics data in *Chapter 6*.

- **Output plugins (load phase)**: The load phase is carried out by output plugins, which are responsible for sending the processed data to its final destination, such as a time series database or a file.

> **Note**
>
> There is no reason to use all types of plugins (inputs, processors, aggregators, and outputs) in your Telegraf setup. Feel free to select only what is essential for your specific needs. For a full list of available plugins, please refer to the Telegraf plugins directory (`https://docs.influxdata.com/telegraf/v1/plugins/`).

This flexible architecture allows you to adapt it to many different use cases and pick the right component in each case. Thus, the next question to solve is, *How do we configure Telegraf to collect network device metrics?* The answer comes in the form of a configuration file. We'll analyze this next.

Telegraf configuration

Telegraf reads its instructions from a configuration file written in **Tom's Obvious Minimal Language (TOML)** syntax. The syntax has grown in popularity mainly due to its clarity and human-readability. You can read more about it here: `https://toml.io/en/`.

Let's break down the core components of the Telegraf configuration:

```
1    [global_tags]
2      dc = "us-west-1"
3      env = "production"
4
5    [agent]
6      interval = "10s"
7
8    [[inputs.net_response]]
9      protocol = "tcp"
10     address = "ceos-01:22"
11
12   [[outputs.file]]
13     files = ["stdout"]
14
15
16   [[outputs.prometheus_client]]
17     listen = ":9077"
18     metric_version = 2
```

Global Tags

Agent Settings

Plugins

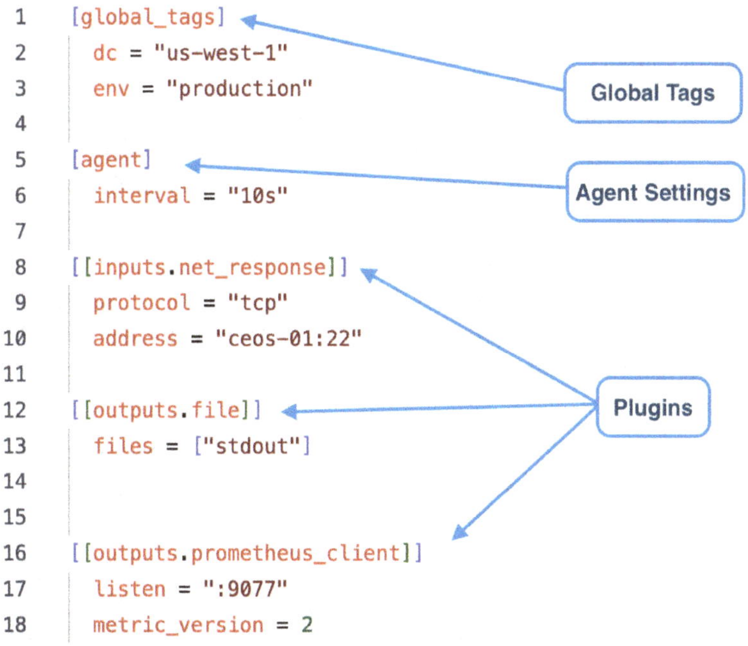

Figure 5.3 – Telegraf configuration taxonomy

Naturally, the Telegraf plugins (that is, input, processors, aggregators, and output) will have their place in the configuration file, with a section that follows the `[[<plugin_type>.<plugin_name>]]` naming convention and contains specific parameters. We will dive deeper into them in the subsequent sections while exploring practical examples.

However, there are still two sections of the Telegraf configuration that we haven't explored yet:

- **Global tags**: These consist of key-value pairs that are attached to every metric the Telegraf agent collects.

- **Agent settings**: These encompass the configurations for the agent's behavior. For instance, this could set the default interval for metric collection across plugins, adjust logging to the debug level, determine the flush interval for all outputs, or even designate the agent's hostname when generating a metric – a recommended practice if the agent operates within a container as the default identifier would otherwise be the container ID.

For more information about the Telegraf configuration file, please see their GitHub page: `https://github.com/influxdata/telegraf/blob/master/docs/CONFIGURATION.md`.

Configuration tips

The Telegraf configuration is the skeleton of how you're going to monitor your devices. The following are some aspects you must consider:

- According to the TOML syntax, each configuration parameter falls within a specific *section*. Notably, even the Telegraf agent settings reside under the `agent` section. *Always be mindful of the section you're in* when you're adjusting or setting configuration parameters.

- *Pay close attention to the usage of single and double brackets* in the sections. A single bracket indicates a unique section (meaning duplicates are not permitted), such as the `[global_tags]` section shown earlier. Conversely, sections with double brackets, such as `[[inputs.net_response]]`, can appear multiple times. In programming terminology, think of a single bracket as representing a table or `dictionary`, while double brackets symbolize an array, or `list`, of such tables.

- The `inputs` and `outputs` plugins run in parallel in Telegraf, whereas `processor` plugins are sequential. So, *order matters for processor plugins* when you're defining them in the configuration file.

- *Telegraf has default values* for its configuration of the agent and plugins; anything that's not explicitly set will be set to a default value. You can find more information about their defaults in each plugin's documentation. For example, the default values for the Telegraf SNMP inputs plugin can be found in the following README file: `https://github.com/influxdata/telegraf/blob/master/plugins/inputs/snmp/README.md`.

Now, it's time to start with some examples! We'll start by monitoring our network devices using SNMP. But to be able to do this, we must prepare our lab environment.

Setting up the lab environment

This lab will show you how to gather metrics and logs from network devices using observability tools. *Figure 5.4* shows the components we'll be using in our lab:

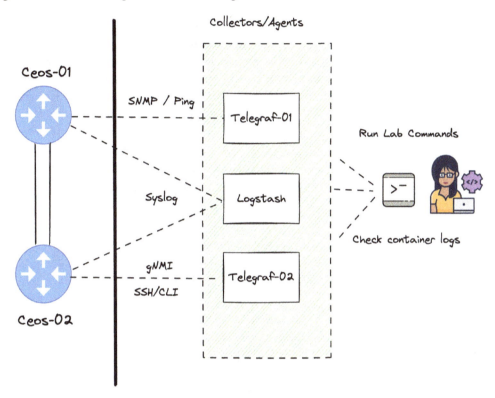

Figure 5.4 – Chapter 5 lab topology (Created using the icons from https://icons8.com)

In this lab, we're going to focus on gathering metrics from two devices, ceos-01 and ceos-02, by setting up and using Telegraf instances. Here's how we're going to do it:

- The telegraf-01 agent will collect metrics from ceos-01 using SNMP and will also perform and report ping results

- The telegraf-02 agent will gather metrics from ceos-02 using gNMI and will also collect data through CLI over SSH

Later in this lab, we will shift our focus to collecting logs. There, we will set up a Logstash instance to listen for and report any syslog messages from both ceos-01 and ceos-02.

The lab environment is designed to provide a flexible and valuable platform for learning and experimentation. Here's an overview of its architecture:

- **Docker containers**: Each lab component is deployed as a Docker container (`https://www.docker.com/resources/what-container/`). Docker packages applications and their dependencies into containers that can run consistently across different environments. Each service in the lab, such as Telegraf and Logstash, is configured with volumes to persist data and map their configuration files. These services have an entry point or command to initiate execution and perform their specific functions, such as collecting SNMP metrics or listening to syslog events. These services are managed by a Docker Compose file specific to each chapter's scenario and help define and run these multi-container Docker applications. If you want to learn more about Docker and Docker Compose, please visit their official documentation (`https://docs.docker.com/`).

- **Containerlab**: To manage the network device portion of the lab environment, we'll be using Containerlab (`https://containerlab.dev/`). This tool facilitates the creation and orchestration of networking devices using containers, making it simpler to integrate with the observability stack.

- **Hosting machine**: The lab environment is hosted on Linux-based machines, which are well-suited for running services such as Docker and Containerlab. These machines can be hosted locally (on your personal computer) or in various cloud environments, such as AWS, Azure, Google Cloud, or DigitalOcean. For more information about installing the services on the host machine, or for an example of setting up a host machine in DigitalOcean, please refer to the *Appendix A*.

- **Forked git repository**: All the configuration pieces for this lab environment are hosted in the network observability repository (`https://github.com/network-observability/network-observability-lab`). By forking this repository, you can create and update the configurations for the services in your lab scenario.

- **netobs tool**: To simplify the process of managing services, including implementing configuration updates and setting up/tearing down lab scenarios, we've developed the `netobs` tool. This tool streamlines the process of running the examples in this book, allowing you to focus on learning rather than managing infrastructure.

To get started with the lab in this chapter, you'll need to set up your lab machine, clone your forked repository, and install the `netobs` tool. Please refer to the *Appendix A* for step-by-step instructions. The *Appendix A* will guide you through preparing your lab machine, whether you're hosting it locally or deploying it on DigitalOcean, and installing the necessary tools to ensure everything is set up correctly.

Once your setup is ready and you're connected to your lab machine via a terminal application or SSH, navigate to the root location of your forked repository by typing `cd ~/<path-to-git-project>/network-observability-lab/` in your terminal. Replace `<path-to-git-project>` with the actual path where you cloned your fork of the repository. This will set you up at the root location of the Git repository, so that you can start working with this lab's materials.

The setup for this chapter's lab is located in the `~/<path-to-git-project>/network-observability-lab/chapters/ch5` folder. In this folder, you will find the Docker Compose file that sets up all the services for this chapter. Additionally, you will find folders containing the configuration files for each service. Please refer to the following figure for guidance:

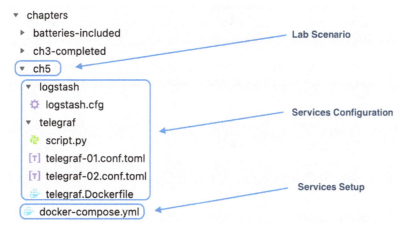

Figure 5.5 – Chapter 5 lab configuration layout

Now, let's initiate the lab scenario for this chapter. On your lab machine, run the `netobs lab prepare --scenario ch5` command to set everything up. Here is a part of what you should expect to see as output:

```
# Omitted rest of the terminal output ...
 ✓ Container netobs-
telegraf-02-1    Started                                          0.1s
 ✓ Container netobs-
logstash-1       Started                                          0.1s
 ✓ Container netobs-
telegraf-01-1    Started                                          0.1s
[21:16:25] Successfully ran: start stack
          Lab environment prepared for scenario: ch5
```

When you run this command, you may notice that the services encounter errors and shut down. This is expected since they haven't been configured yet.

> **Note**
>
> This lab scenario includes a completed version that you can refer to. It is located in the GitHub repository you cloned, under `network-observability-lab/chapters/<chapter-number>-completed`. This folder contains all the configurations that will be used in the examples throughout this chapter, providing a comprehensive guide to ensure you have everything set up correctly.

Great! Now, let's dive into setting up Telegraf.

Telegraf SNMP input plugin

Telegraf includes an SNMP input plugin that utilizes polling from the protocol to collect metrics from SNMP agents on target devices. This plugin offers mechanisms to gather data both from **individual object identifiers** (**OIDs**) and complete SNMP tables.

The following figure shows a zoomed-in portion of `telegraf-01`. We are going to use this for SNMP data collection, as well to generate ping probes toward `ceos-01`:

Figure 5.6 – Telegraf-01 metric collection (Created using the icons from https://icons8.com)

For our current task, we'll configure `telegraf-01` so that it connects to `ceos-01` using SNMP version 2. We're going to collect the system's uptime information as well as the counters and statuses of the interfaces.

We will break down the configuration analysis into three sections: agent settings, input plugin, and output plugin. In this initial example, we will not explore any processor functionalities. These will be covered in *Chapter 6*, where we'll discuss metric data normalization in detail.

Agent settings

The following is an example configuration for the agent settings:

```
[agent]
# Set hostname (defaults to container ID)
hostname = "telegraf-01"
```

Here, we're specifying the hostname of the Telegraf instance statically as it performs data collection because, by default, the metrics are tagged with the container ID. We will cover the use of tags later in this chapter.

SNMP input plugin

The following is an example configuration for configuring the SNMP input plugin so that it can collect metrics from `ceos-01`:

```
[[inputs.snmp]]
  # SNMP targets
  agents = ["ceos-01"]
  # SNMP version (v2)
  version = 2
  # SNMP community string (see example.env)
  community = "${SNMP_COMMUNITY}"
  # Polling interval, timeout, retries
  interval = "60s"
  timeout = "10s"
  retries = 3
```

This initial configuration snippet sets up the SNMP plugin in Telegraf so that it can collect data from multiple targets. In this case, we have only selected `ceos-01`. It includes essential settings for communication, such as the SNMP version, community, polling intervals, timeouts, and retries.

You might have noticed that the SNMP community parameter uses the `${SNMP_COMMUNITY}` value. This approach is used because the parameter might contain sensitive information, and Telegraf's configuration can directly interpret this value from an environment variable. For more on environment variables, see `https://en.wikipedia.org/wiki/Environment_variable`.

Next, we must specify the metrics we want to collect from `ceos-01`:

```
# Uptime Metric (1)
[[inputs.snmp.field]]
  # Metric name override to "uptime"
  name = "uptime"
  # SNMP OID RFC1213-MIB::sysUpTime.0
  oid = "1.3.6.1.2.1.1.3.0"

# Interface Metrics (2)
[[inputs.snmp.table]]
  # Metrics name
  name = "interface"

# Retrieve specific field from the table
[[inputs.snmp.table.field]]
  # Field name override
  name = "name"
  # SNMP OID for Interface Name IF-MIB::ifDescr
```

```
  oid = "1.3.6.1.2.1.2.2.1.2"
  # Set Interface name as tag (3)
  is_tag = true

# Collect specific fields from table walk
[[inputs.snmp.table.field]]
  name = "in_octets"
  # SNMP OID IF-MIB::ifHCInOctets
  oid = "1.3.6.1.2.1.31.1.1.1.6"

[[inputs.snmp.table.field]]
  name = "out_octets"
  # SNMP OID IF-MIB::ifHCOutOctets
  oid = "1.3.6.1.2.1.31.1.1.1.10"
```

Let's break down the configuration snippet section:

- **Uptime metric** (`inputs.snmp.field`): This section configures an SNMP Get operation to fetch the system's uptime. The `oid` value specified as `RFC1213-MIB::sysUpTime.0` points to a standardized location in an SNMP-managed device that reports the system uptime. This metric is then labeled as `uptime` within Telegraf's data outputs.

- **Interface metrics** (`inputs.snmp.table`): This configuration sets up an SNMP GetNext operation, which is primarily used to collect sequences of data, such as lists or tables. In this case, it's focused on the interfaces table of the device. SNMP GetNext helps in fetching multiple entries; it's particularly useful for retrieving all data from a table without knowing all the specific OIDs. For those unfamiliar with SNMP operations such as Get and GetNext, a more detailed explanation is available in *Chapter 3*.

- **Interface name as a tag** (`inputs.snmp.table.field`): Here, the `ifDescr` OID from `IF-MIB`, which stands for **interface description**, is used to fetch the name of each interface. This name is then set as a `tag` value instead of a regular metric value. Tags in Telegraf are key-value pairs that serve to provide additional descriptive information about the metric, which can be extremely useful for filtering and querying data later. For instance, it might tag metrics with `name = Ethernet2`, associating this interface name with all related data that's been collected. This makes it easier to understand and analyze the metrics in the context of the specific network interface they pertain to.

> **Note**
>
> Normally, the `inputs.snmp.table` configuration in Telegraf performs a *walk* operation on the specified SNMP table, which means it retrieves all the entries within that table. However, when you specify `.field`, Telegraf is instructed to fetch only the data for that particular field. This selective data collection can help you focus on the most relevant metrics for your needs, as we saw in the previous configuration snippet.

Once we've set up how the data is collected, the next step is to determine how we can view and use this data. For this purpose, we need to configure output plugins in Telegraf. Output plugins are responsible for sending the data to a destination where it can be stored, visualized, or further analyzed, depending on your observability infrastructure setup.

File output plugin

The `outputs.file` plugin in Telegraf is particularly valuable for debugging purposes. It allows us to direct the metrics we have collected to the **standard output** (**stdout**) of our terminal. This capability is incredibly helpful when you want to verify and monitor the real-time data being collected without needing to store it permanently first. By observing the data directly as it flows, you can quickly spot any issues or confirm that everything is functioning as expected.

Here's an example of the configuration:

```
# Outputs metrics to stdout
[[outputs.file]]
  files = ["stdout"]
```

> **Note**
>
> In *Chapter 7*, we'll learn how to use the output plugin to store the data in a time series database.

Now, it's time to test out the configuration we've set up. First, take the Telegraf configuration snippets for `telegraf-01` and save them. You will place these in your forked repository at `network-observability-lab/chapters/ch5/telegraf/telegraf-01.conf.toml` on your lab machine.

After saving the configuration file, apply the changes to the `telegraf-01` Docker container in your environment by running the `netobs lab update telegraf-01 --scenario ch5` command at the root of the forked repository. This command will restart the service and apply your new configuration, ensuring that everything is set up correctly and running with the latest settings.

> **Note**
>
> Under the hood, the `netobs lab update <service>` command is stopping, removing, and starting `<service>` again. This ensures the changes have been applied to the container service.

Great! Now, let's see what's going on with our `telegraf-01` instance. To verify that it's collecting data as expected, we will check the logs. Use the `netobs docker logs telegraf-01 --follow --scenario ch5` command to view the logs in real time. This command allows you to follow the log output live, helping you monitor the data collection process directly:

```
# Omitted output...
netobs-telegraf-01-1  | 2023-10-09T15:57:42Z I! Starting Telegraf
1.23.2
```

```
netobs-telegraf-01-1  | 2023-10-09T15:57:42Z I! Loaded inputs: snmp
netobs-telegraf-01-1  | 2023-10-09T15:57:42Z I! Loaded aggregators:
netobs-telegraf-01-1  | 2023-10-09T15:57:42Z I! Loaded processors:
netobs-telegraf-01-1  | 2023-10-09T15:57:42Z I! Loaded outputs: file
netobs-telegraf-01-1  | 2023-10-09T15:57:42Z I! Tags enabled:
host=telegraf-01
netobs-telegraf-01-1  | 2023-10-09T15:57:42Z I! [agent] Config:
Interval:10s, Quiet:false, Hostname:"telegraf-01", Flush Interval:10s
netobs-telegraf-01-1  | snmp,agent_host=ceos-01,host=telegraf-01
uptime=8167240i 1696867080000000000
netobs-telegraf-01-1  | interface,agent_host=ceos-01,host=telegraf-
01,name=Management0 in_octets=2731203i,out_octets=2529183i
1696867080000000000
# Omitted output...
```

> **Note**
>
> Use *Ctrl* + *C* to quit the current terminal session so that you can watch the log's output.

Looking at the last two highlighted lines in the preceding code, you will see the metrics we are collecting from ceos-01. These metrics are presented in a specific format known as the **InfluxDB line protocol** format. We'll take a moment to break down and understand this format as it will help us better understand the metrics we'll gather in the upcoming sections.

InfluxDB line protocol

The **InfluxDB line protocol** is a text format that's used within Telegraf to represent all the data points of a metric. Let's check the following figure to break it down:

Figure 5.7 – Influx line protocol syntax

Let's take a closer look at the syntax:

- **Measurement**: This is the name of the measurement being collected. In this example, a BGP metric is being collected.

- **Tags**: These are key-value pairs, delimited by commas, that are used to better describe the metric. They are optional but usually hold information about the source of the metric and characteristics that distinguish it from others. In this example, we have the BGP neighbor address and ASN where this metric is collected from.

- **Fields**: These are *mandatory* key-value pairs that hold the name of the field for the measurement with its respective value. They are the values we wish to collect. They are separated from the tags by whitespace but are comma-delimited between multiple fieldsets.

- **Timestamp**: The time when the measurement was collected. This value is also separated by a whitespace.

Let's conduct a test and dissect the example shown with the `telegraf-01` `uptime` metric. We're receiving the following data:

```
snmp,agent_host=ceos-01,host=telegraf-01 uptime=8167240i
1696867080000000000
```

Now, let's break down this metric:

- The `snmp` value is the measurement's name

- The `agent_host=ceos-01` and `host=telegraf-01` values are the tags.

- The `uptime=8167240i` value is the time the device has been up. These values are represented as timeticks. In this case, they represent 4 days, 17 hours, and 26 minutes. Check out `https://mapmaking.fr/tick/` if you wish to do your own calculations.

- The timestamp value for the metric is `1696867080000000000` and is represented in nanoseconds.

Keep this reference handy for the upcoming examples and whenever you need a quick reminder about the syntax of the metrics. Moving forward with our setup, we'll configure `telegraf-01` so that we can conduct ping tests on `ceos-01`. This approach not only verifies that the device is reachable but also illustrates how Telegraf is reliable for synthetic monitoring.

Telegraf synthetic monitoring input plugins

Telegraf is equipped with a wide range of input plugins. Some of these plugins are used for synthetic monitoring, a technique that's used to simulate and test various network operations to ensure they're functioning correctly. This method helps predict potential failures before they affect users. Here are some of the notable plugins Telegraf offers for these purposes:

- `[http_response]`: This plugin crafts HTTP packets and provides the resulting HTTP code of the response and the overall latency of the requests. It can also match the response with a particular regular expression.

- `[net_response]`: This plugin crafts TCP and UDP packets and checks the connection state and response time.

- [ping]: The good old ICMP ping plugin executes the system ping command and reports the results. This is the one we are going to configure for our test.

- [dns_query]: This plugin collects DNS query times and their statuses. It is particularly useful for collecting the general metrics of internal DNS servers in an organization.

Remember that you also have the option to create your own synthetic monitoring plugins in Telegraf. You can do this by using input plugins such as exec and execd, which allow you to run scripts you've developed yourself. Alternatively, you might consider building your own Telegraf input plugin and tailoring it exactly to your monitoring needs.

> **Note**
>
> The following is a good blog post about writing your own Telegraf plugin: https://www.influxdata.com/blog/collecting-data-from-chess-com-writing-your-own-telegraf-plugin/. Give it a whirl!

Continuing with our telegraf-01 setup, let's configure it so that it uses the ping plugin to check the reachability of the device. We'll do this by performing a ping from telegraf-01 to ceos-01. Notice that this is only the input plugin configuration; the agent settings and output are missing because we added them in the previous section:

```
[[inputs.ping]]
# Interval time between ping commands
interval = "10s"
# Number of ping packets to send per interval
count = 3
# Time to wait between sending ping packets in seconds
ping_interval = 1.0
# Time to wait for a ping response in seconds
timeout = 5.0
# Hosts to send ping packets to
urls = ["ceos-01"]
```

Here, you can observe the settings for the ping operations. The interval between consecutive ping commands is set to 10 seconds. Each ping command sends out three probes, with a 1-second interval between each probe. If a response isn't received within 5 seconds, the ping is considered failed. Although you can specify multiple targets in the urls parameter, for this example, we have chosen to target only ceos-01.

Now, it's time to test the ping configuration. Follow these steps to save and apply the configuration:

1. Take the Telegraf configuration snippets for telegraf-01 and save them to network-observability-lab/chapters/ch5/telegraf/telegraf-01.conf.toml on your lab machine.

2. Apply the changes by running the `netobs lab update telegraf-01 --scenario ch5` command.

3. Check the logs in real time by running the `netobs docker logs telegraf-01 --follow --scenario ch5` command.

The following output shows an example of how the ping metric might appear in the Telegraf logs output:

```
# Omitted output...
netobs-telegraf-01-1  | ping,host=telegraf-01,url=ceos-01
ttl=64i,minimum_response_ms=0.117,maximum_response_ms=0.146,standard_
deviation_ms=0.012,result_code=0i,packets_received=3i,percent_
packet_loss=0,packets_transmitted=3i,average_response_ms=0.133
1697984672000000000
```

From the preceding output, we can see the packet loss percentage, the number of packets that have been received and transmitted, the average response time, and more. We can also see that our ping probes have been successful, as indicated by `result_code=0`. The following table provides a legend of the result codes taken from the `ping` plugin referenced previously:

Result Code	Meaning
0	Success
1	No such host
2	Ping error

Table 5.1 – Ping result codes

Great! Now, let's shift gears to metric collection. Here, we'll be using a streaming telemetry mechanism known as gNMI.

Telegraf gNMI input plugin

Telegraf allows you to collect telemetry data using the gNMI protocol thanks to its gNMI input plugin. We will use this plugin in our network lab to collect data about network interfaces from the ceos-02 device, specifically focusing on traffic usage and the interface's status. This data collection process will be carried out by `telegraf-02` using the gNMI protocol:

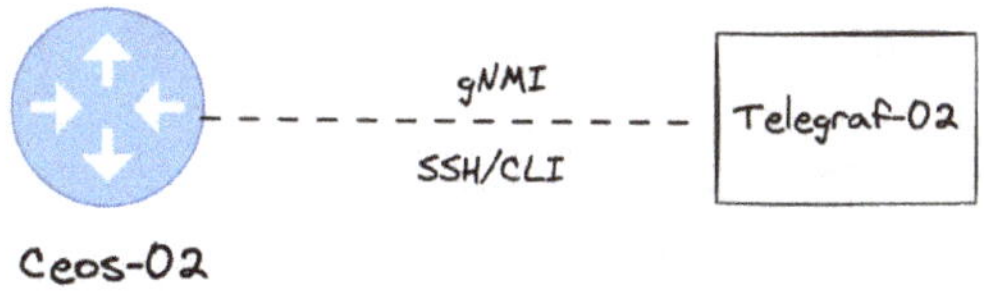

Figure 5.8 – Telegraf-02 metric collection (Created using the icons from https://icons8.com)

Here's an example of the configuration, which includes the agent settings and output plugin (the same as in the previous example), along with the gNMI input plugin:

```
[agent]
  # Set hostname (defaults to container ID)
  hostname = "telegraf-02"

[[inputs.gnmi]]
  # gNMI targets (port 50051 on ceos devices)
  addresses = ["ceos-02:50051"]
  # Credentials for device connection
  username = "${NETWORK_AGENT_USER}"
  password = "${NETWORK_AGENT_PASSWORD}"
  # Retries in case of failure
  redial = "20s"
```

This initial configuration snippet sets up the gNMI plugin in Telegraf so that it can collect data from multiple targets. In this example, we're focusing on ceos-02 and using port 50051, which is the gNMI port that's been configured on the device. The configuration includes the necessary settings for gNMI communication, such as the username and password values, as well as a redial parameter to attempt reconnection if any issues arise:

```
# gNMI subscriptions (1)
[[inputs.gnmi.subscription]]
  # Metric namespace name
  name = "interface"
  # YANG path for interface counters
  path = "/interfaces/interface/state/counters"
  # Subscription ("target_defined", "sample", "on_change") (2)
  subscription_mode = "sample"
  # Interval for each sample
  sample_interval = "10s"

[[inputs.gnmi.subscription]]
  name = "interface"
  # Path to collect interface oper-status
  path = "/interfaces/interface/state/oper-status"
  subscription_mode = "sample"
  sample_interval = "10s"

# Outputs metrics to stdout (3)
[[outputs.file]]
  files = ["stdout"]
```

Let's dive deeper into the gNMI subscription configuration:

- **gNMI subscriptions** (`inputs.gnmi.subscription`): The gNMI input plugin is configured with different subscriptions for `ceos-02`. In this example, we are targeting two endpoints specified by the `path` parameter. These endpoints are for interface counters and can be used to calculate data usage over time, as well as gather interface statuses.

- **Subscription method** (`subscription_mode`): The subscription method we're using is `sample`, so it collects a data sample every 10 seconds. There are other subscription methods available, such as `target_defined`, where the target device decides how often to send data, and `on_change`, where data is only sent if there is a change in the metric being collected. The plugin states it is based on the Openconfig gNMI specification (`https://github.com/openconfig/reference/blob/master/rpc/gnmi/gnmi-specification.md`).

- **Debugging metrics** (`outputs.file`): Similar to our configuration with `telegraf-01`, we want to debug the metrics we are collecting with gNMI. We're using the `outputs.file` plugin for this purpose. In our lab setup, this allows us to see the metrics being collected in real time through the `stdout` stream. This is helpful for verifying that the data collection process is working correctly.

Now, it's time to test the configuration for `telegraf-02`. Just as you did with the SNMP setup previously, follow these steps to save and apply the configuration:

1. Take the Telegraf configuration snippets for `telegraf-02` and save them to `network-observability-lab/chapters/ch5/telegraf/telegraf-02.conf.toml` on your lab machine.

2. Apply the changes by running the `netobs lab update telegraf-02 --scenario ch5` command.

3. Check the logs in real time by running the `netobs docker logs telegraf-02 --follow --scenario ch5` command.

The following example shows the logs from `telegraf-02` and the metrics it's collecting:

```
# Omitted output...
netobs-telegraf-02-1  | interface,host=telegraf-
02,name=Management0,path=/interfaces/interface/state/
counters,source=ceos-02 in_broadcast_pkts=0i,in_errors=0i,in_
fcs_errors=0i,in_multicast_pkts=0i,out_broadcast_pkts=0i,out_
discards=0i,out_errors=0i,out_multicast_pkts=0i 1696785460550368923
netobs-telegraf-02-1  | interface,host=telegraf-02,name=Management0,so
urce=ceos-02 in_discards=2987i 1696874972041962637
netobs-telegraf-02-1  | interface,host=telegraf-
02,name=Management0,path=/interfaces/interface/state/
counters,source=ceos-02 in_octets=2762060i,in_pkts=36672i,in_
unicast_pkts=36672i,out_octets=2565646i,out_pkts=30635i,out_unicast_
pkts=30635i 1696874999083365281
netobs-telegraf-02-1  | interface,host=telegraf-02,name=Ethernet1,sour
ce=ceos-02 oper_status="UP" 1696785464862879899
```

The first highlighted metric provides details about the interface counters for `Management 0`, while the second one specifies the operational status of the interface. Now, let's directly compare a metric from SNMP to one from gNMI to highlight their similarities and differences:

```
# SNMP collected metric for ceos-01 on interface Ethernet1
 interface,agent_host=ceos-01,host=telegraf-01,name=Ethernet1 in_
octets=1479325i,out_octets=0i 1696867080000000000
```

```
# gNMI collected metric for ceos-02 on interface Ethernet1
interface,host=telegraf-02,name=Ethernet1,path=/interfaces/
interface/state/counters,source=ceos-02 in_multicast_pkts=12042i,in_
octets=1518598i,in_pkts=13558i 1696874992046232160
```

You may have noticed that while metrics from both SNMP and gNMI plugins may differ in terms of their format, they essentially give the same message – for example, the network device, interface name, and the total inbound octets. We'll learn how to normalize these metrics in *Chapter 6*.

Having covered interface counters and status metrics using the gNMI protocol, let's turn our attention to collecting BGP metrics data from `ceos-02`. However, we will approach this differently by extracting the data directly from the CLI output of a BGP command, which provides rich information about BGP peers and states. To achieve this, we will write a script that's designed to fetch and parse this information effectively. Then, we'll integrate this script with Telegraf's `exec` plugin to streamline our data collection process in the observability stack.

Telegraf exec input plugins

The `exec` and `execd` input plugins in Telegraf allow us to use scripts or CLI applications to collect data, greatly expanding Telegraf's capabilities. In our network lab example, we want to collect BGP data over SSH using CLI commands. This approach is useful when traditional methods such as SNMP or gNMI don't provide the information we need.

Before we proceed, let's specify what BGP data we need to gather from the network devices. Specifically, it might be useful to collect metrics about the prefixes that are received or accepted from BGP neighbors, and perhaps also the BGP session state. We can obtain this information by running the `show ip bgp summary` command on `ceos-02`, like so:

```
> ssh netobs@ceos-02
(netobs@ceos-02) Password: netobs123
ceos-02>enable
ceos-02#show ip bgp summary
BGP summary information for VRF default
Router identifier 10.17.17.2, local AS number 65222
Neighbor Status Codes: m - Under maintenance
  Neighbor V AS   MsgRcvd   MsgSent  InQ OutQ Up/Down State PfxRcd
PfxAcc
```

```
    10.1.2.1 4 65111      5           5     0     0 00:01:11 Estab   1         1
    10.1.7.1 4 65111      0           0     0     0 00:01:12 Idle(NoIf)
ceos-02#
```

Executing the preceding CLI command provides information about the BGP neighbors, including their state and the number of prefixes received and accepted. By collecting this data programmatically and processing it through a parser that interprets this output, we can convert it into structured data. This structured data can then be manipulated to generate the specific metrics we need.

Let's use Python as our scripting language since it has some handy libraries, such as Netmiko and TextFSM, that we can use to parse the CLI output of the command, as we saw in *Chapter 3*. With them, we can collect and extract the necessary BGP metrics. To give you an idea of what type of structured data we can receive, take a look at the following data structure:

```
[
    {
        'router_id': '10.17.17.2',
        'local_as': '65222',
        'vrf': 'default',
        'description': '',
        'bgp_neigh': '10.1.2.1',
        'neigh_as': '65111',
        'msg_rcvd': '30',
        'msg_sent': '30',
        'in_queue': '0',
        'out_queue': '0',
        'up_down': '00:21:49',
        'state': 'Estab',
        'state_pfxrcd': '1',
        'state_pfxacc': '1'
    },
    # Omitted output...
]
```

First, let's begin by setting up a Python script named `script.py` in the Telegraf configuration directory for this chapter. You will find it at `network-observability-lab/chapters/ch5/telegraf/script.py`. Next, we'll write a function within this script that creates a Netmiko client. This client will handle sending commands to the network devices, allowing us to start gathering the necessary data:

```
import os
from netmiko import ConnectHandler

def netmiko_connect(device_type, host):
```

```
"""Connect to a device using Netmiko."""
# Define the device to connect to
device = {
    "device_type": device_type,
    "host": host,
    # Use environment variables for username and password
    "username": os.getenv("NETWORK_AGENT_USER"),
    "password": os.getenv("NETWORK_AGENT_PASSWORD"),
}

# Establish an SSH connection
net_connect = ConnectHandler(**device)

# Return the Netmiko connection object
return net_connect
```

The function will accept parameters for a Netmiko-compatible device_type and the target host address. Along with environment variables for the username and password values, these parameters are used to establish a connection object. This object allows us to interact with network devices directly. Additionally, we need to create another function that's designed to generate metrics formatted in the InfluxDB line protocol format, as discussed in the previous section. Here's what that function looks like:

```
def influx_line_protocol(measurement, tags, fields):
    """Generate an InfluxDB line protocol string."""

    # Construct tags string
    tags_string = ""
    for key, value in tags.items():
        if value is not None:
            if isinstance(value, str):
                value = value.replace(" ", r"\ ")
            tags_string += f",{key}={value}"

    # Construct fields string
    fields_string = ""
    for key, value in fields.items():
        if fields_string:
            fields_string += ","
        if isinstance(value, bool):
            fields_string += (
                f"{key}=true" if value else f"{key}=false"
            )
        elif isinstance(value, int):
```

```
            fields_string += f"{key}={value}i"
        elif isinstance(value, float):
            fields_string += f"{key}={value}"
        elif isinstance(value, str):
            value = value.replace(" ", r"\ ")
            fields_string += f'{key}="{value}"'
        else:
            fields_string += f"{key}={value}"

    return f"{measurement}{tags_string} {fields_string}"
```

The function we developed takes a `measurement` value or metric name, along with a list of `tags` and `fields`, and constructs a string in the InfluxDB line protocol format. This string represents the metric in a way that it can be readily utilized by Telegraf.

Following that, we need to create another small function that helps map the abbreviated state names from BGP session outputs – such as `Estab` from the `show ip bgp summary` command snippet we saw previously – to more descriptive names. In this example, we'll convert `Estab` into `ESTABLISHED` for clarity and consistency in our reports:

```
def convert_state(state):
    """Convert the state to a more readable format."""
    state_mapping = {
        "Estab": "ESTABLISHED",
        "Idle(NoIf)": "IDLE",
        "Idle": "IDLE",
        "Connect": "CONNECT",
        "Active": "ACTIVE",
        "opensent": "OPENSENT",
        "openconfirm": "OPENCONFIRM"
    }
    # Return mapped state or original state in uppercase
    return state_mapping.get(state, state.upper())
```

The `convert_state` function is designed to return a normalized state value from the various possible states that have been extracted from the output of our CLI command. This ensures consistency in how we interpret and display the BGP state data. With this function in place, we are now ready to develop the main function. In this central function, we bring together all the different functionalities we've developed:

```
def main(device_type, host):
    """BGP neighbor data in Influx line protocol format."""
    # Connect to device (1)
    net_connect = netmiko_connect(device_type, host)
```

```
# Execute command on device (2)
output = net_connect.send_command(
    "show ip bgp summary", use_textfsm=True
)

# Process BGP neighbors data (3)
for neighbor in output:
    measurement = "bgp"
    tags = {
        "neighbor": neighbor["bgp_neigh"],
        "neighbor_asn": neighbor["neigh_as"],
        "vrf": neighbor["vrf"],
        "device": host,
    }
    # Get mapped state (4)
    state = convert_state(neighbor["state"])
    fields = {
        "prefixes_received": neighbor["state_pfxrcd"],
        "prefixes_accepted": neighbor["state_pfxacc"],
        "neighbor_state": state,
    }
    # Generate line protocol string (5)
    line_protocol = influx_line_protocol(
        measurement, tags, fields
    )
    # Print line protocol string (6)
    print(line_protocol)
```

Let's take a closer look at each part of the main function:

- **Device connection**: The first step is to create the object to connect and interact with the network device. This is achieved through the netmiko_connect function, which takes device_type and host as parameters to handle the specifics of the connection. For our lab, these would be arista_eos and ceos-01/ceos-02, respectively.

- **Execute the command on the device**: Once connected, the function sends a command to retrieve the BGP summary information. The use of use_textfsm=True indicates that TextFSM is used to parse the command output into a structured format, which simplifies further processing.

- **Iterate over the BGP neighbors and process the data**: Then, the function iterates through each neighbor found in the BGP summary. For each neighbor, it constructs a set of tags (metadata such as neighbor IP, ASN, VRF, and device host) and prepares data fields (such as prefixes received, prefixes accepted, and the neighbor state).

- **Call the convert_state function**: This line involves a custom function, `convert_state`, which normalizes or maps raw state data (such as converting `Estab` into `ESTABLISHED`) to ensure consistency in how BGP session states are represented in the output.

- **Generate the line protocol string**: This step combines the measurement, tags, and fields into a single line protocol string. The `influx_line_protocol` function we developed is used to perform this task.

- **Print the line protocol string**: Finally, the function prints the formatted line protocol string, which can be used directly by Telegraf.

We now have the script logic ready to collect BGP data and convert it into metrics for our Telegraf instance to process. However, we still need to make this script executable and capable of accepting arguments via the CLI of our terminal. This will enable us to collect data from `ceos-02` dynamically. To achieve this, add the following code to your script:

```python
import sys
# <REST OF YOUR SCRIPT>
if __name__ == "__main__":
    # Get the device type and host from the command line
    device_type = sys.argv[1]
    host = sys.argv[2]
    # Run your script
    main(device_type, host)
```

> **Note**
>
> Remember, a complete example of this script is available in this book's GitHub repository. Be sure to check it out if you need a reference or further guidance.

Now that the script is ready, you can save it and even conduct a test to see how it works. For instance, navigate to the script's location by running the `cd chapters/ch5/telegraf` command. Then, execute the script with the `python script.py arista_eos ceos-02` command. The output should look similar to the following:

```
bgp,neighbor=10.1.2.1,neighbor_asn=65111,vrf=default,device=ceos-02
prefixes_received=1,prefixes_accepted=1,neighbor_state=ESTABLISHED
bgp,neighbor=10.1.7.1,neighbor_asn=65111,vrf=default,device=ceos-02
prefixes_received=,prefixes_accepted=,neighbor_state=IDLE
```

Now, it's time to configure Telegraf so that it can run this script and generate the BGP metrics. To do this, use the following configuration snippet for the `inputs.exec` plugin:

```
# Read metrics from commands outputting to stdout
[[inputs.exec]]
  # Commands executing the python script
```

```
commands = [
  "python /etc/telegraf/script.py arista_eos ceos-02",
]
# Interval between script executions
interval = "30s"
# Timeout for each command
timeout = "5s"
# Data format
data_format = "influx"
```

> **Note**
>
> The script's location inside the Telegraf container corresponds to a local reference that's established through a Docker volume mount. This setup is detailed in the `docker-compose.yml` file for this chapter, located at `network-observability-lab/chapters/ch5/docker-compose.yml`.

The `input.exec` plugin executes the commands or scripts listed under the `commands` array and expects them to return the collected data to `stdout`. The `commands` parameter allows you to specify the necessary arguments for running the script. This setup should be familiar from our earlier exercises. In this instance, we're passing the device type and host address as arguments to initiate the script.

Now, it's time to test the configuration for `telegraf-02`. Just as you did with the SNMP setup previously, follow these steps to save and apply the configuration:

1. Take the Telegraf configuration snippets for `telegraf-02` and save them to `network-observability-lab/chapters/ch5/telegraf/telegraf-02.conf.toml` on your lab machine.

2. Make sure you're at the root of your forked repository and apply the changes by running the `netobs lab update telegraf-02 --scenario ch5` command.

3. Check the logs in real time by running the `netobs docker logs telegraf-02 --follow --scenario ch5 | grep bgp` command:

   ```
   # Omitted output...
   netobs-telegraf-02-1  | 2023-10-22T23:25:42Z I! Loading config:
   /etc/telegraf/telegraf.conf
   netobs-telegraf-02-1  | 2023-10-22T23:25:42Z I! Loaded inputs:
   exec gnmi ping
   # Omitted output...
   netobs-telegraf-02-1  | bgp,host=telegraf-
   02,neighbor=10.1.2.1,neighbor_asn=65111,vrf=default prefixes_
   received=1,prefixes_accepted=1 1698017853000000000
   ```

Here, we can see the metrics of BGP prefixes that have been received and accepted from a BGP neighbor of `ceos-02`. This metric can now be passed through the entire Telegraf pipeline.

In conclusion, we have set up `telegraf-01` using SNMP so that it can collect device uptime and interface counters. We also configured it to send ping probes to `ceos-01`, while `telegraf-02` employs gNMI to pull interface data and statuses and uses a custom script to collect BGP metrics information over SSH and CLI. Great job so far! In real-world setups, you would likely configure Telegraf instances with similar configurations, using diverse collection methods to extract only the necessary metrics data.

With metric collection in good shape, it's time to pivot and dive into logs with Logstash.

A look into Logstash

Logstash is a key part of the **Elastic Stack**, also known as the **ELK Stack**. The Elastic Stack, developed by Elastic (`https://www.elastic.co`), is famous for its Elasticsearch search engine, which helps you search, analyze, and visualize large amounts of data quickly. The *stack* includes **Elasticsearch**, **Logstash**, **Kibana**, and **Beats**, each with its own role in handling data.

Logstash is designed for collecting, transforming, and loading data, and it's particularly well-known for processing log events. It uses `inputs`, `filters`, and `outputs` to gather data from different sources, process it, and then send it to various destinations. Let's take a closer look at its overall architecture.

Logstash architecture

Logstash's architecture consists of what is called a **pipeline**. A **Logstash pipeline** is made up of a series of plugins that work together to handle data: input plugins to extract data, filter plugins to transform it, and output plugins to load it. Take a look at the following figure:

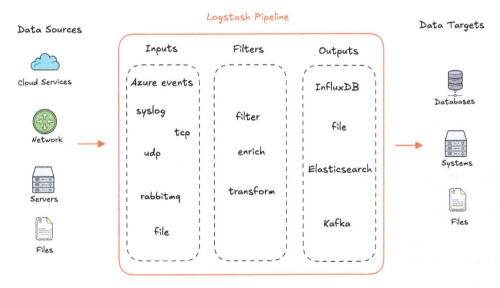

Figure 5.9 – Logstash data pipeline architecture (Created using the icons from https://icons8.com)

Here are a few more details about the phases shown in the preceding figure:

- **Inputs (extract phase)**: In this first step, Logstash uses input plugins to gather data from various sources. A common example is the syslog input plugin, which collects logs from system events. This phase focuses on accumulating all the necessary data for the subsequent stages of processing.

- **Filters (transform phase)**: Once data has been collected, the next task is to refine it using filter plugins. These plugins help you modify, enhance, or remove data as required. A typical use case involves the `grok` filter, which is particularly useful for transforming unstructured syslog data into a structured format. This structuring makes it easier to handle and analyze the data later.

- **Outputs (load phase)**: The final phase uses output plugins to distribute the processed data to various destinations. You can set up these plugins to either display data directly in Logstash's standard output, send it to Elasticsearch, or, as we will explore in upcoming chapters, forward it to Grafana Loki.

> **Note**
>
> For a detailed understanding of Logstash's configuration, its operational mechanics, and a list of available plugins, please refer to the official Logstash documentation (`https://www.elastic.co/guide/en/logstash/current/pipeline.html`).

Understanding the architecture of a Logstash pipeline helps to connect it with the ETL phases we discussed earlier in this book. You might be wondering why someone would choose Logstash for collecting log information when Telegraf can also perform this task. Your company might already be invested in using Logstash for other systems or applications, or perhaps you're already utilizing the Elastic Stack. Additionally, the capabilities of Logstash pipelines could be more appealing, or you might find its syntax for configuring ETL phases more comfortable compared to TOML, which Telegraf uses. There are various reasons why Logstash could be the tool of choice, and we aim to explore different methods of data collection to demonstrate the broad possibilities these tools offer in the field of observability.

To better understand how Logstash is used, we will dive into configuring a Logstash instance so that we can collect logs from our network lab. This will give you a hands-on look at the configuration syntax and how it operates in a real-world scenario.

Logstash syslog input

To begin configuring our Logstash instance, we need to create a Logstash pipeline configuration file. The language that's used to configure these pipeline files is custom-developed by the Logstash developers. It offers programming capabilities similar to those found in Python or Golang, facilitating easier data manipulation within the pipeline.

Let's start by setting up our Logstash configuration. First, create a `logstash.cfg` file at `network-observability-lab/chapters/ch5/logstash/logstash.cfg`. In this file, we'll configure the input settings of our Logstash instance so that we can receive syslog data from our network devices lab (`ceos-01` and `ceos-02`):

```
input {
  # Syslog input plugin
  syslog {
    # Setting a port to receive syslog messages
    port => 1515
    type => syslog
  }
}
```

In this configuration, we specify that Logstash should listen for syslog messages on port `1515` and process the incoming data as syslog data. Next, we must set up an output plugin in Logstash to forward the received data to stdout:

```
output {
  # Outputs to stdout for debugging
  stdout {
    codec => "json"
  }
}
```

The output plugin, `stdout`, has been configured to emit the messages to standard output, and we use the `codec` parameter to format these messages for better readability, specifying JSON as the format. This configuration is enough to handle collecting syslog data from network devices. In *Chapter 6*, we'll dive deeper into how to manipulate and normalize this data using the `filters` plugin.

Now, it's time to test the configuration for `logstash`. Just as you did with the Telegraf setup previously, follow these steps to save and apply the configuration:

1. Take the Logstash configuration snippets for `logstash` and save them to `network-observability-lab/chapters/ch5/logstash/logstash.cfg` on your lab machine.

2. Apply the changes by running the `netobs lab update logstash --scenario ch5` command.

Now that we're all set, let's generate some logs! Let's try connecting to `ceos-02` and turning up the `Ethernet2` interface from another terminal:

```
> ssh netobs@ceos-02
(netobs@ceos-02) Password: netobs123
ceos-02>enable
ceos-02#
```

```
ceos-02# configure terminal
ceos-02(config)# int eth2
ceos-02(config-if-Et2)# no shutdown
```

> **Note**
>
> The network devices in this lab have been pre-configured to direct their logs to the Logstash instance listening on the syslog port. For more details, please check the devices' logging configuration by executing the `show run | include logging` command in one of the devices.

To verify that our Logstash instance is capturing the syslog messages from the network devices, let's check the logs of the Logstash container instance. To do so, run the `netobs docker logs logstash --follow --scenario ch5` command. Here's an example of a syslog message captured by Logstash:

```
netobs-logstash-1  | {"@version":"1","facility":7,"timestamp":"Oct  9
19:44:10","message":"Instance 77: %OSPF-4-OSPF_ADJACENCY_ESTABLISHED:
NGB 10.222.0.1, interface 10.1.7.1 adjacency established\n","severity_
label":"Warning","priority":60,"logsource":"ceos-01","type":"syslog",
"host":"198.51.100.1","facility_label":"network news","severity":4,"@
timestamp":"2023-10-09T19:44:10.000Z","program":"Rib"}[2023-
10-09T19:45:35,086][INFO ][logstash.inputs.syslog    ][main]
[436cd4ed9cb18a2c37aa5746f397d9918b3cd243cb6b8d636651674c52bc3ad6] new
connection {:client=>"198.51.100.1:44950"}
```

> **Note**
>
> In some environments, Logstash may have issues displaying new messages in the Docker logs. To resolve this, Logstash might need to be restarted. We recommend that after generating syslog messages on `ceos-02` by enabling or disabling the interface, you restart the Logstash container and check the logs. Use `netobs docker restart logstash --scenario <chapter> && netobs docker logs logstash --follow --scenario <chapter>` to do so. The syslog messages should appear in the log output once you've restarted.

From the preceding output, you can see that Logstash processes the incoming syslog events and turns them into structured data. This is done by the `syslog` input plugin, which sorts the data much like the regular expression patterns we used for parsing logs in *Chapter 3*. To get a better handle on this, let's take a closer look at one of these log messages shown in JSON format:

```
{
    "@version": "1",
    "facility": 7,
    "timestamp": "Oct  9 19:44:08",
    "message": "%LLDP-5-NEIGHBOR_NEW: LLDP neighbor with chassisId
001c.733e.e8f6 and portId \"Ethernet2\" added on interface
```

```
  Ethernet2\n",
    "severity_label": "Notice",
    "priority": 61,
    "logsource": "ceos-02",
    "type": "syslog",
    "host": "198.51.100.1",
    "facility_label": "network news",
    "severity": 5,
    "@timestamp": "2023-10-09T19:44:08.000Z",
    "program": "Lldp"
}
```

By default, the `syslog` input plugin deciphers the core headers of the message. It extracts essential fields, including the following:

- The `facility` field, with a value of 7, which means the syslog message is coming from the network news subsystem, as per RFC 5424 (`https://datatracker.ietf.org/doc/html/rfc5424`). For more information, please see the section on facility and severity levels tables.

- The `severity` field, with a value of 5, which means the syslog message is of normal severity, as per RFC 5424.

- The `priority` field with a value of 61, which is the result of the `facility` value multiplied by 8 and then adding the `severity` value. Let's do the math to double check: *(7 * 8) + 5 = 61.*

- The `logsource` field, which shows the device emitting the event. In this case, we have `ceos-2`.

- The `program` field, which is the program that sent the event. In this case, we have LLDP.

- The `timestamp` field, which shows the time of the syslog event.

- The `message` field, which shows the actual data that generated the syslog event.

It's important to note that the timestamp from each log event is automatically converted into the standard `@timestamp` field that's used throughout the Logstash pipeline.

With just a minimal configuration on our part, we have successfully managed to collect an event through Logstash, marking a commendable first step in the development of our log collection and processing pipeline.

Summary

In this chapter, we embarked on an in-depth exploration of data collectors within an observability stack, casting a spotlight on Telegraf and Logstash due to their industry recognition and our familiarity with them. We provided a comprehensive view of their internal architecture, how they align with the ETL pipeline for their data operations, and through the help of their plugins ecosystem, demonstrated how they fit within the broader observability landscape.

Through hands-on practice, we covered various configurations for these data collectors so that we can gather information from network devices diversely. For instance, we used Telegraf for data collection via SNMP and gNMI, synthetic monitoring with Ping, and even created a custom script to capture data over SSH and CLI commands. An important focus was also placed on the InfluxDB line protocol, emphasizing its importance in comprehending the metrics that are collected with Telegraf, setting a foundation for the upcoming discussion on data normalization in the next chapter.

Furthermore, we ventured into the workings of Logstash and provided an example of data ingestion and some basic processing regarding syslog-related data. Now that metrics and logs data has been collected in our observability stack, we are well-positioned to advance toward data processing so that we can better meet our monitoring and analytics requirements.

6

Data Distribution and Processing

In the previous chapter, we explored the architecture of data pipelines, understanding their role as the circulatory system for observability data. In this chapter, we'll dive deeper into the specific processes of data distribution and processing. Think of a retail warehouse where online purchases result in items being picked, packaged, labeled with order details, and then distributed to their destinations via conveyor belts:

Figure 6.1 – Warehouse distribution and processing example

Similarly, in telemetry and observability, data collected from metrics, logs, flows, and events in general undergoes a parallel journey:

- **Processing**: Here, similar to packing and labeling in our warehouse analogy, data needs to be normalized into a common format and perhaps enriched with additional details for clarity and context. This step ensures that the data is *package-ready* for its journey through the pipeline.

- **Distribution**: This stage resembles the warehouse's conveyor belts, where the processed data is directed to various backends (that is, databases, message buses, and data lakes), much like packages are sent to different destinations.

Through hands-on practice, we will see how tools such as Telegraf and Logstash are configured to collect network data in diverse ways – be it via SNMP, gNMI, SSH/CLI, or syslog data. Additionally, we will explore the InfluxDB line protocol to understand its significance in comprehending the collected metrics. This knowledge is relevant, especially as we gear up for the next chapter on data normalization and enriching our observability stack for tailored analytics.

This chapter covers the following topics:

- Understanding data normalization
- Enhancing insights with data enrichment
- The scale of the observability data pipeline

Understanding data normalization

Imagine that you have a messy toolbox where all your tools are scattered randomly – screwdrivers mixed with wrenches, nails with bolts, and so on. Whenever you need to fix something, you waste a lot of time finding the right tool. Now, think of data normalization as organizing that toolbox: you separate the screwdrivers, wrenches, nails, and bolts into separate compartments, making it quick and easy to find what you need.

In the context of data, normalization is like organizing that toolbox. It involves structuring and standardizing your data so that it's consistent and easy to manage. This means standardizing the way device names, interface statuses, and other metrics are recorded, which helps in streamlined data management for an observability platform.

Building upon our discussion from previous chapters and practical examples, recall that in *Chapter 4*, within the *Understanding data pipelines* section, we explored the **extract, transform, and load** (ETL) process. Data normalization, a key part of the **transform** stage in the ETL pipeline, plays a key role in ensuring data consistency and usability.

But how is this applied in observability data? Let's focus on Telegraf as an example. Telegraf is part of the **Telegraf, InfluxDB, Chronograf, Kapacitor (TICK)** stack and is widely used for collecting, processing, and writing metrics. Telegraf processes data through an instance with three main stages: input, processor, and output. The input stage collects data from various sources, the processor stage processes and transforms the data, and the output stage sends the data to its final destination. For more information about Telegraf and metrics collection, please refer to *Chapter 5*.

Now, consider the representation of interface counter metrics. The way these metrics are represented when collected via SNMP can differ significantly from their representation via gNMI. For more information on SNMP and gNMI, please refer to *Chapter 3*.

Here's an SNMP metric representation example:

```
interface,device=spine1,interface=Ethernet4 ifHCInOctets=4569765412
```

The following is an example of the same metric collected over gNMI:

```
interface,hostname=spine1,name=eth4 in_octets=4569765412
```

For a more visual representation of these metrics, take a look at the following figure:

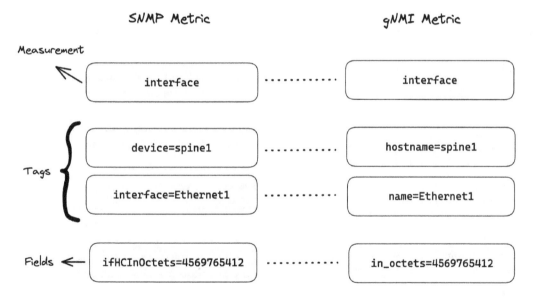

Figure 6.2 – SNMP and gNMI metrics comparison

> **Note**
>
> The metrics that are displayed are in the **InfluxDB line protocol** format. For a deeper understanding of this syntax, please refer to *Chapter 5*. Additionally, a brief recap is provided later in this chapter for your convenience.

Notice the discrepancies: the metric's name (`ifHCInOctets` versus `in_octets`), the device label (`device` versus `hostname`), and the interface label (`interface` versus `name`).

While we, as humans, can often infer similarities between these metrics, a machine might not so readily make this connection – that is, unless we're referring to some particularly savvy AI, but let's not give our algorithms too much credit just yet!

Having highlighted the importance of normalization in harmonizing data from diverse sources, we'll turn our attention to observability data models. These models are important as they provide the necessary structure to enhance normalization. Like a well-structured lecture, a robust data model ensures that our normalized data is not only consistent but also valuable for analytics and monitoring applications.

Observability data models

For the collected data to be properly understood and utilized across various components of our infrastructure, it's essential to normalize it. This means adhering to a schema that clearly defines the fields of the metrics or logs we collect and adapting the data accordingly.

The conversation about data schemas is a substantial one. While some may suggest relying on standards such as gNMI and YANG/OpenConfig, which are meant to be network-centric, the scope of data normalization extends far beyond just network devices. These standards are invaluable in their respective domains, yet modern organizations require an integrated view of their entire infrastructure. This includes network devices, firewalls, servers, Kubernetes clusters, cloud infrastructure, and storage appliances, among many others.

To address this broader need for standardization on data models, vendors have developed ways to manipulate and enforce data schemas for their observability platforms. One such example is the **Elastic Common Schema** (**ECS**), which focuses on standardizing the structure of data ingested into Elasticsearch. ECS provides a common framework that's essential for environments that require data to be normalized from multiple infrastructure components and is particularly useful if you're already utilizing the Elastic Stack. More information about ECS can be found in their official documentation (`https://www.elastic.co/elasticsearch/common-schema`).

The need for normalization has become universal and increasingly relevant in the context of cloud observability and Kubernetes. To address this, the open source and observability community has initiated an industry-wide standardization effort known as **OpenTelemetry**. OpenTelemetry is an open source framework that provides a unified approach to collecting, processing, and exporting telemetry data – such as metrics, logs, and traces – from applications for observability. It offers models that standardize data collection across different systems and platforms. You can explore these models in the official documentation, such as the OpenTelemetry Logs Data Model (`https://opentelemetry.io/docs/specs/otel/logs/data-model/`) and the Metrics Data Model (`https://opentelemetry.io/docs/specs/otel/metrics/data-model/`).

> **Note**
>
> A **data model**, if you are not familiar with the term, is an abstract framework that organizes data elements and standardizes their relationships to each other and to real-world entities. For example, a data model might specify that a data element representing a router includes attributes such as IP address, model, and status, and defines its relationships with other devices in the network.

Regarding data normalization, a key aspect is choosing a data model that aligns with your needs and proves suitable for your end goals, which typically includes creating rich visualizations and providing meaningful alerting. To illustrate this concept of observability data models and to draw inspiration from the SNMP and gNMI data examples already provided, we can propose a data model. This model should contain sufficient data to enable the creation of meaningful alerts and rich visualizations. So, let's take a closer look at our proposed metric model based on our previous examples.

Network interface example data model

In this data model, we represent the network interfaces with specific attributes (labels) and metrics (fields):

- **Measurement**: `interface`
- **Labels**:
  - `device`: The hostname of the device that the interface belongs to
  - `name`: The name of the interface
- **Fields**:
  - `in_octets`: The received (Rx) byte counter, representing the amount of incoming traffic
  - `out_octets`: The transmitted (Tx) byte counter, representing the amount of outgoing traffic

The model helps provide a clearer measurement, labels, and fields for the interface counters metric that more accurately describe the metric.

We don't have to start from scratch or a blank canvas when defining these models. Existing definitions, such as those from **OpenConfig** (`https://github.com/openconfig/public`), can guide us. OpenConfig data models are particularly useful because some attributes are derived from SNMP specifications. This means that if your device only supports SNMP, the extracted data will likely align closely with an attribute in the corresponding OpenConfig YANG model.

While the OpenConfig YANG specification may include more information than what you're looking for, it offers a solid foundation in terms of naming conventions, descriptions, and determining the relevant data to collect.

In the following subsections, we will see how this can be implemented practically using Telegraf.

Breaking down metrics and the data model

To properly manipulate any data, it is important to understand its data model. Let's revisit our lab environment, where Telegraf and Logstash are used to collect metrics and logs, respectively. When data is collected, these tools utilize default data structures: Telegraf employs the InfluxDB line protocol format (refer to *Chapter 5* for an introduction to this protocol) for metrics, while Logstash processes raw syslog messages and presents a JSON-style data structure of the logs.

Having a clear understanding of these data models is important for appropriately applying normalization or enrichment.

Telegraf data normalization

In *Chapter 5*, we examined the InfluxDB line protocol format to understand the data collected from network devices using Telegraf with SNMP and gNMI plugins. Here's a brief recap of the Telegraf data structure in this format:

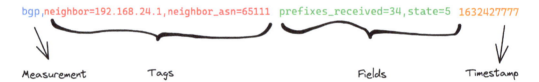

Figure 6.3 – Outline of the InfluxDB line protocol

This breakdown will guide us when we're working with Telegraf processors to normalize data within a Telegraf instance.

In *Chapter 5*, we discussed Telegraf *processors* and *aggregators* as part of the Telegraf architecture. Expanding on this, the processor and aggregator plugins in Telegraf are designed to transform data as it moves through a Telegraf instance. For data normalization, we typically use processor plugins for inline data manipulation; this will be the focus of this chapter. Although we will not cover aggregator plugins in detail here, it is worth mentioning notable ones, such as the `final` aggregator, which only emits the last series collected over a set period. For more information on Telegraf aggregators and to explore the available plugins, refer to their official documentation (`https://docs.influxdata.com/telegraf/v1/plugins/`).

Telegraf processor plugins can be used for various data manipulation tasks. For metrics normalization, some commonly used plugins include `converter`, `enum`, `filter`, `rename`, and `starlark`. These plugins help transform and standardize your data to ensure consistency across your observability data pipeline. For example, the `converter` plugin can change data types, the `enum` plugin can map numeric values to human-readable text, the `filter` plugin can keep or drop certain metrics, the `rename` plugin can change metric or tag names, and the `starlark` plugin allows for advanced custom scripting. We will explore these plugins in detail while covering practical examples later in this chapter. But first, let's set up our lab environment so that we can get some hands-on experience.

Setting up the lab environment

This lab will show you how to process metrics and log data from network devices to normalize the data and enrich it. The following figure shows the components we'll be using in our lab:

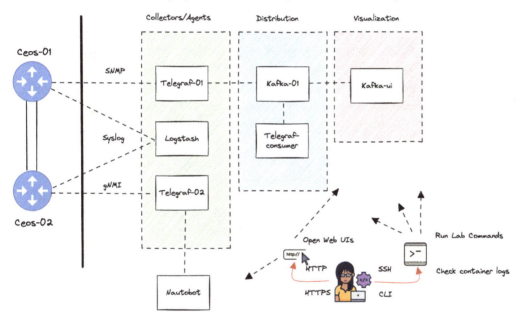

Figure 6.4 – Lab scenario (Created using the icons from https://icons8.com)

In this lab, we will focus on normalizing, enriching, and distributing processed metrics and log data from two devices, `ceos-01` and `ceos-02`.

First, we will focus on Telegraf – not on collecting metrics, which we covered in *Chapter 5*, but on normalizing the metrics we receive from the devices using Telegraf processor plugins. These plugins will help us standardize and format the incoming data to ensure consistency and usability. Similarly, we will configure Logstash filters to further process and refine the syslog information, enhancing its clarity and usefulness.

Next, we will concentrate on enriching the observability data we collect by integrating it with a Source of Truth system that provides additional context about the network. Here, we'll introduce **Nautobot**, a well-known network Source of Truth and automation platform designed to manage network infrastructure. Nautobot offers a centralized repository for network data, including sites, device information, connections, and IP addresses. By leveraging Nautobot, we can enhance our data with valuable details, making it more informative and actionable. In our lab environment, Nautobot will serve as the central hub for the network device information.

Nautobot offers capabilities for accessing data programmatically through its REST API or GraphQL interface. This feature is ideal for showcasing how to enrich data that's been collected from the network via our Telegraf instances by correlating it with the intent information stored in the Source of Truth system. While Nautobot itself is not the focus of this book, understanding some of its characteristics and its GraphQL interface will be beneficial for following along with the examples. For more information, you can visit their documentation (`https://docs.nautobot.com/`) or read *Network Automation with Nautobot* (`https://a.co/d/0b8RbhP`) for a comprehensive review of the system.

Finally, we will discuss scaling the observability data pipeline. We will introduce Kafka, a message broker, to demonstrate how distributed systems can be adapted for an observability data pipeline. Kafka enables the optimized handling of large volumes of data across multiple systems, ensuring reliable and scalable data processing and distribution.

Before diving into the lab activities, ensure your lab machine is set up according to the steps outlined in the *Appendix A*, just as we did in the previous chapter. Make sure you are in the `~/<path-to-git-project>/network-observability-lab/` directory within the project's repository to execute the necessary commands.

Now, let's begin the lab scenario for this chapter. To set everything up on your lab machine, run the `netobs lab prepare --scenario ch6` command.

> **Note**
>
> This lab scenario includes a completed version that you can refer to. It is located in the GitHub repository you cloned, under `network-observability-lab/chapters/<chapter-number>-completed`. This folder contains all the configurations that will be used in the examples throughout this chapter, providing a comprehensive guide to ensure you have everything set up correctly.

Great! Let's start with the data normalization process for our metrics by using Telegraf processor plugins.

Normalizing data with Telegraf processors

Welcome to the beginning of our hands-on lab for *Chapter 6*. Building on the foundational lab setup and Telegraf configurations we explored in *Chapter 5*, where we collected metrics from various network devices via SNMP and gNMI, we're now ready to dive deeper and process the data.

The Telegraf instances, `telegraf-01` and `telegraf-02`, come preconfigured to collect metrics from `ceos-01` and `ceos-02`, respectively.

The `telegraf-01` instance is configured to collect interface metrics information from `ceos-01` using SNMP. You can check the metrics that have been collected by running the `netobs docker logs telegraf-01 --scenario ch6` command. The metrics information is emitted to **standard output** (**stdout**) and is therefore visible in the container logs. The following is an example of a collected metric:

```
# SNMP collected metric for ceos-01 on interface Ethernet1
  interface,agent_host=ceos-01,host=telegraf-01,ifDescr=Ethernet1
ifHCInOctets=1794i,ifHCOutOctets=0i 1704546060000000000
```

The `telegraf-02` instance is set up to collect interface metrics from `ceos-02` using the gNMI protocol. To check the metrics being collected, you need to look at the logs for `telegraf-02`. This can be done by running a command similar to the one used for `telegraf-01`. Viewing the logs will show you the raw metric data being gathered from your device. Here's an example of what a collected metric might look like:

```
# gNMI collected metric for ceos-02 on interface Ethernet1
interface,host=telegraf-02,name=Ethernet1,path=/interfaces/
interface/state/counters,source=ceos-02 in_multicast_pkts=75i,in_
octets=11010i,in_pkts=106i 1704546522961434926
```

If we compare the outputs from the Telegraf instances with the data model we described earlier, we can see discrepancies in some label names and field names. For example, labels such as `source=ceos-02` and `agent_host=ceos-01` are used to depict the device's hostname, but they don't match our standardized naming convention. Similarly, the interface counter fields collected by `telegraf-01` are named `ifHCInOctets`, while our target name is `in_octets`. To address these inconsistencies, our next task is to configure Telegraf processors to normalize the data. This involves adjusting the labels and fields so that they align with our standardized data model.

Let's start by configuring `telegraf-01` so that we can rename SNMP metric tags and fields. Take a look at the following configuration SNMP snippet:

```
# Full Telegraf SNMP configuration from Chapter 5 omitted...
# Only the relevant fields are shown ahead.

    [[inputs.snmp.table.field]]
      # Interface Name - IF-MIB::ifDescr
      oid = "1.3.6.1.2.1.2.2.1.2"
      is_tag = true
      name = "name"

    [[inputs.snmp.table.field]]
      # Inbound Octets - IF-MIB::ifHCInOctets
```

```
    oid = "1.3.6.1.2.1.31.1.1.1.6"
    name = "in_octets"

[[inputs.snmp.table.field]]
    # Outbound Octets - IF-MIB::ifHCOutOctets
    oid = "1.3.6.1.2.1.31.1.1.1.10"
    name = "out_octets"
```

The highlighted section of this snippet shows that the configuration renames tags and fields to make SNMP metrics similar to our desired model. Specifically, it creates the name tag for the interface name and renames ifHCInOctets to in_octets and ifHCOutOctets to out_octets.

We also need to rename the agent_host tag to device on telegraf-01. We can do that by using the Telegraf rename processor:

```
[[processors.rename]]
  # Rename the tag "agent_host" to "device"
  [[processors.rename.replace]]
    tag = "agent_host"
    dest = "device"
```

Let's test this configuration. For this, grab the Telegraf configuration snippets for telegraf-01 and update it under network-observability-lab/chapters/ch6/telegraf/telegraf-01.conf.toml. Then, apply the configuration and check its output with netobs lab update telegraf-01 --scenario ch6 && netobs docker logs telegraf-01 --follow --scenario ch6. Finally, look for the logs concerning the Ethernet1 interface. Here's an example of the expected output:

```
# SNMP Interface counter metric normalized
interface,device=ceos-01,host=telegraf-01,name=Ethernet1 in_
octets=532465i,out_octets=0i 1704576840000000000
```

Great! We're one step closer to having our metrics match our desired data model. Let's continue by working on telegraf-02.

For the gNMI configuration in telegraf-02, we must exclude unnecessary tags in the inputs.gnmi configuration, more specifically the path tag. See the highlighted section in the following configuration snippet:

```
# Full Telegraf gNMI configuration from Chapter 5 omitted...
# Only the relevant fields are shown ahead.

[[inputs.gnmi]]
 addresses = ["ceos-02:50051"]
 username = "${NETWORK_AGENT_USER}"
```

```
password = "${NETWORK_AGENT_PASSWORD}"
redial = "20s"
# Exclude the "path" tag from the output
tagexclude = ["path"]
```

The following Telegraf processor's `rename` snippet is used to rename the device hostname tag from `source` to `device`:

```
[[processors.rename]]
 # Rename the tag "source" to "device"
 [[processors.rename.replace]]
   tag = "source"
   dest = "device"
```

Let's test this data normalization configuration. Take the Telegraf configuration snippets for `telegraf-02` and update them. Apply the configuration with `netobs lab update telegraf-02 --scenario ch6`. Give it a minute and then check the logs by running a command similar to `netobs docker logs telegraf-02 --scenario ch6 --follow | grep Ethernet1`:

```
# gNMI Interface counter metric normalized
interface,device=ceos-02,host=telegraf-02,name=Ethernet1 in_multicast_
pkts=4259i,in_octets=574026i,in_pkts=5323i 1704577652321254630
```

Note that we didn't need to rename the gNMI metric fields as they already matched our desired names.

Great progress! We've successfully normalized metrics from both `telegraf-01` and `telegraf-02`. Now, let's continue the normalization process for log data using Logstash.

Normalizing data with Logstash filters

As we move from Telegraf to Logstash, it's important to understand how Logstash handles data. Logstash is part of the ELK stack and is popular for collecting, processing, and transforming log data from various sources. Logstash processes data through a pipeline with three main stages: `input`, `filter`, and `output`. The `input` stage collects data, the `filter` stage processes and transforms it, and the `output` stage sends the data to its final destination.

Logstash presents syslog information in a structured dataset, and each dataset is called an event. Each event has a set of fields that we can manipulate using Logstash `filters`. This allows us to normalize and enrich our log data, making it consistent and ready for analysis.

In *Chapter 5*, we looked at a sample output showing Logstash's data as a JSON structure. While this format is useful, for better readability and debugging purposes, we should use the `rubydebug` output in our Logstash configuration. This format will make it easier to interpret the log messages and troubleshoot any issues. The following figure will help illustrate this:

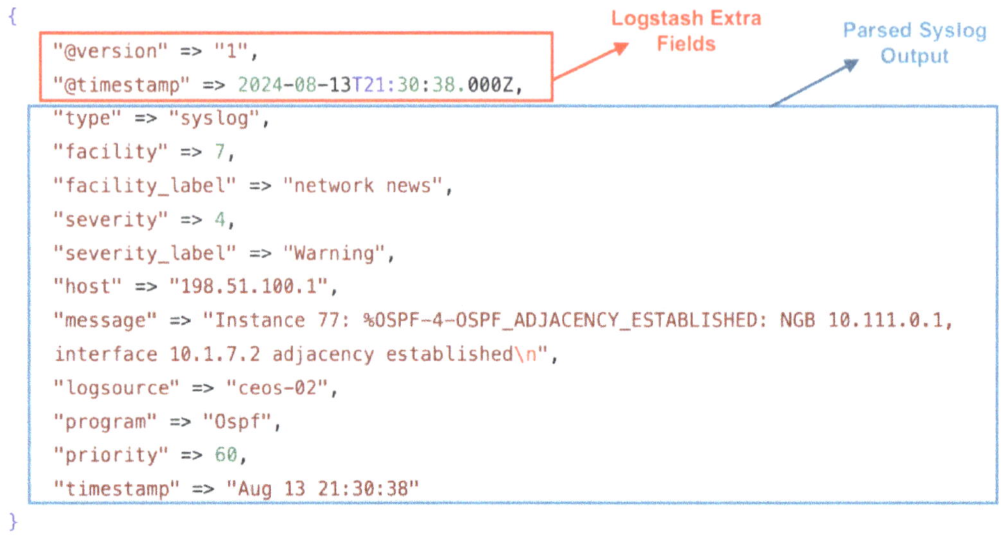

```
{
    "@version" => "1",
    "@timestamp" => 2024-08-13T21:30:38.000Z,
    "type" => "syslog",
    "facility" => 7,
    "facility_label" => "network news",
    "severity" => 4,
    "severity_label" => "Warning",
    "host" => "198.51.100.1",
    "message" => "Instance 77: %OSPF-4-OSPF_ADJACENCY_ESTABLISHED: NGB 10.111.0.1,
    interface 10.1.7.2 adjacency established\n",
    "logsource" => "ceos-02",
    "program" => "Ospf",
    "priority" => 60,
    "timestamp" => "Aug 13 21:30:38"
}
```

Logstash Extra Fields

Parsed Syslog Output

Figure 6.5 – Logstash data structure

This example shows Logstash capturing and parsing a network syslog event. Logstash metadata fields are inherently part of any event it captures, with additional tags (such as `type: syslog`) derived from the specified `input` configuration. The remaining key-value pairs are generated based on the input plugin used, assuming no further processing is applied.

> **Note**
>
> The example Logstash output shown here has been taken from the Logstash instance in the lab environment for this chapter. To explore this yourself, you can run the `netobs docker logs logstash --scenario ch6` command and extract one of the syslog messages.

Our objective now is to normalize the syslog messages so that they match the metrics data model we've been developing throughout this chapter. This alignment is key for accurate data correlation between syslog messages and metric data. For example, in the Logstash data structure, the `logsource` label is similar to the `device` label in Telegraf's metrics.

Logstash provides `filter` plugins (`https://www.elastic.co/guide/en/logstash/current/filter-plugins.html`) to help manipulate the data as it traverses from `input` to `output`. We're going to take advantage of that capability to help normalize attributes in the syslog data structure.

At the moment, in our lab scenario, Logstash is configured to listen for and forward raw syslog messages to stdout. If you need a refresher on the Logstash configuration, please refer to *Chapter 5*.

Let's start by configuring a Logstash filter. The first goal is to rename the `logsource` field to `device` and change the IP address captured as a `host` value to the service name `logstash`, which is the name of the container service:

```
# Full Logstash configuration from Chapter 5 omitted...
# Only the relevant fields are shown ahead.

filter {
    # Filter to mutate the event
    mutate {
        # Rename the logsource field to device
        rename => { "logsource" => "device" }
        # Replace the value of the host field with logstash
        replace => { "host" => "logstash" }
    }
}
```

Add this section to the existing Logstash configuration, typically between the `input` and `output` sections of the configuration file. Save it under `network-observability-lab/chapters/ch6/logstash/logstash.cfg`. Apply the configuration by running the `netobs lab update logstash --scenario ch6` command. Then, wait a minute for Logstash to perform its initialization and be ready to process syslog messages.

Now, let's generate interface flapping events by connecting to `ceos-01` over SSH and turning the `Ethernet1` interface on and off:

```
$ ssh netobs@ceos-01
(netobs@ceos-01) Password: netobs123
ceos-01> enable
ceos-01#
ceos-01#conf t
ceos-01(config)#interface Ethernet1
ceos-01(config-if-Et1)#shutdown
ceos-01(config-if-Et1)#
ceos-01(config-if-Et1)#no shutdown
```

> **Note**
>
> In some environments, Logstash may have issues displaying new messages in the Docker logs. To resolve this, Logstash might need to be restarted. You can restart the Logstash container to refresh its logs output by running the `netobs docker restart logstash --scenario ch6 && netobs docker logs logstash --scenario ch6` command.

An extracted event that's been captured and processed by Logstash should look similar to the following:

```
{
    "@version" => "1",
    "@timestamp" => "2024-01-06T22:34:36.000Z",
    "type" => "syslog",
    "facility" => 7,
    "facility_label" => "network news",
    "severity" => 5,
    "severity_label" => "Notice",
    "host" => "logstash",
    "message" => "%LINEPROTO-5-UPDOWN: Line protocol on Interface
Ethernet1, changed state to down\n",
    "program" => "Ebra",
    "device" => "ceos-01",
    "priority" => 61,
    "timestamp" => "Jan  6 22:34:36"
}
```

From the highlighted fields, we can see that the filter plugins have correctly renamed and replaced some attributes of the syslog data structure. This step is important because it allows us to correlate metrics and log events by referencing the `device` field. We will see this in action later in *Chapter 8*.

This initial step of normalizing syslog data is helpful, but to bridge our logs and metrics for better correlation, they must share matching labels. A closer look at the Logstash output, compared with our Desired Metric Model, reveals we are missing a key piece: the interface name. Referring to the previous snippet, interface syslog messages often contain the interface name within the `message` field:

```
%LINEPROTO-5-UPDOWN: Line protocol on Interface Ethernet1, changed
state to down
```

The `message` field contains rich information that, if extracted, can provide significant value by further categorizing and classifying the syslog messages we receive from network devices. For example, by extracting data from the `message` field, such as the `UPDOWN` field, we can identify messages indicating when an interface changes its operational state. Let's explore how to achieve this with the use of another Logstash filter plugin called `grok`.

Grok filter plugin

The Logstash `grok` filter plugin It's a powerful tool that structures unstructured text data. Using regular expression-based patterns, it transforms text into a more organized format. If a pattern matches, we get our data neatly structured; if not, the data simply moves along unchanged.

Before diving into the `grok` filter configuration, it's important to understand how it operates. Imagine a log message navigating its way through a grok pattern, as depicted in the following diagram:

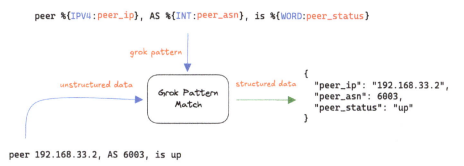

Figure 6.6 – Grok pattern flow

This diagram helps break down the `grok` pattern into more digestible parts:

- Each regular expression pattern in grok is defined with the `%{<regex-pattern>:<resulting-field>}` expression.

- Here, `<regex-pattern>` is a predefined regular expression. When it finds a match in the text data, the result is stored in `<resulting-field>`. Logstash already includes a rich library of such patterns (`https://github.com/logstash-plugins/logstash-patterns-core/blob/main/patterns/ecs-v1/grok-patterns`), but you can also add custom ones as needed.

- Take, for instance, the `%{IPV4:peer_ip}` pattern. If this pattern encounters a string such as `192.168.33.2`, it recognizes it as a valid IPv4 address due to the regex pattern and then saves this IP address under the `peer_ip` field.

Grok Debugger (`https://grokdebugger.com/`) is an incredibly handy tool for testing these regex patterns against your text strings. It helps you see the kind of structured data your `grok` pattern will produce. To give it a try, here is an example you can run on the Grok Debugger website:

1. Copy and paste the following grok pattern into the **Grok Pattern** box on the website. It will try to extract `vendor_facility` and `vendor_facility_process` from the log message; we'll allocate the rest of the message to the `log_message` field:

   ```
   %%{WORD:vendor_facility}-%{INT}-%{WORD:vendor_facility_process}:
   ?%{GREEDYDATA:log_message}
   ```

2. Copy and paste example log lines, such as the following ones, into the **Samples** box on the website. You can add multiple log lines to test against the same pattern by just appending it to the next line.

Here are the example log lines for Log 1:

```
%BGP-5-ADJCHANGE: peer 192.168.33.2 (AS 6003) old state
OpenConfirm event RecvKeepAlive new state Established
```

The following are the example log lines for Log 2:

```
%LINEPROTO-5-UPDOWN: Line protocol on Interface Ethernet1,
changed state to up
```

3. Check the resulting structured data. It should provide an output similar to the following:

```
[
  {
    "vendor_facility": "BGP",
    "vendor_facility_process": "ADJCHANGE",
    "message": "peer 192.168.33.2 (AS 6003) old state
    OpenConfirm event RecvKeepAlive new state Established"
  },
  {
    "vendor_facility": "LINEPROTO",
    "vendor_facility_process": "UPDOWN",
    "message": "Line protocol on Interface Ethernet1, changed
    state to down"
  }
]
```

This tool allows you to see how your patterns will parse log messages, making it easier for you to get them right before you apply them in Logstash. Understanding the structure of a syslog message will help us configure the Logstash filter properly.

Consider `%LINEPROTO-5-UPDOWN: Line protocol on Interface Ethernet1, changed state to down`. Here, `%LINEPROTO-5-UPDOWN` is a standardized message header that provides specific details about the type and severity of the message. Here's a breakdown:

- `LINEPROTO` is the facility or subsystem that generated the message
- `5` is the severity level of the message
- `UPDOWN` is the mnemonic code that gives a quick reference to the specific event

Coming back to our Logstash lab environment, our immediate task with `grok` is to process log messages related to interface events, such as the example mentioned earlier, with the main focus on the `message` field. The following Logstash configuration exemplifies how we can parse this field, extracting details such as the vendor facility and process, and capturing the remaining text in a new field named `log_message`:

```
# Omitted Logstash configuration...
filter {
    # Mutate filter configuration omitted
    grok {
        match => [
            "message", "%%{WORD:vendor_facility}-%{INT}-%{WORD:vendor_
            facility_process}: ?%{GREEDYDATA:log_message}"
        ]
    }
}
```

This setup ensures that successfully parsed messages will now include new fields: `vendor_facility`, `vendor_facility_process`, and `log_message`. If a message doesn't match the `grok` pattern, it will be tagged with `_grokparsefailure` by default. This tag is useful for troubleshooting and refining Grok patterns as it highlights the messages that couldn't be parsed and need further attention.

An example output for parsed interface flapping events would look something like this:

```
{
    "vendor_facility": "LINEPROTO",
    "vendor_facility_process": "UPDOWN",
    "log_message": "Line protocol on Interface Ethernet1, changed
    state to down"
}
```

To further refine our data and extract the interface name, we only process `log_message` when `vendor_facility` equals LINEPROTO and `vendor_facility_process` equals UPDOWN. This ensures we are only dealing with interface flapping events. Let's add another `grok` filter to our configuration for this purpose:

```
# Omitted Logstash configuration...
filter {
    # Mutate filter configuration omitted
    # Grok filter configuration omitted
    if [vendor_facility] == "LINEPROTO" and [vendor_facility_process]
    == "UPDOWN" {
        # Category to identify the event (1)
        mutate {
            add_field => { "event_type" => "interface_status" }
        }
        # Grok the log message to extract the interface and status (2)
        grok {
```

```
                    match => [
                        "log_message", "Line protocol on Interface
                        %{DATA:name}, changed state to %{WORD:interface_
status}"
                    ]
                }
            }
        }
    mutate {
        # Remove the message field (3)
        remove_field => [ "message" ]
    }
}
```

Let's break down this configuration:

1. We add a field, `event_type`, with a value of `interface_status` to categorize these specific messages.

2. Then, we parse `log_message` to extract the name value of the interface and its `interface_status`.

3. Lastly, the original `message` field is removed to avoid data duplication.

Let's activate the new config. First, save the Logstash configuration and then run the `netobs lab update logstash --scenario ch6` command. Wait a minute for Logstash to be ready to receive syslog messages.

Now that our `grok` filter has been configured, let's put it to the test. Repeat the steps we covered earlier to trigger some interface flapping on `ceos-01`, specifically targeting the `Ethernet1` interface. Once done, we can check the logs to see how these events are captured and structured by our new Logstash configuration. To get a snapshot of the results, you can use the `netobs docker logs logstash --scenario ch6 --tail 50 | grep interface_status` command and, if necessary, restart the Logstash instance.

The following is an example of what you might see in the extracted Logstash output message when running these commands:

```
{
  "vendor_facility" => "LINEPROTO",
    "timestamp" => "Jan  7 00:37:42",
    "facility_label" => "network news",
    "vendor_facility_process" => "UPDOWN",
    "name" => "Ethernet1",
    "priority" => 61,
    "@version" => "1",
    "facility" => 7,
    "device" => "ceos-02",
```

```
    "severity_label" => "Notice",
    "event_type" => "interface_status",
    "severity" => 5,
    "program" => "Ebra",
    "@timestamp" => "2024-01-07T00:37:42.000Z",
    "log_message" => "Line protocol on Interface Ethernet1, changed
    state to down\n",
    "host" => "logstash",
    "interface_status" => "down",
    "type" => "syslog"
}
```

Examining the data from the Logstash container's logs output, we can see that the **interface name** field is being successfully captured and parsed. This normalization aligns perfectly with the metric labels from Telegraf and the Desired Metric Model, ensuring consistency across our data. The benefits of this correlation will be demonstrated in *Chapter 8* when we correlate metrics and logs within the same visualization.

With this demonstration, we have wrapped up the normalization aspect of our data collectors within the observability stack. Next, we'll dive into the realm of data enrichment to further enhance our observability capabilities.

Enhancing insights with data enrichment

While data collection and normalization lay the groundwork, these raw messages from our devices provide limited insights within their infrastructure context. Imagine how much more meaningful these logs and metrics could be if they included organizational or business context to better describe the events or metrics collected.

Data enrichment involves adding extra metadata that enhances the description of the original data. Take, for instance, the bgp metrics of a router. These metrics become more informative when they're complemented with the router's role in the network topology. Similarly, understanding the significance of interface_out_octets metrics is much easier if we know the router's role, especially if it's a border router connected to multiple service providers. Adding metadata such as the circuit_id value of an interface can clarify which service provider is involved, assisting in decision-making when traffic thresholds are reached.

The benefits of this extra metadata are multifold:

- **Navigation**: It helps sift through the sea of collected metrics, enabling categorization and filtering based on parameters relevant to your organization.

- **Alerting**: You can create alerts based on specific criteria such as traffic utilization for links with a particular role, circuit_id, or provider. This targeted approach is more efficient than broad global alerts or overly specific per-interface/device alerts.

- **Dashboards and reporting**: Enrichment facilitates the creation of meaningful dashboards and reports. Whether it's identifying the top 10 devices with the highest traffic per site or understanding regional top talkers based on a `service` definition label, enriched data makes these insights more accessible.

So, where do we apply this extra metadata? The answer varies. Data can be enriched as it traverses the pipeline and is stored in the database, or it can be enriched at the visualization or alerting level. Take a look at the following figure, which showcases the data enrichment process in the data pipeline:

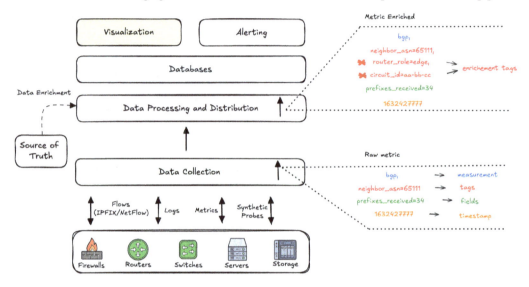

Figure 6.7 – Data enrichment at the data pipeline (Created using the icons from https://icons8.com)

Here, we can see that at the Data Collection layer, an example metric provides only one tag, `neighbor_asn`, which is reported by the device. At the Data Processing layer, the Source of Truth enriches the metric by adding `router_role` and `circuit_id` information.

Typically, key labels that fundamentally describe an event or metric are best introduced during the data pipeline process, as depicted earlier, such as `router_role` or `circuit_id` information. In contrast, data that serves more as decoration (meaning it is not vital for decision-making but adds additional context) is usually added at the visualization and alerting layers, something we'll explore later in this chapter.

A key practice in data enrichment involves identifying the *source* and *intent* of the data. The source reveals where the data originates, while the intent clarifies its purpose.

Expanding on this concept, consider a scenario involving the `ceos-01` device. Basic information such as its hostname might be readily available, but what about its installation location, role in the infrastructure, associated tenant, or the specific rack it occupies? Such details are typically not available directly from the device unless explicitly configured in the device banner, SNMP location, or contact information. This is an example of sourcing enrichment data, where additional context is derived from a designated **source of truth**.

Another aspect of enrichment focuses on the intent of the data. This includes information such as the expected status of the device (for example, provisioning, booting, active, inactive, maintenance, or decommissioning), as well as the status of interfaces, BGP sessions, and more. Such data reflects the intended use or state of the device, providing valuable insights for management and decision-making.

A source of truth in this context is a reliable, primary source of information that accurately reflects the intended state and characteristics of your infrastructure components. It could be a **configuration management database** (**CMDB**), an asset management system, or any other authoritative repository that consistently provides up-to-date and accurate information.

In the upcoming subsections, we'll dive into practical examples of how data enrichment can be integrated into your observability data pipeline.

Data enrichment injection

Data enrichment typically occurs at the processing level of your observability pipeline. Essentially, this process involves embedding additional metadata into the metrics, flows, logs, or events as they move through the observability data pipeline. In this section, we'll explore how to leverage Telegraf and its processor plugin capabilities for data enrichment.

First, we'll introduce two distinct methods of enriching metrics using Telegraf. Let's take a closer look.

Static data enrichment

Static data enrichment is valuable when you already know the source and intent of the data being collected. For instance, using a Telegraf processor plugin such as `regex`, you can match expected events or metrics data using regular expressions and then add descriptive labels. This process involves configuring Telegraf to recognize specific patterns in the incoming data and automatically enriching it with additional context.

Consider the following example. For incoming interface metrics that Telegraf will process, we're matching interface names based on a predefined standard and adding labels to assign roles to the interfaces and the metrics they generate. In our simple topology, `ceos-01` and `ceos-02` are connected via the `Ethernet1` and `Ethernet2` interfaces, so labeling them with `intf_role=peer` is appropriate. Meanwhile, the `Management0` interface is labeled with `intf_role=mgmt`.

We use `telegraf-01` for static data enrichment on `ceos-01` metrics. Here's a snippet of the Telegraf `regex` processor plugin's configuration to achieve this:

```
# Omitted configuration...
[[processors.regex]]
  # Filter metrics that have "interface" name (1)
```

```
namepass = ["interface"]

# Regex processor operation on the metric tags (2)
[[processors.regex.tags]]
  # Name of the tag to match
  key = "name"
  # Regex pattern to match
  pattern = "^Ethernet.*$"
  # Name of the new tag
  result_key = "intf_role"
  # Replacement string
  replacement = "peer"

[[processors.regex.tags]]
  # Name of the tag to match
  key = "name"
  # Regex pattern to match
  pattern = "^Management.*$"
  # Name of the new tag
  result_key = "intf_role"
  # Replacement string
  replacement = "mgmt"
```

If we break down the Telegraf processor configuration, it would look something similar to this:

- **Filtering metrics**: This `regex` processor is configured to process only the metrics with the `interface` measurement name.

- **Matching and adding tags based on the regex pattern**: After filtering for interface-related metrics from `ceos-01`, we match the interface name against a `regex` pattern. If there's a match, an `intf_role` tag is created, and the `replacement` string is added to it as its value.

Now, let's test this configuration to see what the processor can do on `telegraf-01`. Save the new configuration on `telegraf-01` and apply the changes by running `netobs lab update telegraf-01 --scenario ch6`.

Next, verify that the metrics have the new tags applied. Check the logs and filter them for interface metrics by running `netobs docker logs telegraf-01 --scenario ch6 --follow | grep intf_role`. Here are some example lines from the output:

```
interface,device=ceos-02,host=telegraf-02,intf_
role=mgmt,name=Management0 oper_status="UP" 1704546027244306447
interface,device=ceos-02,host=telegraf-02,intf_
role=peer,name=Ethernet2 oper_status="UP" 1704579519330105542
interface,device=ceos-2,host=telegraf-02,intf_role=peer,name=Ethernet1
oper_status="UP" 1704588365783693058
```

Notice how the metrics now include the new label, `intf_role`. Just adding this label can significantly enhance querying, visualization, or alerting capabilities.

It's important to recognize that implementing this method at a larger scale necessitates robust automation and sophisticated infrastructure orchestration. The process involves using scripts and templates to dynamically generate Telegraf configurations with the appropriate regex processor settings tailored to the specific devices being monitored. The complexity of this task increases substantially with the scale and uniqueness (*snowflake* nature) of your network or infrastructure. Essentially, as your environment grows in size and diversity, the process of template generation and rendering becomes more intricate, involving more data and a higher number of conditional elements to manage.

While the design and maintenance of such an orchestration system can be challenging, particularly in environments featuring a wide range of vendors and standards, teams proficient in using template engines for network configurations will likely find this approach within their capabilities. This familiarity can make managing even the most complex and varied network environments more navigable and streamlined. We will revisit this topic in *Chapter 10*, where we'll discuss observability platforms in real-world architectures.

Dynamic data enrichment

Switching gears, let's explore a dynamic approach to data enrichment that involves sending live queries to external data sources to add contextual information to our collected observability data. But what do we mean by *live queries*?

In contrast to static data enrichment, where the additional contextual information is predefined in a static configuration file, the dynamic approach does not rely on such static configurations. Instead, it involves querying an external data source in real time as the metric passes through the processor or enrichment phase. This means the data is automatically enriched with the most up-to-date information available, providing more accurate and timely context.

For instance, instead of relying on a static mapping file to enrich network metrics with device roles or locations, the dynamic approach would query a live database or API to fetch this information as needed. This ensures that any recent changes in the network environment, such as newly added devices or updated configurations, are immediately reflected in the enriched data.

A prime plugin that can help exemplify this concept is the Telegraf `execd` processor plugin. This plugin allows Telegraf to run external programs as separate processes, giving you the ability to manipulate or enrich the metrics data traversing it as you see fit. It's particularly useful for cases where data requires real-time updates or complex processing.

We will demonstrate this with `telegraf-02` and Nautobot. Nautobot is used as a network Source of Truth tool and keeps track of network devices and their properties. It's a rich source of contextual data that can be tapped into for enrichment purposes.

First, you must ensure you have connectivity and can access the Nautobot instance that was spawned when you started the lab for this chapter. Using the IP address of your lab machine, log in via a browser at `http://<machine-ipaddress>:8080`. If you're utilizing the default settings, your login credentials will be as follows:

- **Username**: admin

- **Password**: nautobot123

You'll be taken to a page that looks similar to the following:

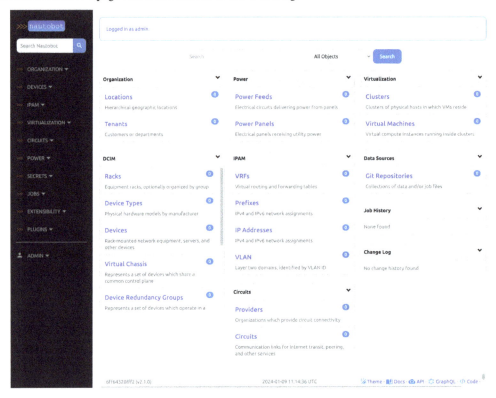

Figure 6.8 – Nautobot home page

Upon your initial visit to the Nautobot home page, you'll notice that it doesn't contain any data yet. To populate it with your network topology data, you have two options: manual entry or using the `netobs utils load-nautobot` command, which will automatically populate our network lab data.

Here's a brief look at what happens when you use the `netobs utils` command:

```
> netobs utils load-nautobot
[11:16:09] Loading Nautobot data from topology file: containerlab/lab.
```

```
yml
        Reading containerlab topology file
        Instantiating Nautobot Client
[11:16:10] Created Role: network_device
[11:16:11] Created Manufacturer: Arista
        Created Device Types: Arista cEOS
[11:16:12] Created Location Type: site
        Created Status: lab-active
[11:16:13] Created Location: lab
        Created IPAM Namespace: lab-default
[11:16:14] Created Prefix: 10.1.2.0/24
...Omitted Output...
```

After running this command, your Nautobot instance should display the newly loaded data. To see the details of a specific device interface, navigate to **Devices** | **ceos-01** | **interfaces** | **Ethernet1** via the navbar. There, you'll find that the `Ethernet1` interface is labeled as **peer**, as shown on the **Interface details** page:

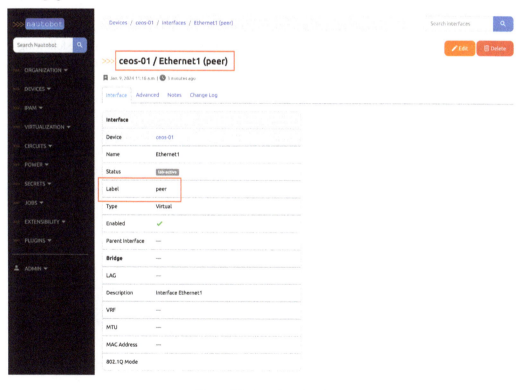

Figure 6.9 – Interface details

Now that our external source, Nautobot, is ready and contains valuable data, we'll develop a Python script to dynamically query information from Nautobot whenever metrics pass through the Telegraf `execd` processor. Similar to `telegraf-01`, we aim to enrich the interface metrics collected by `telegraf-02` by adding an `intf_role` tag, using the **Label** attribute from the Nautobot **Interface** details page.

We will start by creating a script at `network-observability-lab/chapters/ch6/telegraf/script.py`. This script will be used by `telegraf-02` and read by its `execd` processor. Let's walk through the key parts of the script and Telegraf configuration to understand how this works.

Retrieving interface data via Nautobot's GraphQL interface

For the first part of the script, we will use **GraphQL** queries. GraphQL is a powerful query language for APIs, known for its flexibility. It allows clients to request exactly the data they need, making it an ideal choice for querying complex data structures such as those found in Nautobot. For more information about Nautobot's GraphQL capabilities, please refer to their official documentation (`https://docs.nautobot.com/projects/core/en/stable/user-guide/feature-guides/graphql/`).

In your browser, navigate to your Nautobot instance GraphQL interface (`http://<machine-ipaddress>:8080/graphql/`) and paste in a snippet, similar to the one shown in the following figure. Here, we must provide a variable to filter the devices (the device name). We'll only ask for the `name`, `id`, and `interfaces` data:

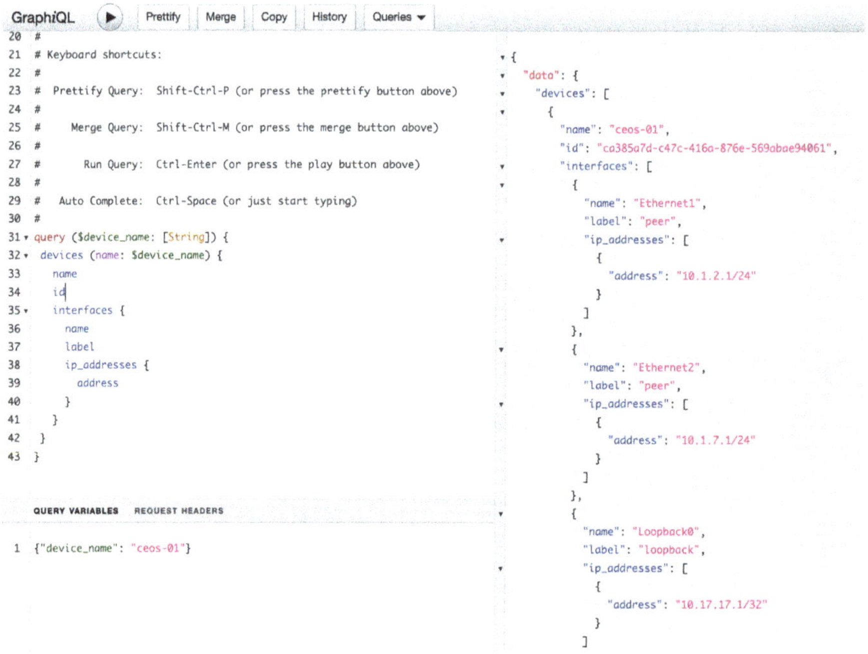

Figure 6.10 – Nautobot's GraphQL interface

On the right-hand side of the preceding figure, we can see the result of the GraphQL query, including the `label` value of each interface, which we need to populate an `intf_role` tag on the collected metrics. Now, let's translate this entire operation into Python code so that we can retrieve the interface role data from Nautobot. We will start by defining the Nautobot connection parameters and the GraphQL query string:

```python
import os

# Define the Nautobot URL and API token
NAUTOBOT_URL = "http://nautobot:8080"
NAUTOBOT_SUPERUSER_API_TOKEN = os.getenv(
    "NAUTOBOT_SUPERUSER_API_TOKEN", ""
)

# Nautobot GraphQL query to retrieve device data
NAUTOBOT_DEVICE_GQL = """
query ($device_name: [String]) {
  devices (name: $device_name) {
    name
    id
    interfaces {
      name
      label
      ip_addresses {
        address
      }
    }
  }
}
"""
```

The Nautobot connection information is obtained through environment variables, while the GraphQL query focuses on requesting interface properties information for a given device.

Next, let's create a function that will perform a GraphQL query for Nautobot and return the relevant results:

```python
import requests
import sys
from typing import Optional
from requests.exceptions import ConnectionError

def get_device_data(device_name) -> Optional[dict]:
    """Retrieve device data from Nautobot GraphQL API."""
    # Retrieve the device data from Nautobot GraphQL API
```

```
try:
    response = requests.post(
        url=f"{NAUTOBOT_URL}/api/graphql/",
        headers={"Authorization": f"Token {NAUTOBOT_SUPERUSER_API_
        TOKEN}"},
        json={
            "query": NAUTOBOT_DEVICE_GQL,
            "variables": {"device_name": device_name},
        },
    )
except ConnectionError:
    print(
        "[ERROR] Unable to connect to Nautobot GraphQL API",
        file=sys.stderr,
        flush=True,
    )
    return None

# Return the device data
if response.json()["data"]["devices"]:
    return response.json()["data"]["devices"][0]
else:
    print(
        f"[WARNING] Device `{device_name}` data not found in
        {NAUTOBOT_URL}",
        file=sys.stderr,
        flush=True,
    )
    return None
```

In this function, the `get_device_data` function performs a GraphQL query to Nautobot given a `device` name, retrieving detailed information about the device, its interfaces, and, most importantly, the labels associated with those interfaces. The `label` value, found in the `label` attribute within Nautobot's data model, is used as the value for the `intf_role` tag for the incoming metrics that will be collected by `telegraf-02`.

Parsing incoming metrics

Our next step is to parse the metrics collected by Telegraf into a Python data structure that is both easy to read and manipulate. This is necessary for the script to determine the `device` name for the GraphQL query for Nautobot.

> **Note**
>
> The example Python snippet shown in this section is useful but includes concepts that are outside the scope of this book, such as Python data classes, which are covered extensively in other Python learning resources. If you prefer not to delve into the details, feel free to skip this section. Just make sure you copy this snippet back into the `script.py` file when it's time to test it.

To achieve this, we'll introduce the `InfluxMetric` Python data class, which builds upon the `influx_line_protocol` function from *Chapter 5*. This data class makes the parsed Influx metrics more accessible and straightforward to handle:

```python
from dataclasses import dataclass
from typing import Optional


@dataclass
class InfluxMetric:
    measurement: str
    tags: dict
    fields: dict
    time: Optional[int] = None

    def __str__(self):
        # Construct tags string
        tags_string = ""
        for key, value in self.tags.items():
            if value is not None:
                if isinstance(value, str):
                    value = value.replace(" ", r"\ ")
                tags_string += f",{key}={value}"

        # Construct fields string
        fields_string = ""
        for key, value in self.fields.items():
            if fields_string:
                fields_string += ","
            if isinstance(value, bool):
                fields_string += (
                    f"{key}=true" if value else f"{key}=false"
                )
            elif isinstance(value, int):
                fields_string += f"{key}={value}i"
            elif isinstance(value, float):
                fields_string += f"{key}={value}"
```

```
        elif isinstance(value, str):
            value = value.replace(" ", r"\ ")
            fields_string += f'{key}="{value}"'
        else:
            fields_string += f"{key}={value}"

    return (
        f"{self.measurement}{tags_string} {fields_string} {self.
        time}"
        if self.time
        else f"{self.measurement}{tags_string} {fields_string}"
    )
```

The `InfluxMetric` class provides a clear structure for storing and processing metric data, including the measurement, tags, fields, and time. The `__str__` method is particularly useful as it converts the data class instance back into an InfluxDB line protocol string. This is essential for the output of our processor as it allows Telegraf to continue processing the metrics without any issues.

With this data class, accessing attributes such as `InfluxMetric.tags` is straightforward, offering a complete view of all tags associated with a metric.

We'll learn how to apply these capabilities next. Here, all these functions and data classes will be brought together so that we can inject the interface role data into the corresponding metrics passing through the processor.

Injecting the enriched data

Now, we'll integrate all the components we developed in the previous steps inside the `script.py` file and create a cohesive `main` function that uses them. This function orchestrates the entire process of enriching the data, so let's take a closer look at what it looks like:

```
import fileinput
import jmespath


def main():
    # Read Telegraf metrics from stdin (1)
    for line in fileinput.input():
        # Parse the line into an InfluxMetric object (2)
        influx_metric = InfluxMetric(**parse_line(line))

        # Extract the device name from tags
        device_name = influx_metric.tags.get("device")

        # Retrieve device data from Nautobot (3)
        device_data = get_device_data(device_name)
```

```
        # JMESPath expression to extract interface data
        jpath = f"interfaces[?name=='{influx_metric.tags['name']}'].
        label"

        # Extract the interface label from device data (4)
        intf_role = jmespath.search(jpath, device_data)[0]

        # Add interface role to tags (5)
        influx_metric.tags["intf_role"] = intf_role

        # Print the line protocol string (6)
        print(influx_metric, flush=True)
```

Let's break down this code snippet:

- **Read metrics**: It begins by reading collected metrics from STDIN (using the `fileinput.input()` method). These are the metrics that are generated by various Telegraf input plugins.

- **Parse metrics**: The incoming metrics are then parsed into an `InfluxMetric` data class. This step simplifies manipulation by converting metrics into a more accessible Python data structure. The `parse_line` function is included in this chapter's `script.py` file, facilitating the testing process.

- **Perform a GraphQL query**: Next, the script uses the `device` name that was extracted from `influx_metric` to query Nautobot for detailed device data. The resulting value is the response from the GraphQL query.

- **Filter with JMESPath**: Once the GraphQL query returns data from Nautobot, JMESPath is employed to drill down to the specific interface level.

- **Extract and inject data**: The script extracts `intf_role` from the Nautobot data and appends it to `influx_metric.tags`. This step is the core of the data enrichment process.

- **Return the enhanced metric**: The enriched metric, now in the InfluxDB line protocol format, is printed and ready for further processing by other Telegraf plugins.

To complete the setup, make this script executable by appending the following at the end of `script.py`:

```
if __name__ == "__main__":
    # Start the processor
    main()
```

This `main` function ties together the steps of reading, parsing, querying, filtering, and enriching the data, showcasing a practical application of dynamic enrichment in an observability data pipeline. Now, let's move on to configuring Telegraf so that it can utilize this script.

Telegraf processor configuration

In this section, we will set up `telegraf-02` so that it can leverage our newly created script. We will configure the `execd` processor plugin so that we can use the script, enabling dynamic querying of Nautobot for interface role data. Let's take a look at the following configuration snippet:

```
# Omitted configuration...

# Telegraf Processor execd for dynamic data enrichment
[[processors.execd]]
 # Filter for "interface" metrics
 namepass = ["interface"]
 # Run the Nautobot data enrichment script
 command = ["python", "/etc/telegraf/script.py"]
 # If script quits unexpectedly, wait 10 seconds before restarting
 restart_delay = "10s"
```

> **Note**
>
> The script's location inside the Telegraf container corresponds to a local reference path that's established through a Docker volume mount. This setup is detailed in the `docker-compose.yml` file for this chapter, located at `network-observability-lab/chapters/ch6/docker-compose.yml`.

Now, it's time to test the changes. Save the Telegraf processor configuration at `telegraf-02` and apply the changes by running the `netobs lab update telegraf-02 --scenario ch6` command.

Check the logs in real time by running `netobs docker logs telegraf-02 --follow --scenario ch6 | grep intf_role`. You'll see that the metrics include the dynamically added `intf_role` label, directly pulled from Nautobot, our Source of Truth:

```
interface,device=ceos-02,host=telegraf-02,intf_
role=peer,name=Ethernet1 in_unicast_pkts=179i 1704801719230029610
interface,device=ceos-02,host=telegraf-02,intf_
role=peer,name=Ethernet1 in_broadcast_pkts=1i 1704796859118000342
```

This dynamic enrichment means that any updates that are made to Nautobot's interface labels are automatically reflected in the corresponding metrics. For instance, if you change the `label` value of the `ceos-02` device's `Ethernet1` interface to `my-custom-label` in Nautobot, this change will be automatically captured in Telegraf's metrics!

To try this out:

1. Log in to your Nautobot instance and navigate to the `Ethernet1` interface of `ceos-02`.

2. Update the label to `my-custom-label` and save your changes.

3. Then, to verify the impact of this change, examine the logs of your `telegraf-02` instance. You can use a command similar to `netobs docker logs telegraf-02 --tail=20 --follow --scenario ch6 | grep Ethernet1` to do so. This should show you that the `intf_role` label in the Telegraf metrics now matches the new label you set in Nautobot.

What we have developed so far is a method to dynamically enrich our collected metrics, which is highly valuable in small-scale or well-defined environments. However, as the volume of collected metrics grows, more advanced techniques, such as caching, become necessary to maintain optimal performance. Let's explore this topic further.

Mastering the dynamics of data enrichment

When integrating dynamic data enrichment into your observability pipeline, it's important to think ahead about scaling this solution properly. As beneficial as real-time updates can be, they bring unique challenges, especially as your network expands and diversifies. Two key considerations stand out in ensuring that this system scales reliably – caching and handling failure scenarios:

*   **Caching**: In larger infrastructure environments, where data points are numerous and queries are frequent, relying solely on real-time queries can strain your external data sources, such as Nautobot. Excessive querying, particularly with complex GraphQL operations, can lead to performance bottlenecks. Implementing a strategic caching mechanism can help mitigate this risk. Storing frequently accessed data locally reduces the dependency on continuous external queries.

*   **Failure scenarios**: It's also vital to plan for times when your external data sources may not be available. Imagine a situation where Nautobot goes offline – perhaps during a critical monitoring period. In such a scenario, your dynamic enrichment process could be compromised, leading to incomplete or missing data enrichment, such as the `intf_role` labels. To counter this, it's essential to have contingency plans. This could involve fallback mechanisms that ensure data enrichment continues, albeit in a limited capacity. For example, you might store a temporary snapshot of essential data to maintain continuity until the external source is restored.

By giving due consideration to these aspects, you can ensure that your dynamic data enrichment strategy not only adds value in real time but also remains robust and dependable, regardless of the scale and complexity of your observability environment. Additionally, there are other methods you can implement to enrich data without the need to inject tags into metrics as they traverse the pipeline. We'll explore these alternative approaches next.

Data enrichment at query time

When it comes to enhancing the insights that you've gained from our observability data, another strategy is to call external sources of data enrichment at different stages of the observability platform.

Info metrics

The Prometheus ecosystem introduced a concept to help decorate and enrich existing captured metrics by adding a type of metric called an **info metric**. This robust concept allows external systems, such as a source of truth, to contribute additional contextual information about the systems being monitored, even if they're not the primary targets. These info metrics are unique in that they provide additional descriptive context to your primary data. Think of them as enriching layers that add depth to the basic metric data, offering a fuller picture of each data point's significance and background. The following figure provides an overview of this concept:

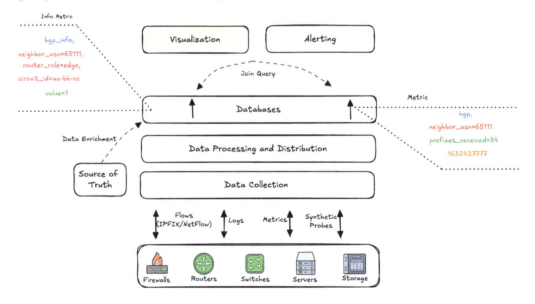

Figure 6.11 – Data enrichment with info metrics (Created using the icons from https://icons8.com)

Here, we can see two metrics: the bgp metric, which indicates the number of prefixes received and includes neighbor_asn, and an info metric called bgp_info, which is injected externally and carries data from the source of truth. The bgp_info metric has a constant value of 1 (indicating it is always on) and provides additional contextual information, such as router_role and circuit_id.

While the specifics of Prometheus info metrics will be covered in more detail in *Chapter 7*, it's important to understand that they enrich data by joining the metrics at the moment of the query. This approach allows for greater flexibility in selecting the contextual data you want to use for visualizations, alerts, or reports with the drawback of more complex queries.

Directly querying external data sources

Another approach, which can be highly specific to the tools or applications you're using, involves leveraging the ability of a data visualization or alerting engine to query and join data from two completely different systems. For instance, you might query the metrics information of devices from Prometheus while simultaneously querying additional contextual data from a source of truth such as Nautobot using GraphQL. This method allows you to dynamically and comprehensively integrate diverse data sources, enhancing the depth and accuracy of your visualizations and alerts. The following diagram illustrates this concept in detail:

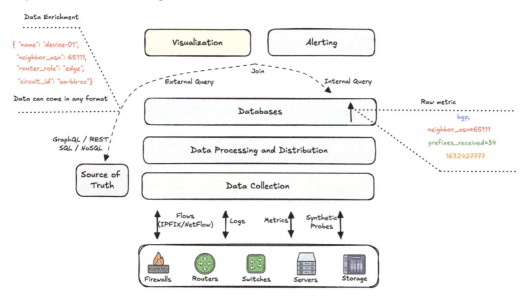

Figure 6.12 – Data enrichment with external queries (Created using the icons from https://icons8.com)

As you can see, this capability requires an application that can support multiple querying languages and technologies to connect to different systems and provide data manipulation capabilities to present all the correlated information at the application level. An example of a tool that can perform these operations is **Grafana**.

Grafana is an open source platform for monitoring and observability that allows you to visualize, analyze, and alert on metrics and logs from various data sources through customizable dashboards. Grafana leverages its querying capabilities through various plugins. These plugins enable dynamic data queries, pulling in additional context and enriching the visualizations you create. For instance, a Grafana dashboard can be configured to not only display real-time metrics but also to overlay this data with contextual information from an external system via its data source mechanism.

This capability is illustrated in the previous figure, where data is queried from the database layer to collect the BGP metric information. Additionally, an external query to a source of truth retrieves contextual information (for example, using a GraphQL query to access contextual information from Nautobot).

Later in this book, we'll explore how Prometheus and Grafana can be utilized together to achieve data enrichment at the time of querying. This will include leveraging Prometheus info metrics and utilizing Grafana's data source plugin ecosystem to enrich and enhance the data that's displayed in your observability dashboards.

The scale of the observability data pipeline

In managing observability for large or complex environments, scaling the data pipeline is often a critical challenge. This necessity arises from various demands, such as the expansive size of the monitored infrastructure, the integration requirements with other systems, the need for high system availability, and sophisticated infrastructure management. As the volume of data grows and the complexity of integrations increases, ensuring that the data pipeline can handle the load without performance degradation becomes crucial.

To address these challenges, introducing message brokers and buses into the observability data pipeline becomes essential. Message brokers and buses facilitate the efficient handling and routing of large volumes of data between various components of the observability stack. They enable asynchronous communication, decoupling data producers and consumers and ensuring that data can be processed and analyzed in a scalable and reliable manner.

In this section, we'll explain why message brokers are important for reaching these goals and improving your observability strategy. We'll also go through a lab so that you know how to use message brokers in your observability pipeline.

Why message brokers/buses matter in observability

First, we need to understand what a **message broker** or bus is. Think of these systems as postal distributors that are good at handling real-time data and routing it to multiple destinations. In IT infrastructure, they are heavily used to help deliver messages and events across a multitude of systems that may need to process them in different manners and/or purposes.

Thanks to this nature, they provide the much-needed flexibility and integration required for an observability data pipeline framework.

In this book, we will touch upon various established tools and systems within this domain, with a particular focus on **Apache Kafka**. We aim to illustrate their functionalities and advantages, particularly in the context of observability.

Kafka

Apache Kafka (`https://kafka.apache.org/`) is a robust, distributed streaming platform that's designed to handle high volumes of data. At its core, Kafka's architecture is built to ensure resiliency and scalability in data management. Let's take a closer look at its key components and functionalities:

- **Distributed architecture**: Kafka operates across a cluster of servers, known as brokers. This distributed nature is fundamental to its high availability and fault tolerance. Data within Kafka is partitioned across these brokers, ensuring that even if one part of the system fails, the rest can continue to function seamlessly.

- **Fault tolerance and scalability**: By replicating data across multiple brokers, Kafka provides a fault-tolerant system. This replication not only safeguards against data loss in the event of a broker failure but also allows the system to scale horizontally. As demand increases, additional brokers can be added to the cluster to handle more data without sacrificing performance.

- **Publish-subscribe model**: Kafka utilizes a publish-subscribe messaging model. Data streams are categorized into topics, which act as channels where data is published. Producers are the clients that publish data to these topics, while consumers are the clients that subscribe to and read data from them. This model facilitates data distribution and consumption.

- **Data persistence**: Kafka is designed for durable data storage. Records (messages) are persisted on disk, ensuring that data is not lost even if the system restarts or encounters an issue. This persistence mechanism also supports Kafka's ability to replay or rewind data streams, which is important for various data processing and analytics tasks.

- **Cluster replication**: Replicating data across the Kafka cluster not only contributes to fault tolerance but also enhances data accessibility. Consumers can read from any broker in the cluster, enabling load balancing and ensuring uninterrupted access to data streams.

The following figure showcases Kafka's architecture:

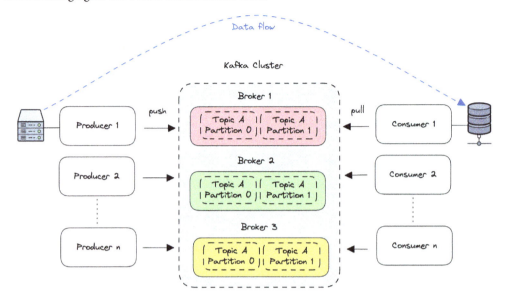

Figure 6.13 – Kafka architecture (Created using the icons from https://icons8.com)

Kafka, with its distributed architecture, typically relies on services such as **ZooKeeper** for cluster management tasks such as state maintenance, leader elections for partitions, and service synchronization. ZooKeeper acts as a centralized system, handling configuration, naming, synchronization, and group services, all of which are important for Kafka's broker management and ensuring consistent operations across the cluster.

The architecture of Kafka, which separates producers and consumers across different data topics, introduces significant scalability and flexibility to observability data pipelines. Imagine an observability stack where collectors send normalized and enriched data to databases. Incorporating a message distribution layer, such as Kafka, enables various solutions to tap into this data flow before it's processed or stored. This design not only enables real-time data connectivity but also offers an efficient way for different systems to interact with and leverage the data being collected.

Integrating a Kafka layer boosts resilience and flexibility in observability systems. However, this advantage comes with the complexity of managing an additional system. Kafka requires a setup with multiple brokers and, traditionally, a Zookeeper instance. This layer needs proper monitoring and internal visibility. In environments where multiple teams own their data collection and aim to leverage a centralized, organization-wide database or SaaS solution for time series data, a message broker solution becomes essential. It offers the necessary reliability, resiliency, and flexibility for these teams' data production and consumption needs. When considering Kafka or another messaging broker solution for your observability stack, it's important to assess your specific needs and infrastructure capabilities, ensuring that they align with your organizational objectives and technical requirements.

Practical example – Kafka in action

To bring our Kafka discussion to life, let's dive into a practical example. In our *Chapter 6* lab, we have a Kafka broker instance functioning as our central message bus. This setup allows us to route telemetry data from one Telegraf instance to another, showcasing Kafka's capabilities in real-time data handling. Additionally, we have integrated `kafka-ui` (`https://github.com/provectus/kafka-ui`), a user-friendly, open source web interface for Kafka, to give us visual insights into the data movements within our Kafka broker:

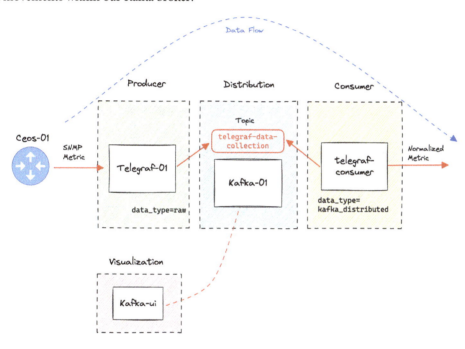

Figure 6.14 – Metrics data flow with Kafka (Created using the icons from https://icons8.com)

Here, we'll initiate our setup by configuring `telegraf-01` so that it can send metrics to a Kafka topic named `telegraf-data-collection`. Additionally, under the `inputs.snmp` configuration, we'll introduce a new tag named `data_type` with a value of `raw`. This tag will help us track the stage the data is in within our pipeline:

```
# Omitted configuration...
[[inputs.snmp]]
agents = ["ceos-01"]
# ... Output omitted for brevity ...
```

```
# Tag to identify the stage of the data pipeline
[inputs.snmp.tags]
  data_type = "raw"


# ... Output omitted for brevity …

# Send the metrics to Kafka
[[outputs.kafka]]
 brokers = ["kafka-01:29092"]
 topic = "telegraf-data-collection"
 routing_tag = "host"
```

Now, update the configuration in `telegraf-01` and apply it using the `netobs lab update telegraf-01 --scenario ch6` command.

Verify the output by examining the logs with `netobs docker logs telegraf-01 --tail=70 --follow --scenario ch6 | grep data_type`. You should see an output similar to the following:

```
interface,data_type=raw,device=ceos-01,host=telegraf-01,intf_
role=peer,name=Ethernet1 in_octets=2739624i,out_octets=0i
1705039140000000000
interface,data_type=raw,device=ceos-01,host=telegraf-
01,intf_role=peer,name=Ethernet2 in_octets=0i,out_octets=0i
1705039140000000000
```

To visually track the data through Kafka, you can browse the `kafka-ui` service deployed in the lab. Navigate to `http://<machine-ip-address>:9080` to access Kafka UI. There, you'll find the `telegraf-pipeline` cluster pre-configured from the lab's `docker-compose` setup:

Figure 6.15 – The UI for Apache Kafka home page

Explore the `telegraf-data-collection` topic under **Topics**. Then, in the **Messages** tab, observe the incoming messages:

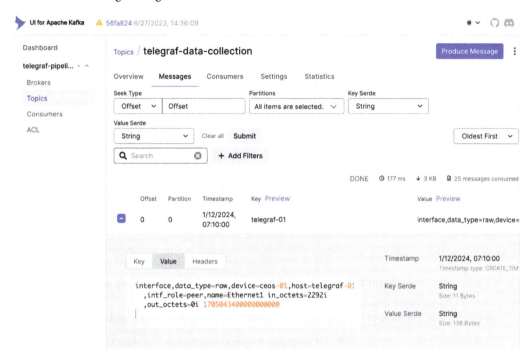

Figure 6.16 – The Kafka UI Topics page

Now that we've confirmed message flow in the `telegraf-data-collection` topic via Kafka UI, our next step is to configure a `telegraf-consumer` instance to consume data from this topic. To track and differentiate the metrics at this stage, we'll introduce a `kafka-distributed` label value to the `data_type` tag. This specific tagging helps us to identify metrics that have been processed by `telegraf-consumer` and originated from the Kafka broker. Take a look at the following Telegraf configuration snippet:

```
# Consume Kafka data (1)
[[inputs.kafka_consumer]]
  brokers = ["kafka-01:29092"]
  topics = ["telegraf-data-collection"]

  # Adding Kafka tag specifying the kafka broker
  [inputs.kafka_consumer.tags]
    kafka_pipeline = "kafka-01"

# Update the data_type tag (2)
```

```
[[processors.regex]]
 namepass = ["interface"]

 [[processors.regex.tagpass]]
   data_type = "raw"

 [[processors.regex.tags]]
   key = "data_type"
   # Regex pattern to match
   pattern = "^raw$"
   # Replacement string
   replacement = "kafka_distributed"

# Send metrics to stdout (3)
[[outputs.file]]
 files = ["stdout"]
```

Let's break down this configuration:

- **Consume metrics from Kafka**: The configuration starts by consuming metrics from the `telegraf-data-collection` Kafka topic on a Kafka broker at `kafka-01:29092`. It also adds a tag called `kafka_pipeline` to indicate the source Kafka broker.

- **Update the data_type tag**: Next, it uses the `regex` processor to update the `data_type` tag. Specifically, it looks for metrics with a `data_type` tag set to `raw` and changes this tag to `kafka_distributed`, signifying that the data has progressed to a new stage in the processing pipeline.

- **Send metrics to stdout**: Finally, the configuration sends the processed metrics to the console, allowing them to be displayed or processed further as needed.

To test this configuration, save the configuration under `network-observability-lab/chapters/ch6/telegraf/telegraf-consumer.conf.toml` and run the `netobs lab update telegraf-consumer --scenario ch6` command.

We should be able to see `telegraf-consumer` connected to the `telegraf-data-collection` topic in the Kafka UI at `http://<machine-ip-address>:9080/ui/clusters/telegraf-pipeline/consumer-groups/telegraf_metrics_consumers`:

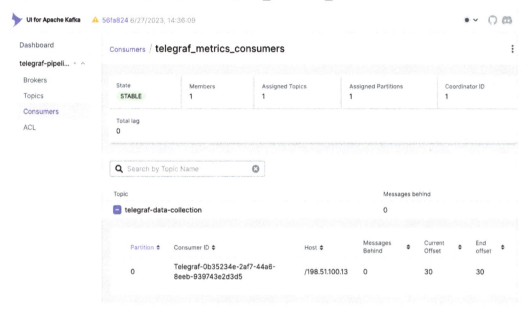

Figure 6.17 – The Kafka UI Consumers page

For our metrics, we should be able to see them being consumed by this Telegraf instance in the logs with a tag of `data_type=kafka_distributed`. This tag is indicative of metrics that have been processed through the Kafka broker.

Verify the output by examining the logs with `netobs docker logs telegraf-consumer --tail=70 --follow --scenario ch6`. You should see an output similar to the following:

```
interface,data_type=kafka_distributed,device=ceos-01,host=telegraf-
01,intf_role=peer,kafka_pipeline=kafka-01,name=Ethernet1 in_
octets=2292i,out_octets=0i 1705043400000000000
interface,data_type=kafka_distributed,device=ceos-01,host=telegraf-
01,intf_role=peer,kafka_pipeline=kafka-01,name=Ethernet2 in_
octets=0i,out_octets=0i 170504340000000000
```

This log output confirms the successful transmission of metrics: originating from `ceos-01`, collected by `telegraf-01`, routed through a Kafka broker, and finally consumed by `telegraf-consumer`.

From here on, the capabilities are endless; you can have the `telegraf-consumer` instance report this data back to another backend, perform more data manipulation and processing, or even use it to debug messages and look at the data traversing the pipeline in real time.

While Kafka is a popular choice, it is not the only message bus platform. Others, such as **Neural Autonomic Transport System** (**NATS**) (https://nats.io/) and **Message Queueing Telemetry Transport** (**MQTT**) (https://mqtt.org/), operate on a publisher and subscriber model too. Each one comes with different characteristics:

- NATS is characterized by providing low-latency and lightweight messaging capabilities. This is useful for status changes or alerts.

- MQTT is characterized by minimal bandwidth usage. This is useful for IoT scenarios where bandwidth is limited.

Analyzing these tools is outside the scope of this book. However, the key takeaway from discussing message buses and brokers, considering Kafka and the Telegraf consumer as examples, is the ability to leverage highly available and scalable data distribution systems. These systems are necessary when scale, reliability, and flexibility are essential. By publishing metrics from the Data Collection layer – whether processed or unprocessed – and enabling real-time consumption by various services or teams, you gain significant flexibility. This approach allows different teams or services to utilize the data, enhancing the overall observability pipeline and its operational value.

Summary

In this chapter, we highlighted the significance of data normalization, a practice that's particularly fundamental in multi-vendor environments with diverse data sources. The need for consistent data representation remarks the relevance of initiatives such as OpenTelemetry and the Elastic Common Schema for observability data models. Telegraf, along with its range of inputs and processor plugins, plays a vital role in this area, especially in the network industry. We delved into how Telegraf achieves data normalization through its processor plugins, and we ran practical labs that guided us in shaping metrics to fit a hypothetical desired metric model.

Similarly, we manipulated log messages with Logstash and introduced filter plugins to help us normalize this type of data. We did a deeper dive into the Grok filter plugin, which provides us with an excellent way to get structured data from the text data of log messages.

Next, we delved into data enrichment, where we concentrated on utilizing Telegraf and its processor plugins, such as `regex` and `execd`, to enhance data as it flows through the system. We also explored dynamic enrichment techniques by employing a custom script and leveraging external data sources such as Nautobot for more sophisticated data augmentation.

We wrapped up this chapter by talking about the scale of the data observability pipeline and its distribution layer. We introduced message brokers/buses as a great way to provide data distribution and resiliency while placing a deeper focus on Apache Kafka. We even ran a practical lab around Kafka brokers and Telegraf producers and consumers to help drive home the value and flexibility a message broker/bus layer can give to your observability stack.

In the next chapter, we'll discuss data persistence. There, we'll uncover various data storage solutions and explore Prometheus, our chosen backend for this book, in detail so that we can explore its capabilities and how we can integrate it within our observability framework.

7

Data Storage Solutions for Network Observability

At the heart of any observability platform lies its databases – the repositories where all data converges. Just as a well-designed network ensures optimal data flow, well-structured databases ensure seamless data storage and retrieval, allowing for historical analysis and real-time decision-making. In this chapter, we will provide an overview of various database types and concepts. We will briefly explain what these databases are, their main differences, and how they're used in observability platforms. The goal is to give you a clear understanding of the most relevant databases at a high level, without delving into overly technical specifics.

Alongside this, we'll look at practical examples using tools like **Prometheus** and **Grafana Loki**, which are popular in network observability solutions. You'll learn about their architecture and main components, with a focus on how data is written to and read from these systems. This practical approach will help you understand how to integrate databases into observability tools in a flexible way, providing you with valuable insights for managing and monitoring network systems.

The chapter covers the following topics:

- Databases for observability
- A look into Prometheus **time series database** (TSDB)
- A look at Grafana Loki
- Persistence tips and best practices

Databases for observability

When we discuss databases, we're venturing into a territory that is as vast as it is varied. But our mission here is specific – we are narrowing our focus to database types that are not just good but exceptional when it comes to observability. These databases distinguish themselves not only by their features but also by their ability to meet the unique demands of monitoring and analysis in real-time environments.

First things first – let's outline the key features of observability databases:

- **Time series proficiency**: Essential for observability, these databases are great at handling data with timestamps. They need to excel at storing, querying, and analyzing time-based data.

- **Real-time performance**: The ability to handle data in real time is essential. These systems must be strong enough to manage large amounts of data quickly, ensuring that data is both taken in and retrieved smoothly.

- **Metric handling capacity**: Observability produces many metrics with various labels to describe them. The ideal database should handle this variety well, offering proper storage, compaction, and fast retrieval of numeric data.

- **Structured data compatibility**: Beyond metrics, observability includes structured data such as logs and network flows (after proper processing). These databases should handle this data well, supporting a variety of labels and values for thorough analysis.

- **Data life cycle**: Observability tools generate a lot of data, so managing this data properly is essential. The database should have strategies for data rollover, aggregation/rollup, and deletion to maintain performance and scalability.

- **Powerful API for data operations**: A strong and versatile API is vital for data operations such as aggregation, filtering, and searching through large datasets.

- **Schema flexibility**: Flexibility in data storage is key during the development and testing phases of observability solutions. Using a schema-less approach to store observability data allows us to quickly adapt as our data and requirements change, keeping our process agile and responsive. However, when moving to production, having a structured schema becomes important. It ensures data integrity, readability, and consistency, which are key for accurate dashboards and alerts.

It sounds like a tall order, and it is. But the good news is that there are databases out there built with these very challenges in mind.

What ties all these databases together is their proficiency with time series data, a cornerstone of robust observability. Let's dive a bit deeper into what sets time series databases apart and why they're pivotal for observability.

Time series databases

A TSDB is a powerful tool designed specifically for managing data points that are linked to time. Imagine a vast sea of data points, each with an exact timestamp. That's where a TSDB excels. It's built to handle the heavy work of storing large amounts of time-stamped data and ensuring that you can insert and retrieve this data fast and reliably.

TSDBs handle time series data that is not only large in volume but also consistently structured. This uniformity allows TSDBs to be more optimized and quicker than general-purpose databases. One reason for this optimization is the use of specialized **compression algorithms** such as Gorilla compression (`https://www.vldb.org/pvldb/vol8/p1816-teller.pdf`), developed by Facebook. Gorilla's methods, such as delta-of-delta encoding for timestamps and **eXclusive-OR** (**XOR**) operations for values, help TSDBs such as Prometheus and **InfluxDB** save significant storage space. These techniques can shrink data to a fraction of its original size, making long-term storage more cost-effective and data retrieval much faster.

Moreover, TSDBs are good at managing data over time. They can regularly delete or downsample old data, which is not common in general databases that keep data indefinitely. This feature helps keep storage manageable without losing important data.

Downsampling involves reducing the granularity of data – converting high-frequency data into a more compact form while retaining essential information. This technique is particularly useful for preserving historical data that remains accessible and easy to query over long periods, all while minimizing storage requirements. One common implementation is to create summary versions of data once it surpasses a specific age threshold. The TSDB then optimizes query performance and storage use for these summarized datasets, ensuring that long-term data remains both compact to store and quick to retrieve.

As we prepare to explore different TSDBs, it's important to understand two key ideas – dimensionality and cardinality. These concepts are central to how these databases work, affecting how they store and retrieve data. Understanding them helps us choose the right database for our needs:

- **Dimensionality** refers to the attributes or tags that describe each piece of data in a time series. Think of it like the details on a name tag. In a database tracking server metrics, these tags might include the server's ID, its role, or its model. More tags mean richer data analysis and better insights. However, there is a trade-off – more tags add complexity to storing and querying data. It's like trying to find a book in a library where every book fits into many categories.

- **Cardinality** refers to the number of unique values within each tag. Using the library analogy, if one of your tags is the genre of a book, the cardinality is the total number of unique genres available. In the context of monitoring routers, if you have a tag for the router name, the cardinality is the count of all the different routers you monitor. However, high cardinality increases storage needs and query complexity. This is because each unique tag combination creates a new entry in your database, requiring storage and indexing. Imagine a library where every single book is its own genre – that's the challenge that high cardinality presents.

So, what does all this mean for TSDBs? Let's break this down:

- **Storage and performance**: High cardinality and dimensionality can cause exponential growth in data volume. This growth can strain your storage capacity and slow down queries as a database processes an ever-increasing number of data points.

- **Query complexity**: More tags and unique values make queries more complicated. This complexity can slow down access to real-time data and insights.

- **Design considerations**: Managing a high-cardinality system well requires smart design choices. This means using good indexing strategies and data compression to keep a database productive and fast.

- **Scalability challenges**: To handle high dimensionality and cardinality, TSDBs need to scale smartly. This means not only storing more data but also having enough processing power to keep things running smoothly.

Matching databases with observability needs

Navigating through the myriad of databases for your observability platform means understanding what job needs doing. It's all about the data – what kind, how it's stored, and how it's accessed. Let's break it down by use case.

> **Note**
> Check *Chapter 3* to review the different types of data observability brings.

Metrics

For tracking application or system metrics, time series data is the principal actor, and the TSDBs are your go-to. Here's a quick look at some notable options:

- **Prometheus**: A popular choice for many, Prometheus is an open source TSDB that is widely used and supported by a strong community. It collects and stores metrics, offers powerful querying, and is highly customizable. Prometheus is also the base for many other databases and tools, making it very versatile.

- **Thanos**: Built on Prometheus, Thanos is made for large-scale environments that need distributed, high-availability setups. It adds features such as long-term storage, global querying across multiple Prometheus servers, and better reliability. Thanos is perfect for organizations that need to scale their monitoring while staying robust and resourceful.

- **Mimir**: Another option based on Prometheus, Mimir is supported by **Grafana Labs**. It's similar to Thanos, offering scale and high availability for large environments that need efficient metrics storage.

- **InfluxDB**: InfluxDB is known for its strong performance in monitoring applications. It's good at processing data optimally. The release of InfluxDB 3.0 brought big improvements, with a columnar data structure based on Apache Arrow, which helps with faster data ingestion and querying in large environments.

- **Promscale**: Created by Timescale, this tool is designed to handle metrics and trace data for Prometheus and Jaeger. Built on **PostgreSQL** and **TimescaleDB**, it allows you to query metrics using **PromQL** (**Prometheus Query Language**) and standard SQL, making it easy to work with observability data.

- **Elasticsearch (Elastic Observability)**: Known for logs and analytics, **Elasticsearch** is now also becoming popular for handling metrics. With its focus on observability and time series data, it provides strong capabilities to monitor and analyze metrics.

Logs

Examining log data is not just about reading text records; it's also about finding the stories they tell about your systems. The key is to use good search, filter, and parsing techniques, especially when the data is just text that needs to be decoded, as with syslog data (which includes fields such as source, priority, and facility – see RFC 5424 at `https://datatracker.ietf.org/doc/html/rfc5424` for more details).

These logs have valuable insights waiting to be found. To uncover these insights, you need databases that are not only strong but also good at handling and managing text data.

Here is a closer look at some standout tools in the log management realm:

- **Elasticsearch**: Known for its fast log search and analysis, Elasticsearch is great at handling large sets of log data. Its powerful querying capabilities make it especially useful for engineers working on observability.

- **Grafana Loki**: Created by the team at Grafana Labs, **Loki** is inspired by Prometheus and works smoothly with Grafana. It focuses on storing and querying logs using **LogQL**, a language inspired by Prometheus' PromQL (which we will use in the following sections).

- **Splunk**: A robust platform for log aggregation and analysis, celebrated for its sophisticated data handling. It offers swift log visibility, requiring minimal effort from the team – a significant advantage when engaging with less technical users.

Traces

In the realm of applications and software development, tracing involves recording the path and timing of transactions as they move through a system, much like tracking a package through various checkpoints. A trace is essentially a map of a single transaction or request as it travels through various parts of an application. It provides a timeline of all the steps involved in processing that request. Traces play an important role in observability, which is about understanding what is happening on the applications that your systems host. By examining traces, you can see how different services interact, spot where issues arise, and understand the overall health and performance of your application.

To handle and store this tracing data, there are specialized backends designed to properly manage and query traces.

- **Jaeger**: Jaeger is an open source tool for end-to-end distributed tracing. It is used to monitor and troubleshoot microservices-based distributed systems. With Jaeger, you can track how requests move through various services, identify performance issues, and analyze latency problems.

- **Zipkin**: Zipkin is another popular tool for distributed tracing. It's known for being easy to use and works with various programming languages and transport protocols.

- **Elastic APM**: Part of the Elastic Stack, Elastic APM provides **application performance monitoring** (**APM**) capabilities, including distributed tracing. It helps you monitor the performance of your applications and services, giving you insights into latency issues, error rates, and overall system health. Elastic APM integrates well with other tools in the Elastic Stack, such as Elasticsearch and Kibana.

- **Grafana Tempo**: Grafana Tempo is a high-volume, minimal-dependency distributed tracing system. It is designed to integrate seamlessly with the Grafana observability stack. Tempo allows you to collect, store, and query traces from your applications, helping you understand their performance and behavior. Its integration with Grafana makes it easy to visualize and analyze trace data alongside metrics and logs.

Packet flow data (NetFlow, IPFIX, and sFlow)

Packet flow data from network devices is complex. It has many detailed attributes (high dimensionality) and lots of different values for each attribute, such as source and destination IP addresses (high cardinality). This data also comes in copious amounts (heavy volumes). Managing this kind of data requires specialized databases.

Columnar-table databases are becoming popular for this task. They store data in columns instead of rows, which makes them very efficient for certain operations. For example, if you need to scan a large dataset or perform calculations across many records but only need to look at a few attributes, columnar-table databases handle this very well. This makes them good candidates for dealing with complex and large-scale data from network devices. Here are some example databases that are used for this use case:

- **InfluxDB**: InfluxDB is great for handling detailed and large volumes of data. It supports analysis and storage, making it easier to work with complex network data.

- **Elasticsearch**: Elasticsearch adapts well to packet flow data analysis. Its powerful search capabilities allow you to gain quick insights into network traffic patterns.

- **ClickHouse**: ClickHouse is known for its high-performance query capabilities on large datasets. This makes it ideal for analyzing complex network traffic and packet flows, allowing you to process and query vast amounts of data quickly.

- **Apache Druid**: Apache Druid is designed for real-time data ingestion and analytics. It is particularly effective for immediate analysis of network flow data, giving you quick insights into what is happening in your network right now.

- **Grafana Loki**: While primarily focused on logs, Grafana Loki can also be applied to packet flow data where log-like analysis is useful. It helps you analyze packet flow data similar to how you would analyze logs.

The world of monitoring and observability is full of tools and solutions, each offering something unique. Big names such as Splunk, InfluxDB, Elastic, and Grafana Labs provide all-in-one packages ready to tackle these challenges.

The databases we have discussed so far are just the beginning. The observability field includes many other specialized databases for handling time series data, such as **VictoriaMetrics**. Each of these databases has its own strengths and weaknesses, making each one suitable for different use cases. Some might excel in performance at scale, others in data retention and query efficiency, or in providing detailed data analysis tools. The key is to match the specific needs of your network environment with the capabilities of these databases to find the best solution for your observability needs.

In this book, for hands-on experience, we've chosen Prometheus for its excellent features to manage time series data, powerful query language (PromQL), and ample learning resources. For logs, we're using Grafana Loki, which offers a query setup similar to Prometheus with LogQL.

Next, we'll dive into how these tools work, focusing on how they manage data – both in terms of storing it and retrieving it.

A look into Prometheus TSDB

Let's dive in on Prometheus, a name that has become synonymous with metrics monitoring and TSDB standards, especially among system administrators and DevOps teams.

Conceived by the SoundCloud team in 2012, Prometheus was inspired by Google's **Borgmon** monitoring project, bringing powerful metrics and scalability into the open source world. Its talent for handling large-scale data input and rapid data access, combined with an easy-to-deploy nature in cloud-native environments, quickly cemented its status as a go-to tool, particularly for those navigating the Kubernetes landscape.

However, what truly sets Prometheus apart is its strong commitment to community and open source principles. This dedication hasn't gone unnoticed, earning it a spot in the **Cloud Native Computing Foundation** (CNCF) and the distinguished award of being a Graduated project since 2018 (`https://www.cncf.io/projects/prometheus/`).

Now, let's dive into the nuts and bolts of Prometheus, exploring the architecture and components that define its approach to data monitoring.

Prometheus architecture

Prometheus is built with specific components, each designed for a part of the TSDB system. The following is a diagram from their official documentation (`https://prometheus.io/docs/introduction/overview/`) that shows the main parts:

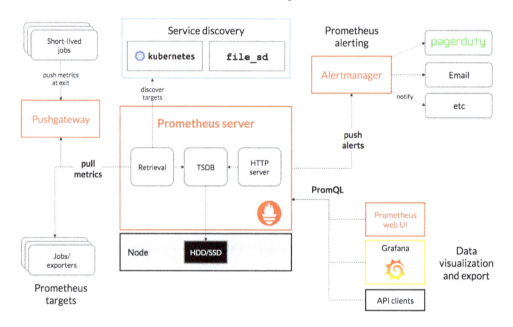

Figure 7.1 – Prometheus architecture (source: https://prometheus.io/docs/introduction/overview/, produced via CC BY 4.0, Attribution 4.0 International)

Next, we'll break down these components and their roles in Prometheus.

Prometheus server

At the core of everything, the Prometheus Server acts as the system's heartbeat. It's all about gathering and sharing data, managing the TSDB for storage, and running an HTTP server for queries and management tasks.

Data retrieval

Prometheus collects data by *scraping* metrics. If you're new to Prometheus, think of scraping as the process where Prometheus regularly checks (or *scrapes*) specified sources at fixed intervals to gather the latest data. These sources can be applications or services that have been set up (or *instrumented*) to expose their current state in a format that Prometheus understands.

There are two main ways Prometheus gets this data:

- **Direct scraping from instrumented jobs (pull mode)**: This is when services or applications are configured to expose their metrics directly to Prometheus. Imagine Prometheus as a visitor that periodically checks a designated spot where services leave updates about their health and performance.

- **Using the Pushgateway for indirect sources (push mode)**: Some applications can't be scraped directly, perhaps because they are short-lived tasks or cannot expose metrics due to their environment. For these situations, Prometheus uses a **pushgateway** – similar to a drop box where these applications can deposit their metrics. Prometheus then collects these metrics from the pushgateway later.

Prometheus is quite flexible in finding what to monitor. It can be told explicitly through its configuration (i.e., statically defined) where to look, or it can dynamically discover targets as they appear online in environments such as Kubernetes, keeping monitoring seamless and up to date.

We'll dive deeper into how all this works and how to set it up in the next section in the *Writing to Prometheus TSDB* section.

TSDB storage

The real action happens in the TSDB, where data lives. Here's how it brings a bit of magic to data management:

- **Writing and reading to disk**: Prometheus's TSDB is designed to write this time-stamped data to disk (such as HDDs or SSDs). This means it can save vast amounts of data in a way that makes it quick to retrieve or "read" when you need to analyze your metrics.

- **Remote storage options**: Besides saving data locally, Prometheus's TSDB can also integrate with remote storage solutions (e.g., Thanos, Grafana Mimir, and InfluxDB). This flexibility allows you to scale your storage needs or integrate with your existing data management systems.

- **The nuts and bolts of storage**: For those curious about the inner workings, the TSDB employs several techniques to handle data effectively. This includes the following:

  - **The TSDB format**: The specific way data is formatted for storage, ensuring compactness and speed. For more information, see the format on GitHub (`https://github.com/prometheus/prometheus/tree/main/tsdb/docs/format`).

- Write-Ahead Log (WAL): A system that records data first to a *log* before it's written to the database, ensuring no data loss even if something unexpected happens. The *Prometheus TSDB (Part 2): WAL and Checkpoint* article (`https://ganeshvernekar.com/blog/prometheus-tsdb-wal-and-checkpoint/`) provides a good overview of how a WAL works for Prometheus.

- Blocks and checkpoints: Methods for organizing data in chunks (blocks) and creating regular save points (checkpoints) for proper data management and recovery.

The HTTP server

The HTTP server acts as a bridge between your data and the insights you can glean from it. Here's how it simplifies interaction with Prometheus:

- Querying with PromQL: Prometheus introduced its own query language, PromQL, which you can use through the HTTP Server to ask questions about your data. Whether you're looking into trends, anomalies, or the current state of your systems, PromQL lets you dive deep into your metrics with precision and flexibility. We will go into more detail about querying Prometheus data in the *Reading from Prometheus TSDB (PromQL)* subsection.

- Dynamic configuration updates: Beyond data queries, the HTTP Server plays a key role in managing Prometheus itself. It allows for dynamic updates to your monitoring setup without needing to restart Prometheus. This means you can change how and what you're monitoring *on the fly*, adapting quickly to new requirements. For more information on this feature, see the official documentation (`https://prometheus.io/docs/prometheus/latest/management_api/#reload`).

- Health checks: The HTTP Server also provides endpoints to check the health and status of your Prometheus server.

Alerts

Alerts are your early warning system in Prometheus, designed to notify you when something isn't quite right. Here's how the alerting process works:

- Triggering alerts: Prometheus constantly evaluates the alerting rules you've set up. When certain conditions are met (such as a metric reaching a critical threshold), Prometheus flags it and triggers an alert.

- Alertmanager takes over: Once an alert is triggered, it's sent to a component called the **Alertmanager**. Think of the Alertmanager as a control center for all your alerts. It's where the management of these notifications happens. It can do the following:

  - Organize: Group similar alerts together to avoid flooding your inbox or chat with too many messages

- **Mute**: Silence alerts that aren't immediately actionable or relevant

- **Forward**: Send the important alerts to the right places, whether that's your email, Slack, or any other notification channel that you use to stay informed

This setup ensures that you are alerted to potential issues promptly but without unnecessary noise. We will dive deeper into how to set up and manage alerts in Prometheus in *Chapter 9*.

Exploring data

Prometheus provides a web interface that acts as a window into the collected metrics, configured alert rules, system status, and more. Here is why it's a valuable tool:

- **Instant access to metrics**: The web interface, through the expressions browser (`https://prometheus.io/docs/visualization/browser/`), allows you to quickly view and explore the metrics that Prometheus has collected. It's like having a dashboard where you can see the data stored in the TSDB at a glance.

- **Querying made easy**: Besides just viewing metrics, you can use this interface to run PromQL queries. This means you can ask specific questions about your data, filter for certain conditions, or even aggregate data points across time, all through a user-friendly web interface.

- **Configuration and status insights**: The interface doesn't stop at data. It also gives you a peek into how Prometheus is configured and its operational status. This is useful for troubleshooting or making sure that Prometheus monitors exactly what you intend it to.

Prometheus data model

At the very heart of the heart (the Prometheus server), we have the Prometheus data model. This model outlines how metrics are structured and stored within the TSDB.

Figure 7.2 – The Prometheus data model

Here's a breakdown of the model:

- **Metric name**: The specific identifier for a measurement. It's more precise than a general measurement name that you might find in other formats, such as InfluxDB's line protocol.

- **Metric labels**: Similar to tags in other systems such as Telegraf, labels are key-value pairs attached to metrics. They add detail and context, helping you distinguish between different measurements or instances of a metric.

- **Samples or values**: These are the actual data points. Each sample includes a numerical value and a timestamp, marking when the measurement was taken.

Every time series in Prometheus is uniquely identified by its metric name, combined with its labels, making each one distinct. Here's what three different metrics might look like:

```
interface_in_octets {intf_role="mgmt", name="Management0"} 47747 @ 1632427777

interface_in_octets {intf_role="peer", name="Ethernet1"}  24566 @ 1632427777

interface_in_octets {name="Ethernet2"}                    12345 @ 1632427777
```

Figure 7.3 – A Prometheus metrics example

From the Influx line protocol to the Prometheus data model

In this chapter, our lab involves collecting metrics from Telegraf and storing them in Prometheus. Before we dive into practical examples with Prometheus and PromQL, it's important to understand the mapping and comparison between these two models. So far, we have analyzed network device metrics collected with Telegraf in *Chapters 5* and *6* and how they are represented using the **Influx Line Protocol**. If you haven't noticed this, take a moment to review those chapters for a deeper understanding. Now, let's explore how these metrics translate into Prometheus metrics. This understanding will be important for our upcoming lab activities, where we will work hands-on with Prometheus and see how it handles and processes metrics:

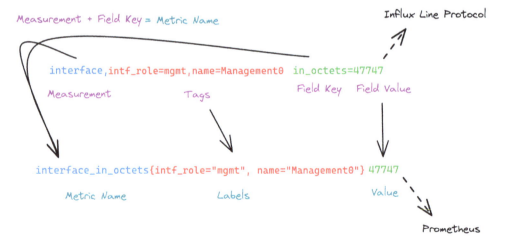

Figure 7.4 – The Influx line protocol to Prometheus format

Let's break down the preceding comparison figure:

- **Measurement name to metric namespace**: In Influx, the measurement name serves a similar purpose to the metric namespace in Prometheus, organizing metrics into groups.

- **Tags to labels**: Both Influx tags and Prometheus labels describe metrics further, adding context and dimensions for filtering and querying.

- **Fields to metrics**: In Influx, fields hold data values. In Prometheus, these become individual metrics. Telegraf's `outputs.prometheus_client` turns each Influx field into a Prometheus metric, combining the Influx measurement name and field name, separated by an underscore.

This conversion process ensures that detailed Telegraf metrics are neatly packaged for Prometheus.

Metric types

Prometheus organizes the data it collects into four main types of metrics, each suited for different purposes:

- **Counters**: Imagine a step counter; it only goes up. Prometheus counters are similar and are used to track things that increase over time, such as `interface_in_octets`.

- **Gauges**: Think of a gauge as a thermometer; the temperature can go up or down. Gauges in Prometheus work the same way, tracking values that can change in either direction, such as the amount of free memory in a system (i.e., `memory_used`).

- **Histograms**: Imagine you're looking at a chart showing how long it takes data packets to travel across your network. This is what a histogram in Prometheus might help you visualize. It categorizes these travel times into buckets, such as how many packets take less than 10 ms, between 10 ms and 50 ms, or more than 50 ms to reach their destination.

- **Summaries**: Summaries are like getting a detailed report on your network's packet loss rates. Instead of just an average, summaries provide insights into the range of packet loss experiences. They might show you that 50% of the time, your packet loss is below 0.5%, but in the worst 10% of cases, it jumps to 2%. This detailed view helps pinpoint problems in network segments that might not be clear when looking at average data alone.

In addition to these standard metrics in Prometheus, there are metrics designed to provide more context and data to other metrics. These are known as **info metrics**. Let's take a closer look at info metrics next.

Prometheus info metrics

Prometheus info metrics, often suffixed with `_info`, are a special type of metric used to represent static, descriptive information about an entity. Unlike traditional metrics that represent numerical data changing over time, info metrics convey metadata or state information that doesn't usually change, or changes infrequently.

> **Note**
>
> This is another method of **data enrichment** for the metrics in the overall observability solution; please refer to *Chapter 6* for a more in-depth view of the topic.

Here are some of the most common benefits of using info metrics:

- **Representing static attributes**: Info metrics can be used to expose static labels (key-value pairs) that provide more descriptive information about an entity. For instance, `device_info{rack="rack-012", row="17", cage="cage-17"}` could represent information about a device's physical location in a data center.

- **Joining with other metrics**: They are often used in conjunction with other Prometheus metrics to add context or labels to the numerical data. This is especially useful when you want to correlate metrics with static attributes. For example, let's take a look at the `topk(5, sot_device_info{role="border"} * on(device) group_right() interface_out_utilization_ratio{})` query, where, without diving too deep into the query, it matches metrics coming from your **SoT** (**Source of Truth**) as device information, using the `border` label against the `interface_out_utilization_ratio` live metrics captured from the network device. The resulting metrics will be matched by the device label and enriched with labels coming from your SoT. And if you were wondering, we will discuss all the components of this query later in this chapter.

- **Configuration and version tracking**: They can be used to track configurations, versions, or deployment information, which is particularly useful in observing changes over time in your infrastructure or applications. For example, imagine having a metric such as `device_info{config_version="v2024-01-07-prod"}`, where the label indicates the configuration version deployed on the device.

Here are some important considerations to keep in mind when using info metrics:

- **Cardinality issues**: While info metrics are valuable, they can increase metric cardinality if not used carefully. High cardinality can impact the performance of the Prometheus server.

- **Update frequency**: Since info metrics are for static or rarely changing information, they should not be used for frequently changing data.

As an example of an info metric, let's take a look at this metric in Prometheus:

```
go_info{job="telegraf", version="go1.20.5"}    1
```

This `go_info` metric provides additional context and data through its labels. In this example, the value is always 1, indicating presence, while the labels (`instance`, `job`, and `version`) provide valuable information about the Golang version for the Prometheus instance, the type of job that was scraped, and the source of the data. The key value of info metrics lies in these labels, which offer detailed context to understand and analyze other metrics.

Writing to Prometheus TSDB

Prometheus has a unique way of collecting data, known as **scraping**. It reaches out and pulls in data from systems or services ready to share their metrics. These sources just need to expose their metrics over an HTTP endpoint in the format Prometheus understands.

But what if a system prefers to send its data rather than wait to be asked? That's where the **pushgateway** comes into play. Think of it as an intermediary. Systems send (or *push*) their metrics to the pushgateway, and then Prometheus swings by to collect (or *scrape*) them later.

There are many external systems with valuable metrics that don't natively support Prometheus. This is where **exporters** come in – they act as translators, exposing these external system metrics in a format that Prometheus can understand and scrape. A popular example is the node exporter (`https://github.com/prometheus/node_exporter`), which shares insights into hardware and OS metrics for Unix-like systems.

Let's explore this process by diving into our practical scenario for this chapter.

Setup lab environment

In this lab, we will focus on storing and querying the metrics and log information we collected and enriched from the network devices in previous chapters, primarily using Prometheus and Loki:

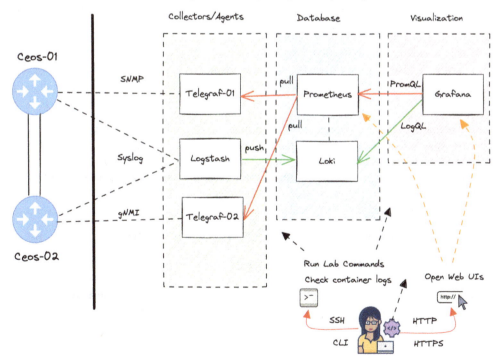

Figure 7.5 – A lab scenario (Created using the icons from https://icons8.com)

We're setting up our lab to enable the Prometheus server to start its mission – scraping metrics collected from two network devices, ceos-01 and ceos-02. Our goal is to configure Telegraf on these devices to serve the collected metrics through an HTTP endpoint, making them ready for Prometheus to gather. These metrics, which have been normalized and enriched through our lab exercises in earlier chapters, will then be stored by Prometheus.

Similarly, we will configure Loki to handle log data. Logstash will read and parse the syslog information collected from the network devices and store the resulting data in Loki. Grafana will be briefly introduced to help us read the log and metrics information from these systems, but we will explore its role and functionality in more detail in the next chapter.

You will get a detailed understanding of these tools, such as writing and reading from them, as you encounter their application and significance within the lab exercises.

> **Note**
>
> For this lab, the Telegraf and Logstash collectors are already configured to collect metrics from ceos-01 and ceos-02, with some metrics and logs already normalized and enriched.

Before diving into the lab activities, ensure that your lab machine is set up according to the steps outlined in the *Appendix A*, just as we did in previous chapters. Make sure you are in the ~/<path-to-git-project>/network-observability-lab/ directory within the project's repository to execute the necessary commands.

Now, let's begin the lab scenario for this chapter. To set everything up on your lab machine, run the netobs lab prepare --scenario ch7 command.

> **Note**
>
> This lab scenario includes a completed version that you can refer to. It is located in the Git repository you cloned, under network-observability-lab/chapters/<chapter-number>-completed. This folder contains all the configurations used in the examples throughout that chapter, providing a comprehensive guide to ensure that you have everything set up correctly.

Now, let's start with writing metrics to our Prometheus TSDB using Telegraf.

Configuring Telegraf output for Prometheus

The Telegraf configuration for this chapter has been completed to the point where we are actively collecting and enhancing metrics from the network devices. Now, we need to expose this metric information so that Prometheus can scrape and store it.

To prepare Telegraf's metrics for Prometheus, we will use the `prometheus_client` Telegraf output plugin. This plugin helps us convert the metrics from Telegraf's default format (the Influx line protocol) into a format that Prometheus can understand (the Prometheus data model).

The configuration to expose the metrics for Prometheus needs to be set up for both `telegraf-01` and `telegraf-02`. Let's take a look at the following configuration snippet for `telegraf-01`:

```
# Rest of the telegraf configuration omitted...
[[outputs.prometheus_client]]
  # HTTP port to listen on
  listen = ":9004"
```

In this configuration, we specify the Prometheus HTTP client to expose the metrics on port 9004. By default, the plugin will convert any metrics it processes into the Prometheus data model format. (Refer to the earlier section on converting the Influx line protocol to the Prometheus Data Model if you need a refresher.)

For `telegraf-02`, we will use port 9005. These ports need to be different, since the services reside on the same Docker network:

```
# Rest of the telegraf configuration omitted...
[[outputs.prometheus_client]]
  # HTTP port to listen on
  listen = ":9005"
```

This setup configures each Telegraf instance to broadcast its metrics, making `telegraf-01` accessible on port 9004 and `telegraf-02` on port 9005.

To bring this configuration to life, copy these snippets into the Telegraf configuration directory we set aside for *Chapter 7*. Make sure that each snippet is saved in its corresponding Telegraf instance TOML file (`network-observability-lab/chapters/ch7/telegraf/telegraf-<number>.conf.toml`).

Now, let's apply the configuration by running the `netobs lab update telegraf-01 telegraf-02 --scenario ch7` command. This should load our latest configuration changes.

Once everything is in place, check out the metrics now on display by the Prometheus client plugin. Visit `http://<machine-ip-address>:9004/metrics` and `http://<machine-ip-address>:9005/metrics` in your browser. You'll see pages with the newly exposed metrics, using the Prometheus data format:

```
# HELP bgp_neighbor_state Telegraf collected metric
# TYPE bgp_neighbor_state untyped
bgp_neighbor_state{collection_type="exec",device="ceos-01",host="telegraf-01",neighbor="10.1.2.2",neighbor_asn="65222",site="lab-site-01",vrf="default"} 1
bgp_neighbor_state{collection_type="exec",device="ceos-01",host="telegraf-01",neighbor="10.1.7.2",neighbor_asn="65222",site="lab-site-01",vrf="default"} 4

# HELP bgp_prefixes_accepted Telegraf collected metric
# TYPE bgp_prefixes_accepted untyped
bgp_prefixes_accepted{collection_type="exec",device="ceos-01",host="telegraf-01",neighbor="10.1.2.2",neighbor_asn="65222",site="lab-site-01",vrf="default"} 1

# HELP bgp_prefixes_received Telegraf collected metric
# TYPE bgp_prefixes_received untyped
bgp_prefixes_received{collection_type="exec",device="ceos-01",host="telegraf-01",neighbor="10.1.2.2",neighbor_asn="65222",site="lab-site-01",vrf="default"} 1

# HELP cpu_used Telegraf collected metric
# TYPE cpu_used untyped
cpu_used{collection_type="gnmi",device="ceos-01",host="telegraf-01",name="CPU0",site="lab-site-01"} 3
cpu_used{collection_type="gnmi",device="ceos-01",host="telegraf-01",name="CPU1",site="lab-site-01"} 7
cpu_used{collection_type="gnmi",device="ceos-01",host="telegraf-01",name="CPU2",site="lab-site-01"} 7
cpu_used{collection_type="gnmi",device="ceos-01",host="telegraf-01",name="CPU3",site="lab-site-01"} 30
```

Figure 7.6 – The Telegraf Prometheus client HTTP output

Now, let's explore the Prometheus configuration to scrape these metrics exposed by the Telegraf collectors.

Prometheus scrape jobs

With our Telegraf instances ready to share their metrics, it's time to configure Prometheus to start collecting them. This involves introducing Prometheus to `telegraf-01` and `telegraf-02`, through scrape jobs.

Prometheus configuration is specified in a **YAML (Yet Another Markup Language)** file, where you can define various aspects of its behavior, including scrape jobs (to collect metrics), alert rules (to trigger alerts), alertmanager configuration (to send alerts), and remote write configuration (to forward metrics to compatible TSDB backends such as Mimir or Thanos).

To get Prometheus ready for our lab environment, we will set up a specific job within its configuration to regularly scrape metrics from our Telegraf instances. Here's a preview of what that setup looks like:

```
scrape_configs:
# Scrape job to collect from the Telegraf instances
- job_name: "telegraf"
```

```
  scrape_interval: 15s
  static_configs:
    - targets: ["telegraf-01:9004", "telegraf-02:9005"]

# Scrape job to collect internal Prometheus metrics
- job_name: "prometheus"
  scrape_interval: 15s
  static_configs:
    - targets: ["localhost:9090"]
```

This snippet acts like a map and schedule for Prometheus, instructing it to visit `telegraf-01` and `telegraf-02` every 15 seconds to collect the metrics they're broadcasting. The collected Telegraf metrics are tagged with `job=telegraf` and `instance=<telegraf-address>` labels for easy identification. Additionally, you can create multiple scrape jobs within the configuration. In this case, we also create a scrape job to collect internal Prometheus metrics.

It is time to test this setup; let's replace the existing configuration with the preceding snippet, save this configuration in Prometheus's configuration file (`network-observability-lab/chapters/ch7/prometheus/prometheus.yml`), and apply it with `netobs lab update prometheus --scenario ch7`.

Head over to your Prometheus instance at `http://<machine-ip-address>:9090`, navigate through **Status | Targets**, and you should land on a page that shows both local and Telegraf endpoints. Here, you can find useful stats such as when the last scrape happened and how long it took:

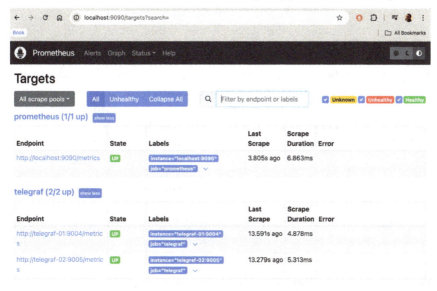

Figure 7.7 – Prometheus target status

Here, you can find useful information such as the configuration of the Telegraf endpoints, the state of those endpoints based on the last time Prometheus scraped the metrics, and other details about the scrape, such as when the last scrape occurred and how long it took.

Now for the moment of truth – let's see those Telegraf metrics in Prometheus. Simply go back to the Prometheus home page and search for `bgp_prefixes_accepted`. If you see data populating, you've successfully bridged Telegraf and Prometheus!

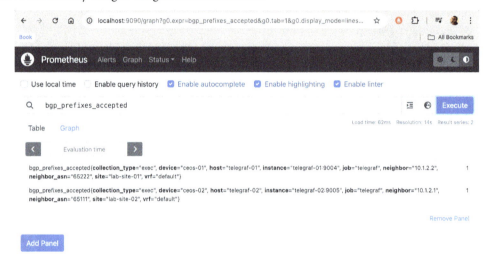

Figure 7.8 – Prometheus network metrics

Dynamic discovery for scrape targets

When setting up the Prometheus configuration, Prometheus offers several methods to determine which targets to scrape. One powerful feature is service discovery, which automates the process of finding and updating the list of targets. This means Prometheus can automatically detect what needs monitoring and adapt as your environment changes. Let's break down how this works, focusing on two particularly user-friendly methods – file-based service discovery and HTTP service discovery.

- **File-based service discovery**: Imagine you have a digital directory where you list all the services you want Prometheus to keep an eye on. This is essentially what file-based service discovery is. You create a file (in the YAML or JSON format) that details all your scrape targets, along with any specific information about them (metadata). Prometheus checks this file regularly for any updates. If you add a new service or remove an old one, Prometheus automatically adjusts its target list based on this file, so there's no need to restart or manually update Prometheus.

- **HTTP service discovery**: This method takes the idea of the file-based approach and puts it online. Instead of a file, you have an HTTP server endpoint that lists your scrape targets in a similar format. Prometheus reaches out to this endpoint to get the latest list of what to monitor.

Like the file-based method, HTTP service discovery allows Prometheus to dynamically adjust its monitoring targets based on the information served by the HTTP endpoint. It's particularly useful when you have an external system managing your service inventory (i.e., NetBox, Nautobot, or Device42) or when you prefer to centralize your configuration in a service that Prometheus can query directly.

The following configuration snippet demonstrates various service discovery methods in Prometheus. Note that this is provided for illustrative purposes only and should not be used in our lab environment:

```
scrape_configs:
  # File-based service discovery
  - job_name: 'file_sd'
    file_sd_configs:
      - files:
          - '/path/to/targets/*.json'

  # HTTP service discovery
  - job_name: 'http_sd'
    http_sd_configs:
      - url: 'http://network-observability-lab/targets'
        refresh_interval: 5m
```

Let's break down this example:

- `file_sd_configs` is used for file-based service discovery, where Prometheus reads target configurations from JSON files located at `/path/to/targets/`

- `http_sd_configs` is used for HTTP service discovery, where Prometheus fetches the target list from a specified URL (`http://network-observability-lab/targets`) every five minutes

While this configuration snippet is just to showcase the capability, the key takeaway is how `file_sd_configs` and `http_sd_configs` are set up. Unlike `static_config`, where targets are manually configured, these methods allow targets to be dynamically discovered and updated, making the setup more flexible and adaptive to changes in your environment.

In modern IT environments, where services can come and go or scale up and down rapidly, manually tracking what needs monitoring is impractical. Dynamic service discovery simplifies this by letting Prometheus adapt to your infrastructure. Whether you are using a simple file or leveraging an HTTP service, integrating Prometheus with service inventories becomes straightforward, enhancing your observability deployment strategy without added complexity.

Fantastic! With Prometheus now integrated into our observability lab and network metrics flowing in, our next task involves querying the data stored in Prometheus's TSDB.

Reading from Prometheus TSDB (PromQL)

Now, let's dive into the *read* aspect of the data stored in the Prometheus server. The focus of this section is on PromQL. This powerful language has become quite popular in the observability realm, with many external systems supporting it (for example, Grafana, which we will explore in the next chapter).

PromQL allows you to query and manipulate your metrics data. You can select specific information and perform various aggregations (e.g., average and sum) on the data returned. You might not have realized it, but we have already executed a couple of queries in our previous examples – to check for metrics coming from our Telegraf instances.

The language is robust and offers a plethora of capabilities to analyze and visualize your metrics data. The following sections will explain these capabilities in detail and provide opportunities to test them in our observability lab.

Running queries in Prometheus

Before diving into the specifics of Prometheus queries, let's ensure that you know how to execute these queries on your Prometheus server. Prometheus provides a user-friendly web interface to run queries, visualize data, and explore metrics in real time. We have seen a glimpse of this in the previous section, so let's continue building more PromQL queries:

1. Navigate to the Prometheus expression browser by entering the URL `http://<machine-ip-address>:9090`. This is where you'll type or paste your PromQL queries.

2. Enter your query in the expression input box. This could be something as simple as `device_uptime{device="ceos-01"}` to check a device's uptime. Hit **Execute** to see the results. For time-based queries, consider switching to the **Graph** view for a visual representation of how metrics change over time.

Figure 7.9 – The Prometheus expression browser

You can also run queries programmatically using Prometheus' API, which we will cover in *Chapter 12*.

Navigating the world of Prometheus and PromQL might feel like stepping into a new realm, but at its core, Prometheus and its querying language simplify the task of monitoring your network infrastructure. Let's dive deeper into PromQL to understand it better, enabling us to create more flexible and powerful queries. We'll focus on data types, selectors, operators, and how they relate to the network metrics we have been collecting.

Prometheus data types demystified

To successfully navigate and use PromQL for your infrastructure metrics, it's important to understand some of the Prometheus data types.

Instant vector

Think of an **instant vector** as a single frame in a movie, capturing a specific moment across multiple series. Each series in this frame records just one data point, all synchronized to the same timestamp. For instance, using the `device_uptime{device="ceos-01"}` query, it fetches the latest uptime for `ceos-01`, providing immediate insight into device availability. By running this query in the Prometheus expression browser (`http://<machine-ip-address>:9090`), you should see results similar to the following:

```
device_uptime{collection_type="snmp", device="ceos-01",
host="telegraf-01", instance="telegraf-01:9004",
job="telegraf"}    2186051
```

The results show the most recent uptime for `ceos-01`, with a value of 2,186,051 time ticks, representing the duration since it booted up.

Range vector

Continuing with the movie frame analogy, imagine extending that single frame into a full scene, capturing a sequence over time. A **range vector** does just that, consisting of multiple data points for each series over a specified period. This is valuable for observing trends or anomalies. For example, the `device_uptime{device="ceos-01"}[1m]` query offers a minute's worth of uptime data, enabling you to spot any recent disruptions. Run this query in the Prometheus expression browser, and you should see results similar to the following:

```
device_uptime{collection_type="snmp", device="ceos-01",
host="telegraf-01", instance="telegraf-01:9004", job="telegraf"}
2204051 @1706118935.855
2204051 @1706118950.855
2210051 @1706118965.852
2210051 @1706118980.852
```

Here, we see multiple entries for the device uptime metrics because the range vector returns multiple data points over the specified time period. This is useful for tracking stability or finding recent issues.

Scalar

Scalars are the simplest form of data in Prometheus, representing a single numerical value without being tied to any specific time series. For example, you can multiply byte counts by a scalar to convert bytes to bits. Using the `interface_out_octets{intf_role="mgmt"} * 8` query, you can perform this conversion. Let's run this query and check the results it returns:

```
{collection_type="snmp", device="ceos-01", host="telegraf-01",
instance="telegraf-01:9004", intf_role="mgmt", job="telegraf",
name="Management0", site="lab-site-01"}    800
```

This example shows the resulting value from the metric collected by `telegraf-01` for interfaces with the role of mgmt. The value of 800 represents the calculated bits, derived from the original byte count of 100.

Fine-tuning your data with selectors

Prometheus's true power shines in its ability to filter and select data with precision using **selectors**. Selectors allow you to drill down into your metrics to extract exactly the information you need.

Starting simply with a metric name, such as `device_uptime`, returns all related time series for that metric. However, you can narrow down the results by adding matchers within curly braces { }. For example, `device_uptime{device="ceos-01"}` focuses specifically on the uptime of the device named `ceos-01`:

```
device_uptime{collection_type="snmp", device="ceos-01",
host="telegraf-01", instance="telegraf-01:9004", job="telegraf",
site="lab-site-01"}    12602543
```

This filtered query gives a focused look at the uptime of `ceos-01`, demonstrating how labels can refine your data search.

Prometheus supports several matching operators to refine your queries further, such as = for exact matches, != for exclusions, =~ for regex matches, and !~ for regex exclusions. Regex matching is especially handy for complex queries across multiple labels, such as fetching discarded packet counts for specific interface types. For example, running a query with a regex selector such as `interface_in_discards_pkts{name=~"Management.+|Ethernet.+"}` allows you to retrieve discarded packet counts for interfaces named `Management` or `Ethernet`, followed by any characters. Let's run this query and check the results:

```
interface_in_discards_pkts{collection_type="gnmi", device="ceos-02",
host="telegraf-02", instance="telegraf-02:9005", intf_role="mgmt",
job="telegraf", name="Management0", site="lab-site-02"}    4541
```

```
interface_in_discards_pkts{collection_type="gnmi", device="ceos-02",
host="telegraf-02", instance="telegraf-02:9005", intf_role="peer",
job="telegraf", name="Ethernet1", site="lab-site-02"}    0
```

The query result shows discarded packet counts for specific interfaces:

- The Management0 interface on ceos-02 at telegraf-02 has 4541 discarded packets
- The Ethernet1 interface on ceos-02 at telegraf-02 has 0 discarded packets

This query uses regex to select metrics for both management and Ethernet interfaces, illustrating the power of pattern matching to gather relevant metrics across similar labels.

Querying with time in mind

Prometheus also offers time-related selectors for range vectors, allowing you to analyze data over a specified period by including a time duration in square brackets. For example, metric_name[5m] retrieves data from the last five minutes. Additionally, the offset modifier shifts the time frame of your query, which is useful for historical comparisons. For instance, metric_name offset 1h shifts data by one hour into the past. The @ modifier sets an exact evaluation time, enabling you to pinpoint data retrieval to a specific moment.

Understanding the temporal aspects of your data is key for in-depth network analysis. Let's run some examples. Let's run this query, interface_in_broadcast_pkts{device="ceos-02", name="Management0"}[5m], and check the results:

```
interface_in_broadcast_pkts{collection_type="gnmi", device="ceos-02",
host="telegraf-02", instance="telegraf-02:9005", intf_role="mgmt",
job="telegraf", name="Management0", site="lab-site-02"}
0 @1707862834.615
0 @1707862849.615
0 @1707862864.615
# Omitted output ...
```

This range vector query provides a view of the interface broadcast packet metrics over a five-minute period for the Management0 interface on ceos-02. It captures multiple data points within this time frame, offering insights into the broadcast packet activity on this interface.

Next, let's delve into Prometheus operators, powerful tools that enable you to perform arithmetic, comparisons, and other operations on your metrics.

Prometheus operators

Stepping deeper into the world of PromQL, let's navigate the essentials of data operations that are important in the realm of monitoring. We'll explore binary, logical, and aggregation operators, along with vector matching, all through hands-on examples from our lab setup.

Binary operators – the arithmetic and comparison tools

PromQL's binary operators are your toolkit for performing mathematical and comparative operations right within Prometheus.

Arithmetic operators

Think of arithmetic operators (+, -, *, /, %, and ^) as your calculators. They are here to help you transform raw data into actionable insights. For example, let's calculate the total memory utilization of a device using the `memory_used` and `memory_available` metrics from our lab environment. To do this, you sum up `memory_used` and `memory_available` to get the total memory capacity, and then divide `memory_used` by this total to get the utilization percentage. The query looks something like the following:

```
(
  memory_used{device="ceos-01"}
  /
  (
    memory_used{device="ceos-01"}
    +
    memory_available{device="ceos-01"}
  )
) * 100
```

By running the query, you should see results similar to the following:

```
{collection_type="gnmi", device="ceos-01", host="telegraf-01",
instance="telegraf-01:9004", job="telegraf", name="Chassis",
site="lab-site-01"}
30.98121479855142
```

In this example, adding the `memory_used` and `memory_available` metrics gives us a total memory capacity. Dividing `memory_used` by this total gives us approximately 0.3098. When multiplied by 100, this results in approximately 31%, indicating that `ceos-01` utilizes about a third of its memory resources.

Comparison operators

Comparison operators (==, !=, >, <, >=, and <=) act as your filters. They help you navigate through data to find the pieces of information that indicate whether something is wrong or whether everything operates as expected. For example, if you want to quickly identify which interfaces on a device are down, you can use these operators to filter out the relevant metrics.

Assuming that an operational status of 1 means that the interface is up and any other value means it is down or in an unintended state, you can use a comparison operator to find interfaces that are not up.

In our lab setup, you can run the following query to identify interfaces that are not in the *up* state – `interface_oper_status != 1`.

This query should return results similar to the following:

```
interface_oper_status{collection_type="gnmi", device="ceos-02",
host="telegraf-02", instance="telegraf-02:9005", intf_role="peer",
job="telegraf", name="Ethernet2", site="lab-site-02"}    2
```

In this example, the query filters out the interfaces that are not up, showing that the `Ethernet2` interface on `ceos-02` has an operational status of 2, indicating that it is down or in an unintended state. This makes it easy to identify issues and perform quick and short audits.

> **Note**
>
> You might not get a response if all of the interfaces of the network device lab are up. In this case, try connecting to a device and shut down some interfaces. And if you are wondering about the operational states available for interfaces, **Border Gateway Protocol** (**BGP**) sessions, and more, processed by the Telegraf instances, check the processor's enum configuration on the Telegraf collectors for this chapter (i.e., `network-observability-lab/chapters/ch7/telegraf/telegraf-01.conf.toml`).

Logical/set operators – getting into data correlation

When you're managing IT infrastructure, not all data points are isolated. You often need to correlate different metrics to get the full picture. Logical/set operators (`and`, `or`, and `unless`) are the glue that binds disparate data points based on conditions you set.

and operator

Operators in PromQL help you find connections between different sets of time series data. For example, the `and` operator links two sets of time series based on shared labels, allowing you to find common ground between two metrics. Suppose you want to correlate high bandwidth usage with specific devices' operational status. You can combine bandwidth metrics with operational status to focus on devices that are both active and experiencing high traffic.

Let's look at the following query example: `interface_in_octets > 100000 and on(device) interface_oper_status == 1`.

This query filters for interfaces with inbound traffic greater than 100,000 octets that are also operational. The results should look similar to the following:

```
interface_in_octets{collection_type="gnmi", device="ceos-02",
host="telegraf-02", instance="telegraf-02:9005", intf_role="mgmt",
job="telegraf", name="Management0", site="lab-site-02"}    426963559
interface_in_octets{collection_type="gnmi", device="ceos-02",
host="telegraf-02", instance="telegraf-02:9005", intf_role="peer",
job="telegraf", name="Ethernet1", site="lab-site-02"}    20004493
# Output omitted...
```

This query helps you pinpoint areas that might be under strain by identifying interfaces with high inbound traffic that are also operational. For more useful measurements, such as the rate of bits per second going through an interface, we will provide an example using Prometheus functions later on.

or operator

Using the `or` operator in PromQL allows you to merge metrics from various sources, eliminating duplicates and creating a comprehensive dataset. This is perfect for compiling data from fragmented pieces. For example, if you need to combine error metrics collected via **Simple Network Management Protocol** (**SNMP**) and **gRPC Network Management Interface** (**gNMI**) for a holistic view, you can use the `or` operator.

Consider the following query:

```
interface_in_errors_pkts{collection_type="snmp"}
or
interface_in_errors_pkts{collection_type="gnmi"}
```

This query combines the inbound error metrics for all interfaces from devices that captured interface counters via SNMP or gNMI. The results in your lab environment should look something like this:

```
interface_in_errors_pkts{collection_type="snmp", device="ceos-01",
host="telegraf-01", instance="telegraf-01:9004", job="telegraf",
name="Loopback1", site="lab-site-01"}    0
interface_in_errors_pkts{collection_type="gnmi", device="ceos-02",
host="telegraf-02", instance="telegraf-02:9005", intf_role="mgmt",
job="telegraf", name="Management0", site="lab-site-02"}    0
# Output omitted...
```

This query helps ensure that you don't miss out on critical error metrics, regardless of the collection method. By merging the data, you get a complete view of inbound errors from both the SNMP and gNMI sources.

unless operator

Sometimes, you need to exclude certain metrics from your analysis. The `unless` operator helps you remove data points that match a specified condition from your dataset. For example, let's check for all BGP neighbors but exclude those with a specific **autonomous system number** (**ASN**). This could be useful to identify BGP neighbors that are active in a network while excluding those belonging to a particular ASN, which might be used for special purposes or maintenance, or those that are expected to be excluded due to specific configurations or policies.

Consider the following PromQL query:

```
bgp_neighbor_state
unless
bgp_neighbor_state{neighbor_asn=~"651.+"}
```

Let's break the query down:

- `bgp_neighbor_state` shows all the BGP neighbor states in the network

- `bgp_neighbor_state{neighbor_asn=~"651.+"}` shows the BGP neighbors with ASN numbers that match the regular expression

- The `unless` operator filters out the BGP neighbors with ASNs that start with `651` from a list of all the BGP neighbors

The results should look something like this in your lab environment:

```
bgp_neighbor_state{collection_type="exec", device="ceos-01",
host="telegraf-01", instance="telegraf-01:9004", job="telegraf",
neighbor="10.1.2.2", neighbor_asn="65222", site="lab-site-01",
vrf="default"}     1
bgp_neighbor_state{collection_type="exec", device="ceos-01",
host="telegraf-01", instance="telegraf-01:9004", job="telegraf",
neighbor="10.1.7.2", neighbor_asn="65222", site="lab-site-01",
vrf="default"}     1
```

In this example, it indicates that the `Management0` and `Ethernet1` interfaces are up but do not have an OSPF neighbor established.

Vector matching – fine-tuning comparisons

Vector matching is a key concept, particularly when analyzing metrics that come in pairs or groups. It's a bit like ensuring you're comparing apples to apples in your network data. So, let's introduce keyword matching capabilities in Prometheus with the `on`, `ignoring`, and `group` modifiers with `group_left` and `group_right`, before diving into more examples:

- `on`: This keyword allows you to specify which labels should match when combining two vectors. For example, by having the `device_uptime` and `interface_admin_status` metrics, the `device_uptime and on(device) interface_admin_status` query will perform a match only if they match the label device and its corresponding value. Go ahead and try it out!

- `ignoring`: Sometimes, you want to do the opposite and tell Prometheus to disregard certain labels. It's like saying, "*Match these up, but ignore the differences in these particular labels.*" This could be useful when you have metrics that are collected differently but need to be correlated, such as CPU usage metrics collected via SNMP versus those collected via another method.

- `group_left` and `group_right`: These modifiers are used when you have a one-to-many or many-to-one relationship. They're like instructions on how to handle a situation where one metric (such as total memory usage on a device) should be considered alongside multiple related metrics (such as memory usage per application). You can also provide a list of labels that should be included in the results, which ensures that the final output has the context you need..

Understanding these capabilities on the PromQL query language will help you understand the examples around vector matching coming up next.

One-to-one matching (comparing equivalent metrics)

One-to-one matching in PromQL is like aligning two different metrics perfectly before comparing or combining them. For example, suppose you want to check whether the traffic going into an interface is higher than the traffic coming out.

Consider the following query:

```
interface_in_octets{device="ceos-02", name="Ethernet1"}
>
interface_out_octets{device="ceos-02", name="Ethernet1"}
```

If the incoming traffic is higher, it will return the evaluated metric – in this case, `interface_in_octets{device="ceos-02", name="Ethernet1"}`. If not, it will not return a value. This means that ONLY metrics with the same set of labels (`device=ceos-02` and `name=Ethernet1`) will be matched.

Many-to-one/one-to-many matching (aggregating related metrics)

This type of matching is about pairing up related data points so that you can compare or compute them in a meaningful way. It's important when you want to match metrics that naturally belong together, such as device metrics and the number of BGP prefixes accepted by a device.

For example, in your network, you might have metrics that show how much memory each device uses (`memory_used`) and another set that shows how many BGP prefixes each device has accepted (`bgp_prefixes_accepted`). The `memory_used` metric gives you one figure per device, while `bgp_prefixes_accepted` could give several metrics per device because it's broken down by the BGP neighbor.

Here's how you can think about it:

- `memory_used`: One metric per device, such as the total memory each router is using
- `bgp_prefixes_accepted`: Several metrics per device, such as how many routes each neighbor has accepted on a router

Now, suppose you want to match these up – you want to see the memory usage related to each BGP prefix count to help you identify whether there's an impact on memory from the routing information scale. But you can't just compare them directly because they don't line up one to one. That's where vector matching comes in – specifically, many-to-one matching.

To properly compare them, you need to follow steps similar to these:

1. Ignore unmatching labels. Since `memory_used` doesn't care about neighbors but `bgp_prefixes_accepted` does, you tell Prometheus to ignore the neighbor-related labels when doing the match.

2. Match up `memory_used` for each device with all the `bgp_prefixes_accepted` by that device.

Let's take a look at the query example:

```
memory_used
* on(device) group_left
sum(bgp_prefixes_accepted) by (device)
```

Let's break down the PromQL query:

- `sum(bgp_prefixes_accepted) by (device)`: This sums up the number of accepted BGP prefixes for each device, grouping the results by device. The `sum` aggregation operation will be explained further in the following section.

- `memory_used * on(device) group_left sum(bgp_prefixes_accepted) by (device)`: This multiplies the memory used by each device with the corresponding sum of accepted BGP prefixes for that device. The `on(device)` clause indicates that the multiplication should align on the `device` label, and `group_left` is used to keep all the labels from the left-hand side operand (`memory_used`) in the result.

The results should look something similar to this in your lab environment:

```
{collection_type="gnmi", device="ceos-01", host="telegraf-01",
instance="telegraf-01:9004", job="telegraf", name="Chassis",
site="lab-site-01"}    3720720384
{collection_type="gnmi", device="ceos-02", host="telegraf-02",
instance="telegraf-02:9005", job="telegraf", name="Chassis",
site="lab-site-02"}    3721461760
```

The result shows you the total memory used next to the total number of BGP prefixes for each device. This way, you can start to see whether there's a correlation between memory usage and the number of prefixes a device is handling. By multiplying memory usage by the number of accepted BGP prefixes, you can get a sense of how resource-intensive each BGP prefix is for a device. This can help in understanding the load that BGP prefixes place on a device's memory.

Aggregation operators

Aggregation operators are tools in Prometheus that help you summarize or consolidate data from numerous sources into a more manageable form. Let's dig into some of these operators with practical lab examples.

sum operator

The `sum` operator adds values from multiple metrics into a single total. For example, you might want to calculate the total number of BGP prefixes accepted across all devices on a site to get a sense of the overall network activity.

Consider the following example – `sum(bgp_prefixes_accepted)`.

This query sums all BGP accepted prefixes across all devices and neighbors. The result would look something like `{} 2`, indicating that there are a total of two BGP prefixes accepted overall in our lab environment.

You might have noticed that the `sum` aggregation function in PromQL removes all labels by default. You can preserve certain labels using `by`, or exclude them using `without`, tailoring the results to your needs.

Let's see the BGP example using `by`. In this case, we want to collect the amount of BGP prefixes accepted per device – `sum(bgp_prefixes_accepted) by (device)`.

The results should look something similar to the following:

```
{device="ceos-01"}    1
{device="ceos-02"}    1
```

The result is the breakdown of the BGP prefixes accepted by each device.

min/max/avg operator

These operators help identify the lowest, highest, or average values from a set of metrics. For example, if you want to find out which device has been running for the shortest time in your network, you can use a query similar to `min(device_uptime)`, returning the lowest value of the device uptime. Go ahead and try it out!

bottomk/topk operators

These operators are your go-to tools for spotlighting the smallest or largest values within your data. Unlike other aggregators, they retain specific details from the original data points, giving you a clearer picture of what's happening.

For example, let's check the CPU cores on the devices with least usage. To accomplish this, we can run the following query – `bottomk(2, cpu_used) by (device)`.

The result might look something like this:

```
cpu_used{collection_type="gnmi", device="ceos-01", host="telegraf-01",
instance="telegraf-01:9004", job="telegraf", name="CPU3", site="lab-
site-01"}    20
cpu_used{collection_type="gnmi", device="ceos-01", host="telegraf-01",
```

```
instance="telegraf-01:9004", job="telegraf", name="CPU2", site="lab-
site-01"}    25
cpu_used{collection_type="gnmi", device="ceos-02", host="telegraf-02",
instance="telegraf-02:9005", job="telegraf", name="CPU2", site="lab-
site-02"}    23
cpu_used{collection_type="gnmi", device="ceos-02", host="telegraf-02",
instance="telegraf-02:9005", job="telegraf", name="CPU0", site="lab-
site-02"}    29
```

This query returns the two lowest CPU usage metrics for each device. However, because we have CPU usage metrics separated by CPU and device, we see a total of four metrics in the result.

The bottomk operator helps identify the smallest values, which is useful to pinpoint devices with minimal resource usage or other low-value metrics. Conversely, the topk operator highlights the largest values, which can be critical for spotting potential bottlenecks or high resource consumption in your network.

count/count_values operators

The count and count_values operators are ideal for determining the frequency of occurrences or how many items meet a certain criterion. For example, they are useful for checking the health of BGP neighbors.

To illustrate, let's count how many BGP neighbors are in an established state on a per-device basis. We can run the following query – count(bgp_neighbor_state == 1) by (device).

The result should look something similar to this:

```
{device="ceos-01"}    1
{device="ceos-02"}    1
```

This result shows that there is one BGP neighbor in an **established** state for each device (ceos-01 and ceos-02).

The count operator simply counts the number of time series that match a specified condition, making it straightforward to get a tally of occurrences. In this example, it counts the number of BGP neighbors with a state of 1 (which we assume represents *established*).

Conversely, the count_values operator can be used when you need to count occurrences of each unique value within a set of time series. This can be particularly useful to get a distribution of states or other categorical data.

There are more aggregator operators that aren't covered in this book, and we strongly suggest checking out the official documentation (https://prometheus.io/docs/prometheus/latest/querying/operators/#aggregation-operators). Among these, we have the calculation of standard deviation (`stddev`) and standard variance (`stdvar`), which are normally used for anomaly detection. Here are some references that cover this topic:

- *How to use Prometheus for anomaly detection in Gitlab* (`https://about.gitlab.com/blog/2019/07/23/anomaly-detection-using-prometheus/`): This article discusses the importance of correctly aggregating data and using statistical methods such as z-scores to identify anomalies

- *Introduction to a Telemetry Stack - Part 4* by David Richey (`https://blog.networktocode.com/post/telemetry-stack-series-part-04/`): This post covers the use of standard deviation and anomaly detection to identify deviations in network telemetry data

Next, we are going to explore Prometheus functions to make our queries more robust.

Prometheus functions

Prometheus functions are, essentially, tools or commands that allow you to manipulate and interpret the metrics data that Prometheus collects from your network. These functions can help you transform raw data into meaningful metrics. Let's categorize some of the functions available:

- **Basic mathematical functions**: These are your go-to tools for direct data manipulation. Need to normalize data or adjust metrics for comparison? Functions such as `abs()`, `ceil()`, `floor()`, and `sqrt()` let you apply basic arithmetic directly to your metrics. For instance, converting negative values to positive with `abs(cpu_used)` ensures that you're only dealing with usable, positive CPU usage figures.

- **Aggregation over time functions**: Imagine that you want to see an overview rather than minute-by-minute updates. Functions such as `avg_over_time()`, `min_over_time()`, `max_over_time()`, and `sum_over_time()` let you aggregate data across specified periods. This is perfect for summarizing average CPU usage over the last hour with `avg_over_time(cpu_used[1h])`, giving you a clear picture of performance trends.

- **Range vector functions**: These functions are all about understanding changes over time. They help you track how metrics evolve, which is important for detecting anomalies or trends. For example, `rate(interface_in_errors_pkts[5m])` can show the rate of errors over a five-minute interval, highlighting potential issues as they occur.

- **Special functions**: Some scenarios require predictive analysis or identifying patterns. Functions such as `predict_linear()` are your crystal balls into future metrics based on past data. Predicting future disk space usage with `predict_linear(disk_space[1h], 4 * 3600)` can help you prevent outages by foreseeing when you'll run out of space.

To better grasp the concept of the Prometheus functions, let's see some detailed examples of their usage when managing your network data.

Network interface traffic

When monitoring network interface traffic, you often work with raw counter metrics, such as bytes or packets transferred, rather than direct bits-per-second measurements. This necessitates some calculations to translate these counters into a more usable metric.

To analyze inbound traffic in bits per second, we utilize the `rate` function in PromQL, which calculates the per-second average rate of increase over a specified time range. Multiplying this rate by 8 converts bytes (octets) to bits, providing the desired bits-per-second metric. For example, by running the `rate(interface_in_octets[5m]) * 8` query, we should have a result similar to the following:

```
{collection_type="gnmi", device="ceos-02", host="telegraf-02",
instance="telegraf-02:9005", intf_role="mgmt", job="telegraf",
name="Management0", site="lab-site-02"}    2976.9935475269026
{collection_type="gnmi", device="ceos-02", host="telegraf-02",
instance="telegraf-02:9005", intf_role="peer", job="telegraf",
name="Ethernet1", site="lab-site-02"}    133.41426701800307
# Omitted output...
```

In this example, the `Management0` interface on the `ceos-02` device has around 2.9 kbps, and `Ethernet1` has around 133 bps.

For these kinds of queries, it is better to see them represented in a time series graph. Luckily, we can check this directly in Prometheus. If you click the **Graph** tab on the Prometheus expression browser page, you should get something similar to the following:

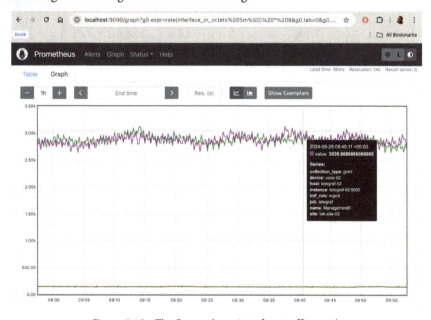

Figure 7.10 – The Prometheus interface traffic graph

Average CPU usage over time

Let's use another example to understand more Prometheus functions. To monitor how CPU usage trends over time or in response to network activity, we use the `avg_over_time` function in Prometheus. This approach helps identify whether CPU load patterns align with expected network behaviors. For a focused analysis, we'll examine the CPU usage of a specific device, `ceos-01`.

First, viewing the raw `cpu_used` metrics for `ceos-01` gives us a baseline of CPU activity. Run the `cpu_used{device="ceos-01"}` query. This query will return the raw CPU usage metrics for `ceos-01`, which we can then analyze over time using the **Graph** tab in the Prometheus expression browser:

Figure 7.11 – Prometheus CPU usage on a device

Now, let's apply `avg_over_time(cpu_used{device="ceos-01"}[1h])`; the resulting graph offers a smoothed overview of CPU usage trends over the specified period:

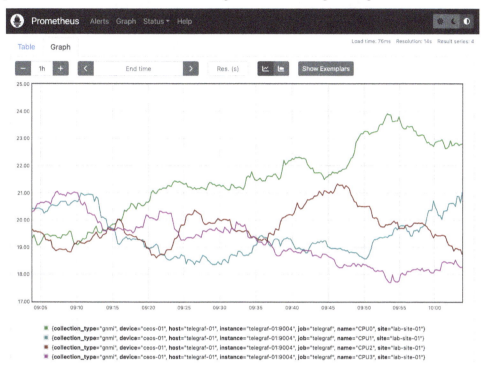

Figure 7.12 – Prometheus average CPU usage on a device

This resulting graph offers a clear visualization of average CPU usage over the last hour. Functions such as `avg_over_time` help represent average calculations of CPU usage, providing insights into CPU load patterns and aiding in identifying any correlations with network activity.

Predicting interface operational status

For our final example using Prometheus functions, our goal is to utilize Prometheus's `predict_linear` function to estimate whether an interface will be up or down in the near future, allowing you to anticipate and address potential issues before they impact network performance.

For this example, we will focus on predicting the status of peer interfaces on the `ceos-02` device over the next hour, based on the last hour's data. For example, take a look at the following query:

```
predict_linear(
  interface_oper_status{device="ceos-02", intf_role="peer"}[1h],
  3600
)
```

Let's break it down:

- **Metric**: `interface_oper_status` – tracks the operational status of network interfaces
- **Function**: `predict_linear(...[1h], 3600)` – applies a linear regression to the operational status of interfaces tagged as `peer` on `ceos-02`, predicting their status one hour into the future
- **Filter**: `{device="ceos-02", intf_role="peer"}` – focuses the prediction on peer interfaces of a specific device to streamline the output

If we apply the preceding query, we will see a result similar to the following, where the `Ethernet1` interface is predicted to be up (a value of `1`) for the next hour while the `Ethernet2` interface is predicted to be down (a value of `2`):

```
{collection_type="gnmi", device="ceos-02", host="telegraf-02",
instance="telegraf-02:9005", intf_role="peer", job="telegraf",
name="Ethernet1", site="lab-site-02"}    1
{collection_type="gnmi", device="ceos-02", host="telegraf-02",
instance="telegraf-02:9005", intf_role="peer", job="telegraf",
name="Ethernet2", site="lab-site-02"}    2
```

These kinds of queries are useful – for example, for proactive maintenance planning.

Diving into PromQL has shown us how to handle, examine, and foresee what's happening in our network with sharp accuracy. Now, we'll step into a new area – Prometheus rules. This is where we get to upgrade how we monitor and manage our network. We will learn how to apply rules directly in Prometheus to perform specific actions on the stored data.

Prometheus rules

Prometheus comes with two powerful tools in its arsenal – alerting and recording rules, both designed to be set up and automatically checked at consistent intervals.

While we'll dive into alerting rules and how to manage alerts more broadly in *Chapter 9*, our current spotlight is on recording rules. This focus will help us streamline complex data into simpler, more accessible metrics for ongoing analysis.

Recording rules

Recording rules are a clever strategy for those looking to streamline their operational tasks. They let us do the math or processes ahead of time on calculations we use often or that are tough to crunch, storing the results as brand-new time series data. This task makes fetching these results faster than recalculating every time you need them. This is especially useful for dashboards, which need to query the same expression repeatedly every time they refresh.

Recording and alerting rules *exist in a rule group*. Rules within a group are run sequentially at regular intervals, all based on the same point in time.

Let's run a practical example with a network traffic overview of a device:

```
groups:
  - name: network_traffic_overview
    rules:
    - record: device:network_traffic_in_bps:rate_2m
      expr: sum(rate(interface_in_octets[2m])) by (device) * 8
    - record: device:network_traffic_out_bps:rate_2m
      expr: sum(rate(interface_out_octets[2m])) by (device) * 8
```

Here's what we are looking at:

- `groups`: Defines a group of rules, named `network_traffic_overview`.

- `record`: Creates new time series with the names `device:network_traffic_in_bps:rate_2m` and `device:network_traffic_out_bps:rate_2m`. The naming follows a structured convention (`level:metric:operations`):

  - `level`: Denotes the aggregation scope and label context of the rule's output

  - `metric`: Refers to the core metric name being recorded

  - `operations`: Lists the applied transformations, starting with the most recent

- `expr`: This aggregates (sums up) the rates calculated by `rate()` for each unique device. This step is important because multiple series (representing different interfaces' metrics) exist for each device.

Let's put this into practice in our lab environment to see the recording rules in action. First, save the preceding recording rules under `network-observability-lab/chapters/ch7/prometheus/rules/recording_rules.yml`. Recording rules are YAML files that Prometheus fetches and reads at start time.

Next, we need to guide Prometheus to these rules in its configuration file, located at `network-observability-lab/chapters/ch7/prometheus/prometheus.yml`. The following is an example snippet to instruct Prometheus on where to fetch and read the rules files:

```
# Omitted configuration...
rule_files:
    - rules/*.yml
```

> **Note**
>
> The file paths are relative to the chapter's folder, and the configuration can be found in the `Docker Compose` file of *Chapter 7*'s scenario.

This will let Prometheus know its rule manager process, where the rules are fetched from.

With the Prometheus configuration saved, we need to reload this service. This can be done in multiple ways. For example, you can use the `reload` function of the Prometheus Management API (see the Prometheus Management API docs at `https://prometheus.io/docs/prometheus/latest/management_api/#reload`), or you can use the `netobs` tool by running the `netobs lab update prometheus --scenario ch7` command.

Now, check the rules configuration in Prometheus by navigating to **Status | Rules**. You should see the new configuration of the recording rules with information on the evaluation time:

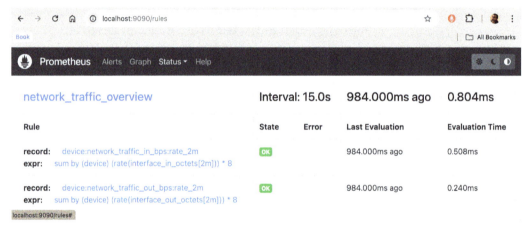

Figure 7.13 – Prometheus recording rules

The new metrics generated from the recording rules should be visible at this stage. Run the `device:network traffic_in_bps:rate_2m` query on the Prometheus expression browser. You should get a result similar to the following:

```
device:network_traffic_in_bps:rate_2m{device="ceos-01"}    3452.93154
device:network_traffic_in_bps:rate_2m{device="ceos-02"}    2998.78095
```

And with that, we have a successful recording rule metric, which we can start using on our network dashboards.

A look at Grafana Loki

Grafana Loki is essentially a toolkit to handle logs coming from systems and applications. It's designed to gather and organize log information in a way that's easy to use and efficient to store. Here's how it works from a high-level standpoint:

- **Collection and indexing**: Loki collects log data and indexes labels associated with log streams, such as application name or environment—similar to how Prometheus handles its data. These labels act as a quick-reference index, eliminating the need to search through the entire text of the logs.

- **Storage**: The actual log messages are compressed and stored in *chunks* –compact pieces of data. These chunks can be kept in various storage solutions, including cloud services such as Amazon S3 or Google GCS, or on local servers. By compressing the logs and keeping the index minimal, Loki ensures compact storage use, reducing both resource consumption and costs.

- **Querying**: Loki's design and architecture are highly inspired by Prometheus, and its query language, LogQL, is similar to **PromQL**. This familiarity allows users to leverage their existing knowledge of the Prometheus query language, making it easier to learn and use.

- **Integration and setup**: We chose Loki because of its native integration with Grafana and simple integration with **Logstash**. Plus, setting up a development instance of Loki is straightforward, making it practical for managing logs in our lab environment.

Let's dive deeper into Loki's architecture next.

Grafana Loki architecture

Loki is a horizontally scalable, highly available, multi-tenant log aggregation system inspired by Prometheus. Loki differs from Prometheus by focusing on logs instead of metrics and collecting logs via push, instead of pull.

As a logging system, Loki does not index the content of the logs but, instead, their metadata as a set of labels for each log stream.

A log stream is a set of logs that share the same labels. Labels help Loki to find a log stream within your data store, so having a quality set of labels is key for precise query execution.

Log data is then compressed and stored in chunks in an object store such as Amazon **Simple Storage Service (S3)** or **Google Cloud Storage (GCS)**, or even, on the filesystem. A small index and highly compressed chunks simplify the operation and significantly lower the cost of Loki.

The following diagram expresses the Loki component architecture:

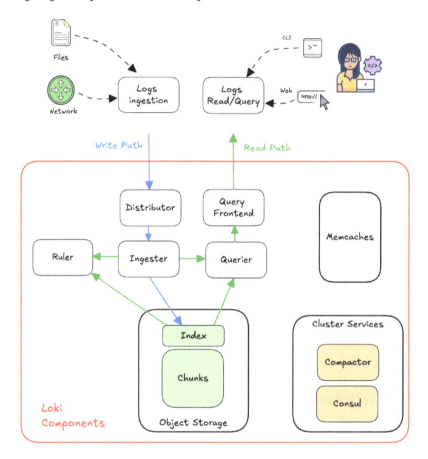

Figure 7.14 – The Grafana Loki component architecture (Created using the icons from https://icons8.com)

Loki's architecture (`https://grafana.com/docs/loki/latest/get-started/components/`) allows users to send, store, and query their log data in a proper manner. This section aims to demystify its components and their interactions.

The write path – the journey of a log entry in Loki

When a log entry arrives at Loki, its first encounter is with the **distributor**, as seen in the Loki component diagram. Imagine the distributor as the front desk of a data hotel, where logs are guests checking in. This component meticulously validates each log entry, ensuring that it adheres to the correct formats and limits. However, it's not just about validation. The distributor also takes care of preprocessing – it organizes and sorts labels within logs for uniformity. Moreover, it plays a role in maintaining system stability by rate-limiting logs based on each tenant's allocation, similar to controlling the flow of guests into a hotel.

> **Note**
>
> In Grafana Loki, a **tenant** is an isolated logical entity used to segregate log data. Each tenant has its own set of logs, configurations, and queries, which allows multiple users or teams to share the same Loki instance without interfering with each other's data. This multitenancy feature is useful for large organizations that need to manage logs from different departments or applications separately.

Once validated, these log entries are then replicated and forwarded to the **ingester**. The ingester acts much like a data steward, managing these logs with care. Here, log streams are transformed into chunks – compressed packets of data, ready for storage. The ingester is vigilant, ensuring that the logs are in the correct chronological order and safeguarding them with a **write-ahead log** (**WAL**) for added reliability (for more information about WALs, see the *Time series databases* subsection). This process prepares the logs for their final destination – long-term storage.

The read path – retrieving insights from logs

Conversely, when users query Loki to retrieve log data, the **Querier** springs into action. This component is akin to a librarian, knowledgeable and skilled in finding the right log entries from a vast repository. It intelligently fetches logs from the ingester for recent data, and for older records, it digs into the long-term storage. To ensure accuracy and eliminate redundancy, the querier deduplicates data, offering a clean and precise set of logs.

Refining this process is the **Query Frontend**. It's the planner and coordinator, optimizing how queries are handled. It queues incoming queries, managing their execution order and ensuring system balance. It breaks larger, more complex queries down into smaller, manageable parts, executed in parallel for speed. Additionally, by caching common query results, the query frontend ensures faster access to frequently sought-after information.

It is important to note that Grafana Loki indexes log entries based on the metadata of the log, particularly its labels and timestamps, rather than the content of the logs themselves. This indexing strategy allows Loki to perform optimized searches and queries without the overhead of full-text indexing, which can be resource-intensive. Think of it as a library catalog that indexes books by title and author instead of scanning every word in every book.

Storage – the final resting place for logs

At the heart of Loki lies its storage capability. Grafana Loki indexes metadata about logs, particularly their labels. The long-term storage backends, such AWS **Simple Storage Service** (**S3**) or **Google Cloud Storage** (**GCS**), serve as the vaults for processed log data. These backends store the log data by using compressed chunks, ensuring that they are readily available for future retrieval and analysis, much like a secure archive preserving valuable records.

Enhancing Loki's capabilities with additional components

To further elevate Loki's functionality, additional components play vital roles. The **Ruler** functions like an alert system, monitoring log patterns and triggering notifications when specific conditions are met. Then there's the **Memcached**, which boosts system performance by caching frequently accessed data and query results, akin to a quick-access memory layer.

A cluster's smooth operation is ensured by services such as the Compactor and Consul. The **Compactor** plays an important role in managing storage by periodically merging smaller data chunks into larger ones, compacting index data, and applying retention policies to remove outdated logs. This process not only optimizes storage usage but also enhances query performance and reduces costs. Meanwhile, **Consul** manages the cluster's state and ensures communication and coordination among services. It helps distribute workloads evenly across the cluster by using the hash ring and enables reliable service discovery, ensuring that different parts of the system can easily find and interact with each other.

> **Note**
>
> The **hash ring** is a consistent hashing mechanism used by Loki to distribute log data evenly across multiple nodes in the cluster. Here's how it works – when log data comes in, it's assigned a hash value based on its metadata. This hash value determines which node in the cluster will handle that piece of data. Think of the hash ring as a circular table with evenly spaced seats (nodes). Each piece of log data gets assigned a seat based on its hash value. This method ensures that the data is spread out evenly, preventing any single node from becoming overwhelmed, which enhances the overall performance and reliability of the cluster.
>
> For a more detailed and in-depth view of Grafana Loki's architecture and concepts, check out its official documentation (`https://grafana.com/docs/loki/latest/`).

The Loki data model

To send data into Loki, you need to format your log entries in a specific JSON payload structure:

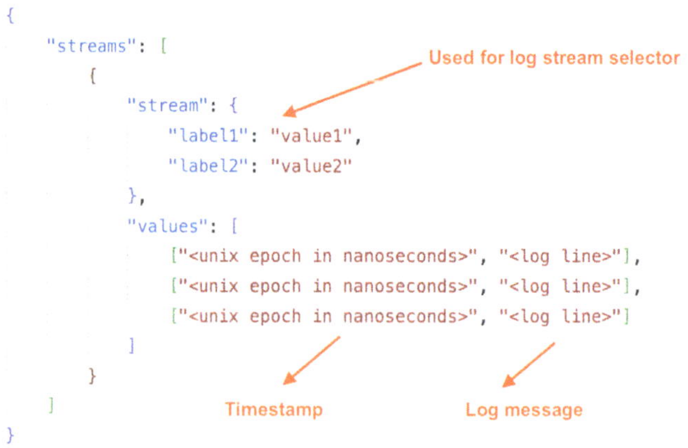

Figure 7.15 – The Loki data model

The payload should include the following fields:

- `stream`: This field represents the log stream or source of the log entry. It is typically used to identify the application or component generating the log.

- `labels`: This field contains key-value pairs that provide additional metadata or labels for the log entry. Labels can be used to filter and group logs in Loki.

- `timestamp`: This field represents the timestamp of the log entry. It should be in the Unix timestamp format or the RFC3339 format.

- `message`: This field contains the actual log message or log content.

Writing to Loki

There are multiple agents and methods available to write data into Loki. Let's list the most popular ones:

- **Promtail**: This is the native agent that ships the contents of local logs directly to Loki. It is usually deployed on the systems or machines that you want to collect log data from. It is a popular choice when running a Kubernetes environment.

- **Grafana Agent**: This is the recommended client when using the Grafana stack, and it can collect metrics and traces as well.

- **Logstash plugin**: This is a Logstash output plugin that ships data to Loki. It is the one we are going to use in our practical lab.

- **fluentd/Fluent Bit/Docker driver**: These are plugins or extensions to processes or log forwarders that allow you to do some processing and send data to Loki.

Let's put this into practice! In the following section, we'll look into capturing logs from network devices, specifically targeting `ceos-01` and `ceos-2`. For a comprehensive view of how these logs traverse our system, consult the lab diagram in *Figure 7.5*. This will give you a visual understanding of the entire log flow process.

Logstash output to Loki

Think of Loki as a destination for your network logs and Logstash as the courier that delivers them. The `logstash-output-loki` plugin is developed to send logs directly to a Loki server.

In our hands-on lab, we have a public Docker image – specifically, `grafana/logstash-output-loki` from Docker Hub, which comes with this plugin ready to go.

We've already set up Logstash to listen for and interpret `syslog` messages from our network devices, `ceos-01` and `ceos-02`. These messages may include information about interface status changes that are already processed and normalized. If you want to know more about the log message parsing and normalization process with Logstash, check out *Chapter 6*.

Now, we're going to add a new destination for these logs in Logstash's configuration file for this chapter's lab – `network-observability-lab/chapters/ch7/logstash/logstash.cfg`. Here's how we tell Logstash to forward logs to Loki:

```
# Omitted Logstash Configuration...
output {
    # Other output omitted...
    loki {
        url => "http://loki:3001/loki/api/v1/push"
    }
}
```

Next, update the configuration file with the new output and run the `netobs lab update logstash --scenario ch7` command to apply the changes.

To see the logs we've sent to Loki, we'll use Grafana, which has been pre-configured to connect to Loki as a data source (see the configuration at `network-observability-lab/chapters/ch7/grafana/datasources.yml`).

We'll explore how to view and analyze these logs in the next section, where we focus on extracting information from Loki.

Reading from Loki (LogQL)

LogQL, inspired by PromQL, is the query language used in Grafana Loki. It's designed to act like a distributed `grep`, allowing users to aggregate and filter log sources. LogQL stands out for its use of labels and operators, enabling users to perform complex log searches and calculations.

In the most basic form, we have **log queries**, which are the basic rules and selectors LogQL has in place to search, parse, extract, or even format log data. Then, we have another level of abstraction called **metric queries**, which extends the log queries to calculate values based on the log results, transforming log data into insightful metrics.

We will begin by understanding log queries and how to read data from logs. Then, we will do a short review of the LogQL operators, which will help us perform more advanced calculations using metric queries on aggregated log data.

Before we delve deeper into LogQL concepts, let's prepare for the practical lab examples. Our first step is to generate log events from our network devices so we can work with real data.

In the lab environment, you can create log events by just accessing the routers (ceos-01 and ceos-02). Shutting down interfaces and bringing them back up is also great for generating interface, **Link Layer Discovery Protocol** (LLDP)- and BGP-related events. Let's see a short example, connecting to ceos-02:

```
# Connect to ceos-02
> ssh netobs@ceos-02
(netobs@ceos-02) Password: netobs123
ceos-02>enable
ceos-02#
ceos-02# configure terminal
ceos-02(config)# interface eth2
ceos-02(config-if-Et2)#shutdown
ceos-02(config-if-Et2)#not shutdown
... repeat ...
ceos-02(config-if-Et2)#shutdown
ceos-02(config-if-Et2)#no shutdown
```

Alternatively, you can execute the following command to automate this process: netobs utils device-interface-flap --device ceos-02 --interface Ethernet2 --count 4 --delay 140.

In the following section, we'll guide you through the process of displaying these log messages in Grafana.

Running LogQL queries

Let's understand how to execute these queries on your Grafana server. Grafana provides a useful web interface for running queries, visualizing data, and exploring metrics and logs in real time, and although we are going to dedicate a chapter on Grafana (*Chapter 8*), this will be an early introduction to visualizing your log messages. Here's how you get started:

1. Navigate to Grafana by entering the http://<machine-ip-address>:3000 URL.

2. Log in using your credentials for the lab (the default user is netobs and the password is netobs123), and you should encounter the **Grafana** home page.

3. In order to access the **Explore** view to visualize the log messages that Loki received, click on the hamburger button in the top-left corner and then select **Explore**.

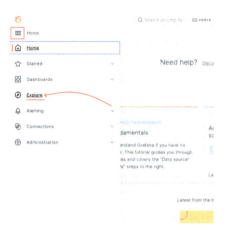

Figure 7.16 – The Grafana home page – Explore

4. Next, select the pre-configured data source of Loki:

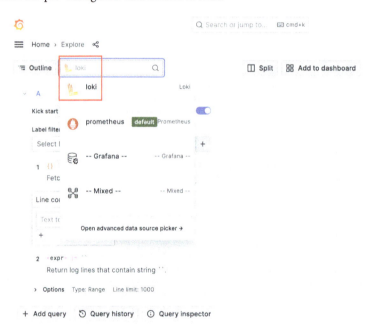

Figure 7.17 – The Grafana Explorer view – the data source

5. For the Loki data source, you will see several options available to you to start exploring your log messages. You will want to enable **Explain query**, select **Builder** mode for your queries, and finally, select **Label browser** to explore the available labels derived from your log messages:

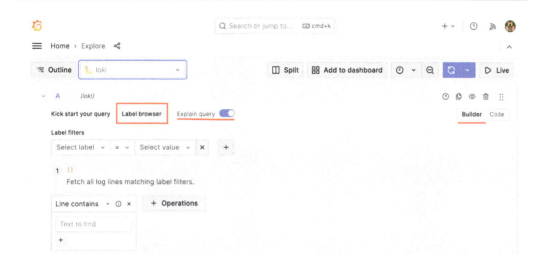

Figure 7.18 – The Grafana Explorer view – the Loki data source settings

6. If the logs from your network devices have arrived successfully—see the previous subsection, *Writing to Loki*—it will present you with labels similar to the following screenshot. Select the labels and the values you would like to use for the log stream selection, and then, click **Show Logs**:

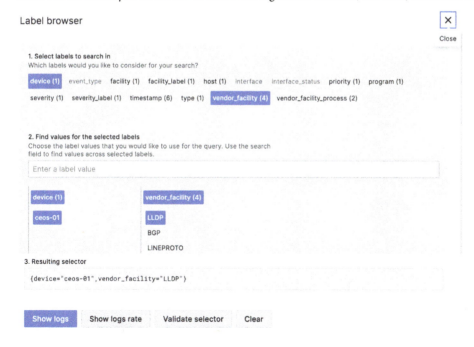

Figure 7.19 – The Grafana Explorer view – the Loki label browser

7. The resulting query will be displayed, along with the **Logs Volume** panel, which is a time series panel showing the volume of logs received over time. Additionally, there will be a dedicated **Logs** panel at the bottom where you can view each log message in detail:

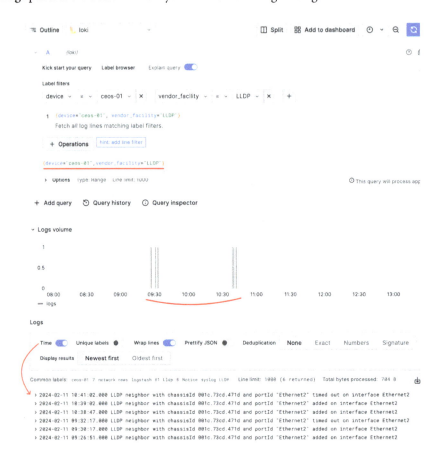

Figure 7.20 – The Grafana Explorer view – Loki logs

8. You can view the labels for each log message by expanding an entry in the **Logs** panel:

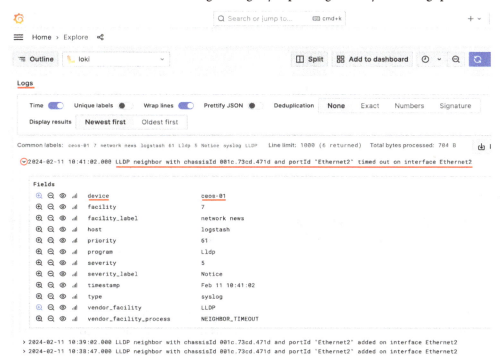

Figure 7.21 – The Grafana Explorer View – the Loki log details

Great! We've got everything set up with Grafana and Loki, and you can already see logs coming in. Moreover, it is possible to interact with Loki programmatically using its API, but we will dive into that approach in *Chapter 12*.

Now, it's time to roll up our sleeves. We are going to not only observe the logs from our lab environment but also demystify LogQL—Loki's query language. We'll walk through it step by step, using examples from our lab environment.

LogQL queries

LogQL queries are the tools you use to dig into your log data within Grafana Loki. They're like a specialized search, homing in on the exact information you need:

Stream Selector   **Log Pipeline (Optional)**

`{ label1="value1", label2="value2" }  <filter operator> <log pipeline expression>`

Figure 7.22 – A LogQL query expression

At the heart of every LogQL query is the **log stream selector**. Think of it as setting the boundaries or the *search area* for the logs you want to analyze. Along with this, you can use a **log pipeline**, although it's optional. A log pipeline allows you to perform a series of operations on the logs you've gathered, such as refining, rearranging, or even reinterpreting data.

Let's understand it by looking at an example. Let's say we want to search for BGP session messages from `ceos-02` that are in the `OpenConfirm` state. We can use a query such as the following:

```
# Example LogQL
{device="ceos-02"} |= ip(`10.1.7.1`) |= `OpenConfirm`
```

Let's break this down and understand the components of a LogQL query expression.

The log stream selector

The log stream selector uses key-value pairs enclosed in `{}` to filter logs based on labels, such as `{device="ceos-02"}`. This filtering is the first step in narrowing down the logs that are relevant to your query, and in this case, we filter the logs coming from the `ceos-02` device.

The log pipeline

After the log stream selector, a log pipeline can be (optionally) applied. This pipeline is a sequence of expressions that process the selected log streams in various ways, such as filtering content, parsing log lines, or mutating labels.

For example, the `|= ip(`10.1.7.1`) |= `OpenConfirm`` pipeline performs the following actions:

1. It filters logs containing the `10.1.7.1` IP address. To do this, it uses a special matching IP address function, `ip("<pattern>")`, that helps properly parse IP addresses, ranges, or even prefixes/**Classless Inter-Domain Routing (CIDR)** specifications.

2. It further filters out messages that contain the BGP state of `OpenConfirm`.

Pipeline expressions

In the world of Grafana Loki, pipeline expressions are like filters and tools that help you pinpoint the log messages you're interested in. In this book, we are going to discuss two types, filter and formatting expressions, but there are many more, which you can check out in their official documentation (`https://grafana.com/docs/loki/latest/query/log_queries/#log-pipeline`).

Filter expressions

Filter expressions are search terms you use to narrow down the logs, acting like keywords or phrases you look for in a conversation. For example, if you're searching for any mention of a router with a BGP session flapping, you can use the **line filter expression**, `|= "old state Established"`. This tells Loki, *"Show me logs that indicate whether a BGP session is starting to go down."*

Let's see this in action. In the Grafana **Explorer** view, select the label filters, `{device="ceos-01",` `vendor_facility="BGP"}`, select + **Operations** | **Line Filters** | **Line Contains**, and then enter the text old state, `Established`. The resulting query is `{device="ceos-01", vendor_` `facility="BGP"} |= `old state Established``:

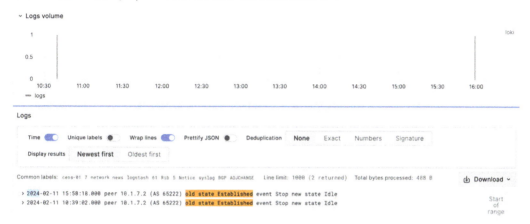

Figure 7.23 – The Grafana Explorer Logs view

The logs returned from this query will be the ones that match your specified criteria. If you want to keep a record of these logs, simply click the **Download** button.

As another example, we can use **label filter expressions**. These work on the metadata tags attached to each log entry, allowing for more refined sorting and selection. For instance, if you notice a severity label marked as 5 in the payload of a BGP log entry and want to filter logs from `ceos-01` with a severity less than 5, you would craft a query such as `{device="ceos-01"} | severity <` 5. This query returns only the logs where the severity is marked lower than 5, helping you quickly identify potential issues that require your attention.

Try it out! See whether it finds log messages with a lower severity code.

Format expressions

These expressions are your formatting toolkit – they let you restructure log lines or labels for clearer analysis. For instance, if you're gathering BGP messages and want to quickly discern which device each message is from, you can use a line format expression, tweaking the message layout to include the device label upfront.

Take the following query:

```
{vendor_facility="BGP"} | pattern `<message>` | line_format `{{.
device}} - {{.message}}`
```

Let's break down the expression:

- **Filter by BGP log messages**: Scopes all the device's log messages to filter only those related to BGP, `{vendor_facility="BGP"}`.

- **Parse log lines**: Adds a **parser expression** that allows the extraction of fields from the log lines, and in this case, the pattern expression captures all the content of the log line and saves it to a new `message` label.

- **Apply line format**: Applies a line format expression to the log line message using a Go template expression (`https://pkg.go.dev/text/template`), where the variables are all the labels available for that log event. The `| line_format `{{.device}} - {{.message}}`` expression will first print out the device name alongside the log line message.

After the parser expression does its job, the log messages are transformed for easy reading. For example, you'll see entries laid out like this – `ceos-02 - peer 10.1.7.1 (AS 65111) old state Established event Stop new state Idle`. It's as if each message has been rewritten to put the key details right up front so that you can quickly see what's happening without digging through the entire log.

Here is an example screenshot of the log messages formatted:

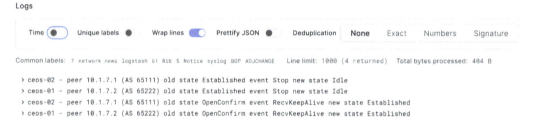

Figure 7.24 – The Grafana Explorer Logs view – the BGP state

Special considerations

Here are some tips when working with pipeline expressions:

- **Dealing with special characters**: When you're working with patterns in your logs that use special characters, opt for backticks (`` ` ``) instead of double quotes. This little trick keeps things simple by skipping the extra step of escaping those characters.

- **Boosting performance**: Want to speed up your log searches? Put your line filter expressions – those specific search terms – at the start of your pipeline. It's like telling the system to focus on the most important task first, which can make everything else go faster.

- **Understanding Regex quirks**: If you're using **Regex** for searching within lines, remember that it might catch patterns anywhere in the text, not just at the beginning or end. This is different from when you're filtering by labels, where Regex sticks strictly to what you specify.

Loki metric queries

Metric queries in LogQL let you turn log data into numerical metrics, such as calculating how many error messages appear over time or identifying which log sources are the busiest. It's about going beyond just reading logs to actually measuring what's happening in your systems. Let's see how they work.

Transforming logs into metrics

Loki metric queries allow you to aggregate log data and turn it into metrics. This is particularly useful when you want to monitor the frequency of certain events or measure the duration of operations within your network infrastructure.

There are two types of metric queries in LogQL – log metrics and derived metrics.

Log metrics

Log metrics are directly derived from log data based on certain aggregation operations, performed over the logs themselves. These operations include counting occurrences, summing numerical values found within log lines, or calculating rates of log entries over time. The primary purpose of log metrics is to transform raw log data into quantifiable metrics that can be monitored over time.

Imagine you're monitoring an application and you want to track the number of error messages logged over time. You could use a query like this:

```
count_over_time({app="example-app", level="error"}[1h])
```

This query uses the `count_over_time()` function to count the number of log entries that match the `app="example-app"` and `level="error"` labels during a one-hour window. This gives you a metric representing the volume of error logs in that period.

Here's another example if you're interested in monitoring the number of **Open Shortest Path First (OSPF)** adjacency teardown events across different network devices over a period of time:

```
sum by(device) (
    count_over_time(
        {vendor_facility_process="OSPF_ADJACENCY_TEARDOWN"}[$__auto]
    )
)
```

This query can be broken down as follows:

- `sum by(device)`: This part of the query aggregates (sums up) the counts by the device label. This means it will provide a separate sum for each unique value of the device label in the log data.

- `count_over_time()`: This function counts the number of log lines that match the specified filter within the given time range.

- {vendor_facility_process="OSPF_ADJACENCY_TEARDOWN"}: This is the filter criteria. It filters the log lines to only those where the vendor_facility_process label has the OSPF_ADJACENCY_TEARDOWN value, which likely indicates a log message generated when an OSPF adjacency is torn down.

- [$__auto]: This is a Grafana-specific variable that automatically adjusts the time range of the query, based on the dashboard's time range selection. It allows the query to be dynamic and adjust to the time range that the user views in Grafana.

Try it out! You might need to log into a device and create multiple interface flaps in order to generate OSPF neighbors' teardown events, or just run the netobs utils device-interface-flap --device ceos-02 --interface Ethernet2 --count 10 --delay 2 command. The result should be similar to the following:

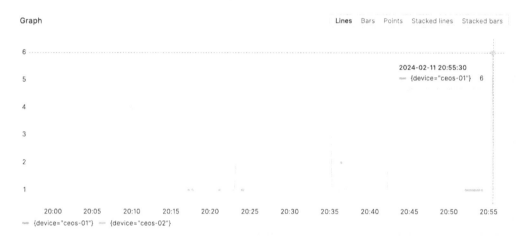

Figure 7.25 – The Grafana Explorer view logs – the metrics query graph

In the previous figure, you can see the maximum of six events captured at one point in time for ceos-01.

Derived metrics

Conversely, derived metrics, involve more complex operations on the log metrics, often combining multiple metric queries using arithmetic operations, comparisons, or other vector operations. Derived metrics are used for more advanced analysis, allowing you to compare metrics, calculate ratios, or perform other mathematical operations on the metrics derived from log data.

Let's say you want to compare the rate of error logs to the rate of all logs for a specific application to understand the proportion of logs that are errors. You could do this with a derived metric query:

```
rate({app="example-app", level="error"}[5m])
/
rate({app="example-app"}[5m])
```

In this query, the numerator, `rate({app="example-app", level="error"}[5m])`, calculates the rate of error logs, while the denominator, `rate({app="example-app"}[5m])`, calculates the rate of all logs for `example-app`. The result is a ratio of error logs to all logs, providing a measure of how many logs are errors, relative to the total volume of logs.

Another example could be to understand the frequency of interface state changes as they relate to all warning events over a given time frame:

```
sum(rate({vendor_facility_process="UPDOWN"}[$__auto]))
/
sum(rate({severity="4"}[$__auto]))
```

Let's break it down:

- `sum(rate({vendor_facility_process="UPDOWN"}[$__auto]))`: This is the numerator of the fraction. It calculates the rate of log entries where the `vendor_facility_process` label is UPDOWN over the time range automatically selected in Grafana. The `rate` function measures how many log entries per unit of time match the filter, and the sum aggregates these rates across all log streams that match the filter. The UPDOWN event is typically associated with an interface-changing state (for example, going from up to down), which could indicate that a network interface has gone offline or been disconnected.

- `sum(rate({severity="4"}[$__auto]))`: This is the denominator of the fraction. It calculates the rate of log entries, with a severity label of 4 over the same time range. Severity 4 is typically considered as a *warning*, which could represent a variety of issues or alerts that are not as critical as errors but still require attention.

- **The division operation – /**: This divides the total rate of UPDOWN events by the total rate of *warning* events. This gives you a ratio of these specific network events (interface state changes) to the total number of warning events in your logs.

If you try the query in the lab environment, the result should be similar to this:

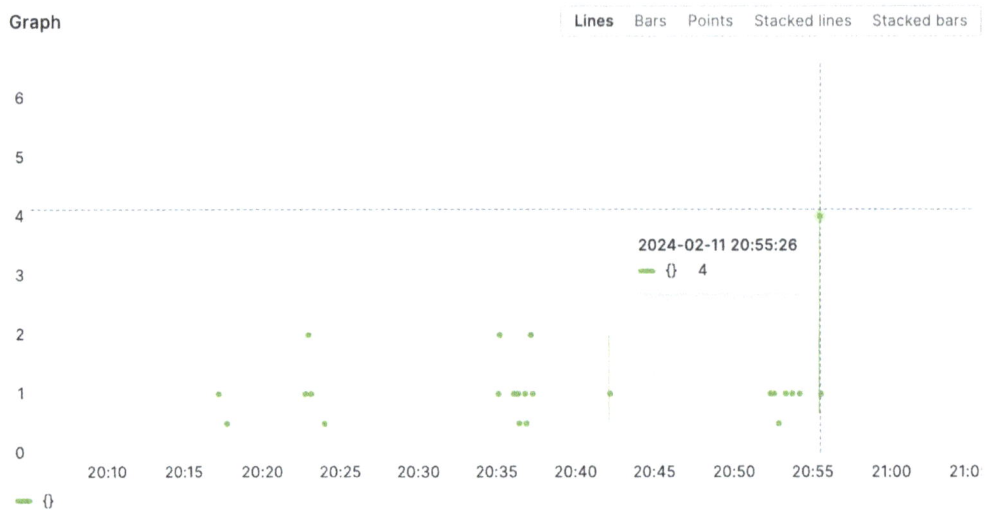

Figure 7.26 – The Grafana Explorer logs view – the derived metrics

Having delved into LogQL metric queries, you may have noticed that we bypassed an in-depth discussion about operators and functions. This is largely because there's a significant overlap with what you might already be familiar with from PromQL, thanks to our earlier coverage of Prometheus. The parallels between these query languages mean that many of the principles and practices transfer over quite neatly. However, LogQL does come with its own set of quirks and additional capabilities. For those finer points and advanced features, turning to the official documentation will be your best bet to fill in the gaps.

The journey through LogQL has been about laying the groundwork that will serve you well as we step into the realm of Loki rules. In this next section, your LogQL and PromQL experience will shine.

Loki rules

In Grafana Loki, there's a component called the **ruler** that watches over your logs continuously. It evaluates specific queries that you've set up and can trigger actions based on what it finds. Think of it as a watchdog for your log data, alerting you when something important happens or keeping track of key metrics over time.

Similarly to Prometheus, you define *rules*, the ruler components find them, and then they use them to know what to look for in the logs.

There are two main types of rules:

- **Alerting rules**: If your logs meet certain conditions (such as too many error messages), an alert is sent out. It's like setting a trap for problems in your logs. We will dive deeper with an example in *Chapter 9*.

- **Recording rules**: These let you keep tabs on specific metrics extracted from your logs, such as the number of requests to your web server. It's a way to summarize your log data as useful stats.

Recording rules

Recording rules in Loki are used to precompute frequently needed or computationally expensive expressions, with their results saved as a new set of time series. This can significantly improve the performance of queries, especially for dashboards that need to query the same expressions repeatedly.

Let's introduce the concept with an example. Imagine you are managing a large infrastructure and would like to closely monitor the rate at which interfaces flap for each device in the network. The following recording rule should help you track interface flap events:

```
groups:
- name: interface_updown_events
  interval: 1m
  rules:
  - record: events:interface_updown_rate:2m
    expr: sum(rate({vendor_facility_process="UPDOWN"}[2m])) by
(device)
```

Let's break down the recording rule:

- `groups`: A collection of rules. Here, we've named the group `interface_updown_events`.

- `name`: The name of the group, `interface_updown_events`, indicates that this group contains rules related to tracking interface flap events.

- `interval`: This specifies how often a rule should be evaluated. We've set it to `1m` (one minute), meaning the rule will run every minute.

- `rules`: The actual list of rules within the group.

- `record`: This is the name of the metric that the rule will generate. We've named it `events:interface_updown_rate:5m`. This follows a naming convention that helps us understand that this metric is a rate of interface-down events over the last five minutes, by job.

- `expr`: The PromQL expression that defines what we're measuring. In this case, `sum(rate({vendor_facility_process="UPDOWN"}[5m])) by (device)` calculates the rate of UPDOWN log events (which we are using to track interfaces going down) over a five-minute window, summing this rate across all devices.

This recording rule is like setting up an automatic counter that keeps track of network interface issues across all devices. It's set to update every minute, summarizing the frequency of interfaces going up or down in a single metric.

Let's test it out! First, save the configuration snippet into the Loki recording rule file located at `network-observability-lab/chapters/ch7/loki/rules/recording_rules.yml`.

Next, we need to configure Loki to read from the `rules` directory, execute the defined recording rules, and push the newly generated metrics toward the Prometheus instance. For this, we will use Loki's `ruler` and `remote_write` features. Here's how to set them up:

```
# Omitted Loki configuration...
ruler:
  # Omitted ruler configuration...
  storage:
    type: local
    local:
      directory: /rules
  remote_write:
    enabled: true
    client:
      url: http://prometheus:9090/api/v1/write
```

Let's update the configuration of Loki with these settings – `network-observability-lab/chapters/ch7/loki/loki-config.yml`.

Finally, the lab is pre-configured to map the `rules` folder into the container and enable the remote-write receiver by default in this chapter's lab. This configuration is done at the Docker Compose level, so there is no need for you to update it. However, if you want to check it out, feel free to view the Loki and Prometheus service definitions in `network-observability-lab/chapters/ch7/docker-compose.yml`.

Now that the Loki config file is updated, the recording rule is set, and the necessary changes are already present in our Docker Compose file for these services, we are ready to test it out. Run the `netobs lab update loki --scenario ch7` command to allow Loki to perform the recording rules evaluation and forward any new metrics to Prometheus.

To generate some of these metrics, we need to simulate an interface flapping so that the recording rule can capture it. Instead of connecting to the network device, as we did in the previous example, let's run a command to facilitate this – `netobs utils device-interface-flap --device ceos-02 --interface Ethernet2 --count 30 --delay 2`.

Give it a few minutes for the recording rules metrics to be generated. In the meantime, you can check Loki's log output to ensure that the rule is being properly evaluated with the `netobs docker logs loki --follow --scenario ch7` command. The logs should show something similar to the following:

```
loki-1  | level=info ts=2024-08-07T08:31:10.771959723Z caller=engine.
go:232 component=ruler evaluation_mode=local org_id=fake
traceID=72185c62ba40e208 msg="executing query" type=instant
query="(sum by (device)(rate({vendor_facility_process=\"UPDOWN\"}
[2m]))..."
```

Then, to see the generated metrics, go to Prometheus at `http://<machine-ip-address>:9090` and search for the `events:interface_updown_rate:2m` metric. You should see something similar to the following:

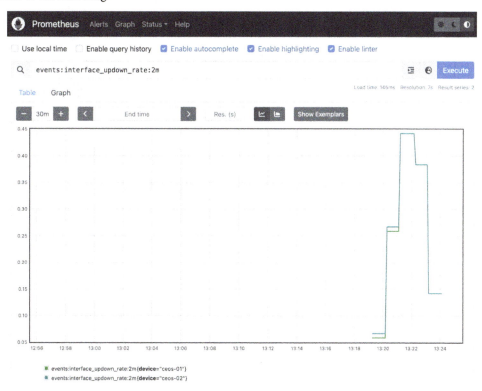

Figure 7.27 – The Loki recording rule metric

A limitation of the events:interface_updown_rate:2m metric is that it becomes unavailable when no interface flap events occur, as shown in *Figure 7.27*. To ensure continuous monitoring and keep device labels present, we can use another metric that is always available. By extracting device labels from this persistent metric and setting their value to 0 in the absence of flap events, we maintain consistent data.

Prometheus provides always-on metrics for devices. To keep the events:interface_updown_rate:2m metric available, even without interface flap events, create a Prometheus recording rule. This rule will ensure each device records a value of 0 when no events are detected, based on an always-on metric such as interface_admin_status. Add the following recording rule configuration to your Prometheus recording rules file at network-observability-lab/chapters/ch7/prometheus/rules/recording_rules.yml:

```
# Omitted configuration...
- name: interface_updown_events
  rules:
  - record: device:interface_updown_rate:2m
    expr: events:interface_updown_rate:2m or (sum by (device)
(interface_admin_status) * 0)
```

This configuration will create a new, continuous metric for interface UPDOWN events, called device:interface_updown_rate:2m:

To update Prometheus with the new configuration, run the netobs lab update prometheus --scenario ch7 command.

Next, generate interface flaps to create events by running the netobs utils device-interface-flap --device ceos-02 --interface Ethernet2 --count 30 --delay 2 command.

The resulting metric, `device:interface_updown_rate:2m`, should now display values similar to the following graph:

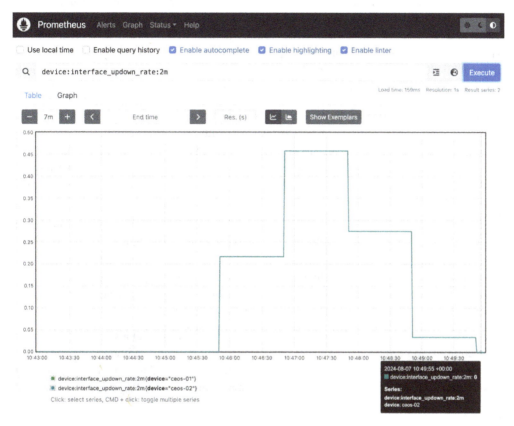

Figure 7.28 – A Prometheus recording rule metric

With this recording rule, you will receive a continuous measure of network stability. Instead of manually sifting through logs for updates, you get a real-time feed of the health of your network's interfaces. This ensures that you can quickly identify and address any network issues.

Persistence tips and best practices

For an observability system, the database layer is central to its architecture. Two key factors to consider for a successful deployment are its performance and scalability. It's important that the system can expand to meet your infrastructure's growing needs. Managing these systems can be challenging and risky if not done properly. Therefore, adopting standardized best practices and automation can mitigate many of these risks, making management less daunting and more efficient.

Performance and scale

As the core of the observability stack, it's vital for the data storage layer to be performant and scalable. It needs to handle substantial amounts of data optimally, managing thousands or even millions of time series and log events. If these systems are slow, users will avoid them and seek other ways to get the information they need. Reliable performance is also important for timely alerts and incident management in critical systems. While this book focuses on the operational aspects of these systems, here are some high-level tips and considerations for TSDBs such as Prometheus:

- **Time series and cardinality**: Each unique combination of a metric name and labels (key-value pairs) in Prometheus creates a new time series. Cardinality refers to the number of these unique combinations. High cardinality means a lot of unique time series, which can strain Prometheus's storage and memory. Understanding time series and cardinality is crucial because they directly affect Prometheus's performance and resource consumption. High cardinality can slow down queries and increase costs.

- **Data retention and rollups**: Data retention policies in Prometheus determine how long data is kept before being deleted. Rollups involve summarizing detailed data into more compact forms over time, although Prometheus primarily relies on raw data and recording rules for precomputed summaries. Setting appropriate retention policies helps manage storage requirements and costs. Knowing how Prometheus handles data summarization (via recording rules) is essential for optimized long-term data analysis without overwhelming storage.

- **Monitoring and alerting**: Prometheus not only collects metrics but also provides alerting capabilities. You can define alert rules based on specific metrics' conditions. For evaluators, understanding the dual role of Prometheus in monitoring and alerting is key. It's not just about collecting data; it's also about acting on it. Prometheus's integrated alerting helps teams respond to issues proactively.

- **Scaling**: As the volume of metrics grows, Prometheus may need to be scaled out. This can involve strategies such as sharding (splitting data across multiple instances) or federation (having a parent Prometheus instance scrape aggregated data from child instances). Knowing how Prometheus scales is important for planning and evaluation. Effective scaling strategies ensure that Prometheus can handle growing data volumes and query loads without degradation in performance.

Let's also highlight some tips and considerations for using Grafana Loki as your log data database:

- **Schema and index management**: Loki uses schemas to define how log data is stored and indexed. The choice of schema affects query performance and storage utilization. Evaluators should understand that Loki's flexibility in managing log data comes with choices that impact performance. Selecting the right schema based on your data and query patterns is important for optimal operation.

- **Data compaction and retention**: Like Prometheus, Loki allows you to set data retention policies. Additionally, Loki compacts log data to reduce its size. For evaluators, it's important to know that Loki can manage the life cycle of log data, from compaction to deletion. These mechanisms help control storage costs and ensure data is manageable.

- **Query performance**: The performance of queries in Loki is influenced by several factors, including the structure of log data, the use of indexes, and query patterns. It's essential for operators integrating log data into the platform to understand these factors, as they directly impact the performance of query execution. By setting appropriate guardrails, operators can ensure that queries run smoothly, delivering fast and responsive log analysis for users.

- **Scaling and high availability**: Loki can be deployed in a microservices architecture, allowing components to be scaled independently for performance and availability. Evaluators should consider Loki's scalability and fault tolerance as critical factors. A properly scaled Loki setup can support high volumes of log data and ensure availability, even in the face of failures.

Automation is your best friend

When managing complex systems such as Prometheus for monitoring and Loki for log management, the volume of data can grow exponentially, and the tasks required to maintain system health can become overwhelming. This is where automation steps in as a critical ally. Automation can handle repetitive tasks, ensure consistency, reduce human error, and free up valuable time for more strategic work.

Systems such as Prometheus and Loki are incredibly powerful tools for observability, but their value hinges on your ability to manage the data they collect. Over time, metrics and logs accumulate, potentially leading to storage issues, slower query responses, and higher operational costs. Here's why keeping things tidy through automation is essential:

- **Purging unnecessary/old data**: Over time, you accumulate data that's no longer relevant for current operations. Automatically identifying and deleting this old or unnecessary data ensures that your storage is optimized, costs are controlled, and system performance is maintained. For someone new to these systems, think of this like cleaning your house regularly; it prevents clutter from building up and makes it easier to find what you need when you need it.

- **Data migration**: As your data storage needs evolve, you might need to move data to different storage solutions (e.g., from local storage to cloud storage for scalability). Automation can manage these migrations smoothly, ensuring data integrity and minimizing downtime. Imagine moving to a new house; automation is like having a team that packs, moves, and unpacks for you, ensuring that everything is exactly where it should be with minimal effort on your part.

- **Management of topics/tables/indexes/schemas**: To keep your monitoring and logging systems running smoothly, it's vital to manage the underlying data structures such as topics (in messaging systems), tables, indexes, and schemas. Automation can handle creation, updates, and maintenance tasks based on predefined rules. This is akin to organizing a library. Without a system to categorize and manage books (data), finding the book you need becomes a time-consuming challenge. Automation ensures that your data library is well-organized and accessible.

- **Efficiency and time savings**: Automation performs tasks faster and more consistently than manual processes, freeing up time to focus on strategic improvements.

- **Scalability**: Automated processes can easily scale with your data and system growth, handling increased loads without additional manual effort.

- **Reliability**: Automation reduces human error, ensuring that critical maintenance tasks are performed accurately and on time.

- **Predictability**: With automation, you can ensure that maintenance tasks are performed regularly, making system performance more predictable and stable.

Summary

In this chapter, we've navigated the essential data storage solutions that underpin effective network observability. Databases like those used by Prometheus and Grafana Loki are the engines of observability platforms, crucial for insightful data analysis. Through practical examples, we've revealed how data flows from collection to storage, and how it's harnessed using powerful querying languages like PromQL and LogQL.

With hands-on exercises in a lab setup that mirrors real-world networks, you've learned how to write data to and query it from these observability tools. You've also seen how automated rules in Prometheus and Loki can streamline your monitoring process, preparing you to proactively address network events.

As we move forward, the next chapter shifts our focus to the art of **data visualization** using **Grafana**. Here, we'll bring the queries and data from this chapter to life, transforming raw metrics and logs into compelling visual narratives that make monitoring not just informative but also intuitive. Get ready to turn numbers and text into actionable insights with the visual power of Grafana.

8

Visualization – Bringing Network Observability to Life

In the previous chapter, we delved into the foundational aspects of data storage using tools such as Prometheus and Grafana Loki to handle time series information such as metrics and logs. So far, we've covered how to collect, store, and query data in preparation for using observability data.

As we shift our focus to **visualization**, we'll not only transition from using data storage to graphically represent data but also emphasize the importance of dashboards. There are several options, but Grafana is an excellent tool in this space as it can transform complex datasets into clear, intuitive visual narratives. This chapter is dedicated to leveraging Grafana to breathe life into your data (most of the concepts can be reused for other tools), turning it into actionable insights for decision-making.

Dashboards are at the heart of this transformation. They are not merely tools for data presentation but vital instruments for storytelling within your network. Whether it's diagnosing an ongoing issue, reporting on the health status of your network, or monitoring key performance metrics, dashboards enable you to tell a compelling story.

We'll explore how to successfully create and utilize dashboards in Grafana, making them both informative and easy to understand. Customizing and optimizing dashboards will be key themes, ensuring they meet the specific needs of your network's observability. By the end of this chapter, you'll have a solid grasp of how to use visualization to enhance your infrastructure management, making complex data accessible and actionable.

This chapter covers the following topics:

- The art of visualization in network observability
- A look into Grafana
- Visualization tips and best practices

The art of visualization in network observability

In the world of network observability, how we present data is as important as the data itself. The shift toward visualizing network metrics and logs isn't just about making things look good; it's about clarity, understanding, and action. Here, we'll break down the essentials of dashboard design into practical principles that every individual should consider, and wrap up this section by discussing various data visualization tools and libraries that enhance data representation.

Data visualization principles

In today's world, we're surrounded by more data than ever before. Think about the health apps on your phone – tracking everything from your steps and heart rate to your sleeping patterns. There's so much information being collected from various sources, and yet, these apps manage to present it all in a way that's easy for us to understand and act upon. They have to make it simple, engaging, and informative, condensing vast amounts of data into digestible insights.

This approach isn't just good for fitness apps; it's essential for network observability too. Just like those health apps help you make sense of your data, network observability tools need to present complex network data clearly and concisely. This ensures that you can quickly grasp what's happening in your infrastructure, make informed decisions, and take action when needed. So, let's dive into the principles that make this possible:

- **Clarity and simplicity**: Dashboards and their visualizations should be straightforward tools, not puzzles that leave you scratching your head trying to figure them out. The key here is simplicity. This means going for a clean and clear approach over a complicated one so that anyone on your team (or the intended audience for the dashboard) can immediately understand what's being shown. Every detail on the dashboard, from each element and panel to the choice of colors, should have a clear reason for being there, all aimed at aiding the decision-making process.

- **Effective use of colors**: Colors in your dashboards play a vital role – they're not just for looks. They should make it easier to tell data apart, spotlight urgent issues, or signal alerts. But it's important to choose colors that are easy on the eyes. Even better, pick a palette that follows the Web Content Accessibility Guidelines (https://www.w3.org/WAI/standards-guidelines/wcag/). This helps make sure that everyone, including those with low vision or color blindness, can easily understand the visuals. And remember, always pair your colors with text descriptions or symbols. This way, you're ensuring that the story you're telling with your data is clear to everyone, no matter how they see your dashboard.

- **Contextual relevance**: Every chart, graph, or metric you include should tell a part of your network's story. Irrelevant data becomes just noise. The goal is to focus on what matters – whether it's pinpointing a bottleneck or showcasing performance improvements in your infrastructure, the data that's displayed must be directly related to the operational goals or issues at hand.

- **User-centered approach**: Remember who you're creating these dashboards for. What works for a senior network architect could be too much for a junior engineer or a manager. Tailoring the detail and complexity to match your audience's needs is key. This way, you're not just making dashboards that are used; you're making them indispensable. By focusing on who will be using the dashboard and what they need from it, you ensure it becomes a valuable tool in their toolkit, not just another piece of software they must navigate.

- **Interactivity**: A snapshot of your network's status can tell you what's happening now, but it might as well be a JPEG if that's all it does. Dashboards that let you interact, zoom in on the details, drill down into specifics, or adjust time frames go from good to great. This kind of interactivity empowers users to explore the data further, giving them the tools or methods to uncover the *why* behind the *what*. This level of engagement turns passive viewers into active investigators.

- **Data hierarchy and layout**: Think of your dashboard like a well-organized storybook. The way you lay out the information sets the stage for the entire narrative. Start things off with the big-ticket items – those critical metrics that demand immediate attention – right up front and center. From there, it should cascade to the finer details that paint the full picture but might not need the spotlight right away. This setup not only makes your dashboard easier on the eyes but also guides anyone looking at it through your infrastructure's tale in a logical, easy-to-follow way. And remember, you don't have to cram every piece of data into a single dashboard. It's perfectly fine to have an overview dashboard that provides a quick health check of your infrastructure and separate, more detailed dashboards that focus on specific aspects, such as device health or traffic analysis. This approach keeps things clean, focused, and, most importantly, useful.

Visualization tools for observability

The visualization tools that are available on the market are designed with specific strengths, whether it's handling real-time data, integrating with various data sources, or offering advanced analytical capabilities. It's common for monitoring systems to offer an integrated visualization component directly tied to their database systems. Examples include **Prometheus**, **Graylog**, and **Jaeger**. However, some tools opt for a more modular approach, offering separate visualization components that support a decoupled architecture. Let's look at a few notable examples:

- **Grafana**: Well known for its versatility, it offers a wide array of integrations, enabling connections to numerous data sources and supporting a vast range of visualizations. Thanks to its plugin framework, users have the flexibility to incorporate new data sources or visualizations crafted by either partners or the Grafana community.

- **Kibana**: As an integral component of the Elastic Stack, Kibana serves as the primary visualization tool for Elasticsearch data. It excels in offering insightful discovery panels that help in exploring stored data, specialized observability applications, and even tracking application performance metrics.

- **Chronograf**: Chronograf is the visualization component of the **InfluxData** platform, designed specifically for time series databases. It simplifies the process of visualizing and querying data stored in **InfluxDB**. It also enables the creation of dashboards and alerts from monitored applications, infrastructure, and general network performance.

There are other tools, not necessarily open source, in the realm of **business intelligence** (**BI**), such as **Power BI** and **Tableau**, that also play a pivotal role but with a broader focus. Power BI and Tableau are well known for their ability to convert data into strategic business insights, serving non-technical users through intuitive interfaces and powerful data visualization capabilities. These tools emphasize ease of use, allowing users to quickly identify trends, patterns, and anomalies within data.

Beyond these application-specific tools, the foundation of modern data visualization lies in specialized software libraries. They offer a wide spectrum of functionalities, from creating interactive web visualizations with **D3.js** (`https://d3js.org/`) and **Plotly** (`https://plotly.com/`) to generating sophisticated statistical charts with **Matplotlib** (`https://matplotlib.org/`) and **Seaborn** (`https://seaborn.pydata.org/`). These libraries empower developers and data scientists to tailor visualizations to their exact requirements.

In this book, we'll focus on using Grafana. We've chosen Grafana because it offers a wide range of visualizations and it can easily connect to various data sources, such as Prometheus, Loki, and Nautobot. This makes it a practical tool for our needs, allowing us to work with diverse types of data in a more straightforward manner.

A look into Grafana

Grafana is an open source software that started as a tool for visualizing time series data. Over time, it has grown to include features for alerting and exploring metrics, logs, and traces from various remote databases or backends. This expansion has turned Grafana into a versatile tool not just for visualization but also for monitoring, analyzing, and, as we'll explore in the next chapter, alerting across various dimensions of data.

One of the strengths of Grafana is its open source framework, which allows it to be integrated with a wide array of external systems, including various databases and **IT service management** (**ITSM**) software. This flexibility makes it an excellent choice for teams looking to customize their observability stack to their specific needs.

This chapter will continue to leverage our lab environment, which was detailed in *Chapters 5 to 7* and further elaborated on in the *Appendix A*. Our lab features the ceos-01 and ceos-02 network devices, configured with BGP and in communication with each other, providing a real-world context for our exercises. You'll be guided through creating a **Device Health** dashboard that's designed to offer network engineers a comprehensive overview by correlating metrics and log data.

The **Device Health** dashboard should serve as a roadmap for this chapter. It's going to show us what Grafana can do, and then do it step by step. You'll get to learn by doing, which means that by the end of this chapter, you'll have made a really useful dashboard for network engineers. This hands-on experience is all about getting you comfortable with Grafana and teaching you how to use it to keep an eye on network health in a practical way:

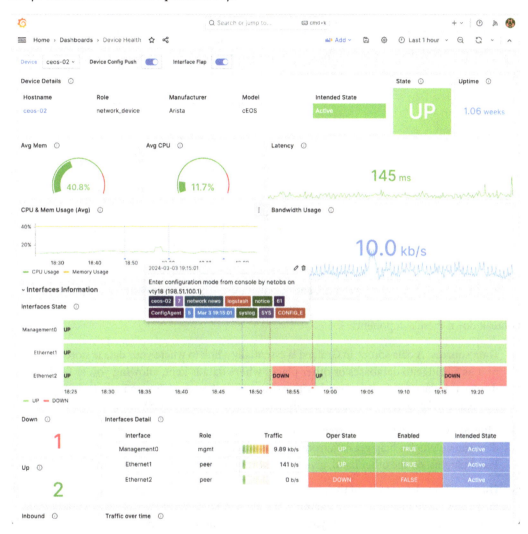

Figure 8.1 – Example dashboard

So, let's dive into how Grafana works behind the scenes. We'll break down its key components to see how it helps us visualize our data, from metrics to logs. Understanding this will show us how we can make the most of Grafana for our data visualization needs.

Architecture

Grafana's heart lies in its dashboards, the canvases where we bring data to life through custom visualizations. These dashboards are vital for keeping a watchful eye on our systems, networks, and applications, presenting observable data in a way that's meaningful for us. But how do they function, and what elements do they consist of? To get a clearer picture, let's take a closer look at the structure and components behind a Grafana dashboard:

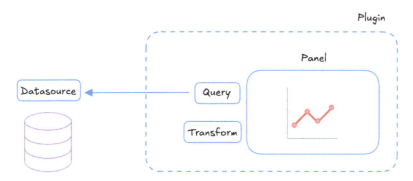

Figure 8.2 – Grafana dashboard components (Created using the icons from https://icons8.com)

Let's break down these components:

- **Data sources**: A data source in Grafana is essentially a link to where your data is stored. Think of it as a bridge to databases or other places where data is kept, allowing you to fetch and display this data. Grafana can work with many different types of data sources, including SQL and NoSQL databases, **time series databases** (**TSDBs**), and even CSV files.

- **Plugins**: A plugin in Grafana is something that brings new features to the platform. For instance, data source plugins help get data from various places and put it in a format Grafana can use. Panel plugins, on the other hand, give you new ways to show that data visually.

- **Queries**: These let us pull out the pieces of data we need from our data sources using different languages. Take **PromQL** as an example: it's the special language we use to grab data from Prometheus or similar sources (if you want to see it in more depth, please refer to *Chapter 7*). With a PromQL query, we can narrow things down to just the data we're interested in. For instance, writing `device_uptime{device="ceos-02"} [2m]` fetches the last 2 minutes of uptime data for the `ceos-02` device.

- **Transformations**: This layer acts like a toolbox that helps us clean up and organize our data just how we need it for our visuals. It lets us combine data pieces, change details, or get rid of what we don't need. It's not always necessary, but sometimes, it's the key to making our data look just right for our charts and graphs. We'll dive into how to use this tool in our lab section, giving you a clear picture of how it can improve our visualizations.

- **Panels**: A panel is where all the magic happens. It takes the data we've gathered, the specific questions we've asked through our queries, and any fine-tuning we've done with transformations, and puts it all together into a clear, informative display. Grafana offers an extensive range of panel options right out of the box, including time series graphs, bar charts, heatmaps, geomaps, and tables, among others. Beyond these built-in panels, the Grafana community plays an important role in enhancing its capabilities. Developers from the Grafana ecosystem contribute plugins that add even more variety, such as **FlowCharting** for intricate diagrams or **Business Calendar** for specialized calendar panels. This collaborative effort has enriched the Grafana plugin library (`https://grafana.com/grafana/plugins/`), so feel free to explore and discover the available options.

Now, it's time to put theory into practice! We're going to roll up our sleeves and construct a network dashboard for our lab environment. Our goal is to adhere as closely as possible to the principles introduced earlier, making our dashboard not just functional but truly useful. Along the way, we'll dive deeper into Grafana's capabilities, exploring how it can enhance our monitoring and visualizations.

So, let's take a closer look at the lab environment for this chapter.

Setting up the lab environment

This chapter is filled with examples to guide you in visualizing the observability data. For a visual representation of the elements we're going to consider, check out the following diagram:

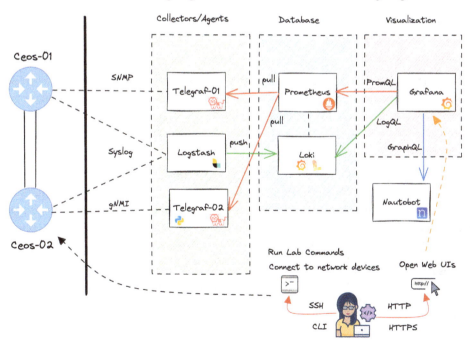

Figure 8.3 – Chapter 8 lab scenario (Created using the icons from https://icons8.com)

You may have noticed that this lab environment is pretty similar to the one in the previous chapter. That's because we're building on top of the configurations and components we set up in previous labs and chapters. Our hands-on activity here focuses on visualizing metrics and logs using Grafana. You'll get the chance to interact with network devices to generate activity, allowing you to witness the real-time impact on your dashboard.

For this lab, tools such as Telegraf, Logstash, Prometheus, Loki, and certain aspects of Grafana have been pre-configured. This setup is tailored to visualize the overall health of the network devices (`ceos-01` and `ceos-02`).

Before diving into the lab activities, ensure your lab machine is set up according to the steps outlined in the *Appendix A*, just as we did in previous chapters. Make sure you're in the `~/<path-to-git-project>/network-observability-lab/` directory within the project's repository to execute the necessary commands.

Now, let's begin the lab scenario for this chapter. To set everything up on your lab machine, run the `netobs lab prepare --scenario ch8` command.

> **Note**
>
> This lab scenario includes a completed version that you can refer to. It is located in the git repository you cloned, under `network-observability-lab/chapters/<chapter-number>-completed`. This folder contains all the configurations that have been used in the examples throughout this chapter, providing a comprehensive guide to ensure you have everything set up correctly.

This configuration includes a **Nautobot** instance, which acts as a supplementary data source for Grafana. If you're not familiar with Nautobot, check out *Chapter 6*, where we use it for data enrichment purposes. Alternatively, take a moment to look through the official Nautobot documentation (`https://nautobot.readthedocs.io/`) for a comprehensive understanding.

You can verify your connection to Nautobot at `http://<machine-ip-address>:8080` using the default credentials (username: `admin`, password: `nautobot123`). The Nautobot service may take a few minutes to start up completely; on smaller machines, it can take up to 10 minutes to go online. You can monitor its progress by checking the logs. To do this, run the `netobs docker logs nautobot --follow --scenario ch8` command. If you only want to see the most recent entries, add `--tail=20` to show just the last 20 lines of the log.

Next, to ensure that Nautobot reflects the current state of your network devices in the lab, run the `netobs utils load-nautobot` command. This will import the necessary device data into Nautobot, synchronizing it with your lab setup.

Accessing Grafana

Let's walk through how to access Grafana in our lab environment. This will give you a preview of the features you can expect to use. Getting into Grafana is the first step in bringing your network data to life.

First things first, connect to Grafana at `http://<machine-ip-address>:3000`. Log in with the lab credentials (username: `netobs`, password: `netobs123`) to land on the Grafana home page:

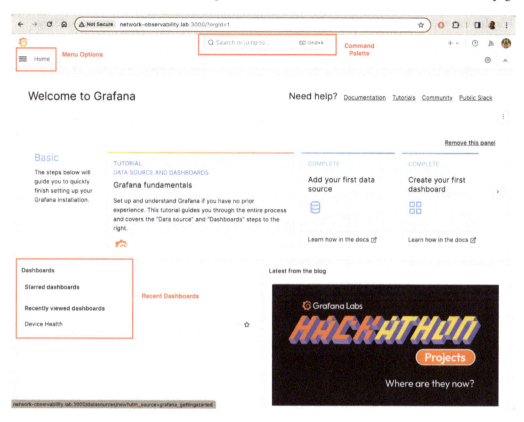

Figure 8.4 – Grafana home page

You will be presented with the Grafana home page, which features menu options in the top-left corner, a command palette at the top center, and a view of recent dashboards in the bottom-left corner.

Connecting to and exploring data sources

With your lab environment set up, you're ready to establish connections to your data sources and delve into the data that's been collected from your network devices. In Grafana, go to **Menu** | **Connections**. Here, you'll see the data sources – **Prometheus**, **Loki**, and **Nautobot** – already listed and ready for you to explore:

> **Note**
>
> For this lab, the data sources have been pre-configured to streamline our focus on building dashboards. This saves us time setting up and lets us dive straight into the visual representation of our data. If you're interested in how the data sources for this lab scenario have been configured, take a look at the `network-observability-lab/chapters/ch8/grafana/datasources.yml` file included with the lab for more detailed information.

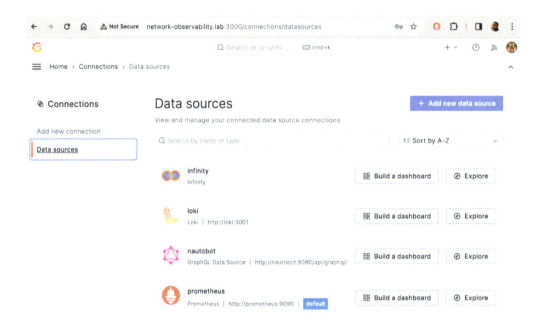

Figure 8.5 – Grafana – Data sources

Now, you're ready to test each connection to ensure they're ready to retrieve data.

Great! To wrap up our Grafana setup, let's take a closer look at our data by conducting some simple queries on our data sources. Navigate back to **Data Source | Connections** and click on the **Explore** button next to each source to begin this exploration:

1. Let's start by executing a straightforward PromQL query on the Prometheus data source to gather BGP neighbor state information. Our objective is to ensure we can successfully pull data from each source. To do this, click on **Explore**, choose **Code** for your query type, and enter `bgp_neighbor_state{device="ceos-01"}` for your PromQL expression. For those new to Prometheus and PromQL, a deeper dive can be found in *Chapter 7*. The outcome in the panel should look similar to the following:

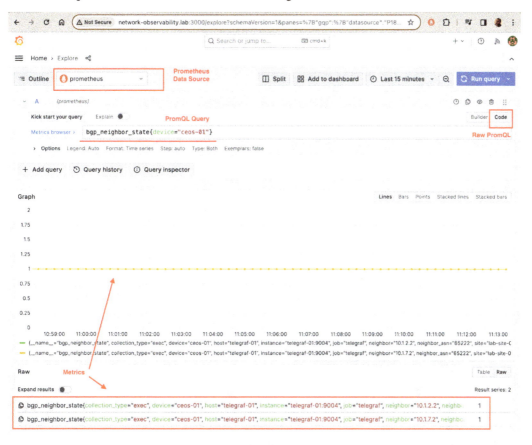

Figure 8.6 – Grafana Prometheus – Explore

2. Next, let's switch our focus to Loki for log searching related to the `ceos-01` device. Use the `{device="ceos-01"} |= ``` query to filter logs for this specific device. Just like with PromQL, if you're looking to brush up on how to utilize LogQL properly, *Chapter 7* provides a comprehensive overview:

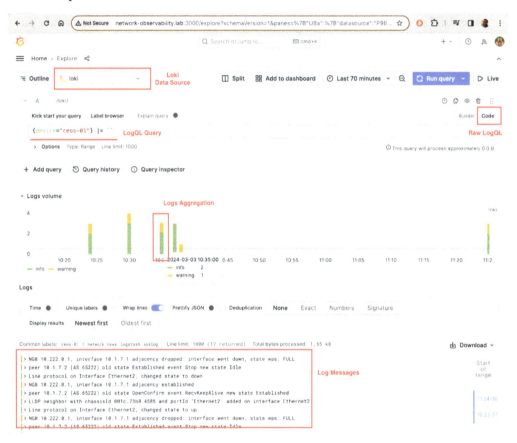

Figure 8.7 – Grafana Loki – Explore

3. Lastly, let's use a GraphQL query within Nautobot to enumerate device names along with their interfaces by using the `{ devices { name interfaces { name } } }` query. If you want a refresher or want to gain a deeper understanding of interfacing Nautobot with GraphQL, *Chapter 6* offers some insights. Consulting the official Nautobot documentation on GraphQL is highly recommended for more comprehensive guidance:

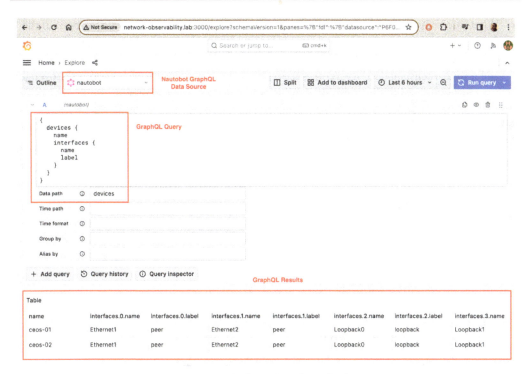

Figure 8.8 – Grafana Nautobot – Explore

With all the groundwork laid, we're well-positioned to start building our Grafana dashboard. Let's dive in and see what we can create!

Creating your first Grafana dashboard

In this section, you will be guided through building out some example dashboards using Grafana. One of the best aspects of using the Grafana dashboards for your visualizations is that you get to customize the dashboard to your liking.

But first, we'll review the **data visualization principles** we mentioned previously. The overarching goal is to create a dashboard that tells a story, has a target audience, and provides good enough information that is clear and simple. So, let's define some of these principles:

- What do we want to visualize? This is the main title and purpose of your dashboard. For instance, we're going to build a **Device Health** dashboard to gain an overall view of the status of a network device. We will use this dashboard to explore Grafana's capabilities.

- Who is the intended audience? This normally falls into the type of *persona* the dashboard is being built for. We'll cover this in more detail in *Chapter 11*, but for now, we're focusing on building out the dashboard for the network operations team.

- What information is critical and informational on the dashboard? As we talked about in the *Data visualization principles* section, the big ticket items of the dashboards should be front and center. For this type of dashboard, they are usually the device information, status (up, down), and resource consumption. Then comes the contextual informational aspects, such as the overall interface status and traffic usage, BGP information, device logs, and more.

With this overview in mind, let's sketch a mockup or wireframe to lay out our ideas visually. This will help ensure we have a clearer direction before we dive into the actual building process:

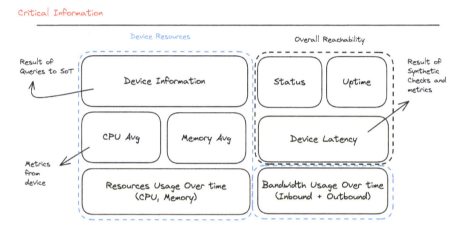

Figure 8.9 – Device critical information wireframe

This is a good start, but we need deeper insights customized to different topics. Let's sketch a wireframe specifically for the device interfaces, their current state, average traffic, and an overall view of inbound and outbound traffic over time:

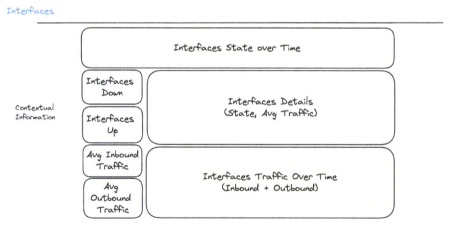

Figure 8.10 – Device interfaces wireframe

Lastly, let's create a wireframe that showcases protocol information, such as BGP sessions, alongside device logs. This will enrich our understanding of *why* there are changes in the protocols on the network:

Figure 8.11 – BGP and logs wireframe

To construct the dashboard panels, we'll tap into three key sources: the metrics database (Prometheus) for the bulk of our metrics data, Loki for event logs from the devices, and an SoT (Nautobot) for intended device insights.

Now, let's get started on the dashboard. Head back to the Grafana home page and click **New dashboard**, under the + icon in the top-right corner. Then, click **Add Visualization** to begin crafting your panels:

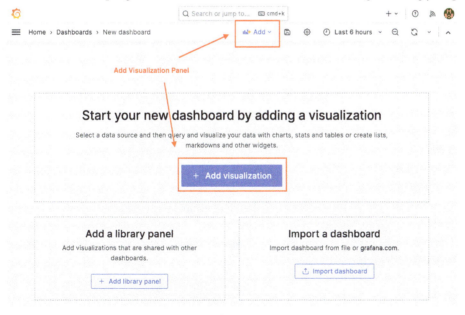

Figure 8.12 – Grafana – New dashboard

Metrics panels

The initial wireframe features various panels derived from metrics collected from the network devices, including reachability latency, overall status, uptime, CPU, and memory usage, among others. So, to kick things off, let's begin creating our new panel by selecting Prometheus as our data source.

You'll be presented with a blank space that's ready to show your visualization. At the top right, there's a panel type selector where you can choose the kind of visualization you're aiming for. Just ahead, you can tweak its settings in the **Panel options** section, and finally, you can specify the query and the transformation you want to apply to your data at the bottom left of the panel canvas. See the following figure for a visual representation:

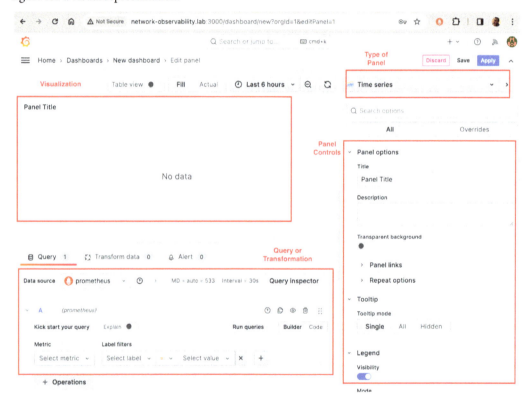

Figure 8.13 – Grafana panel – blank canvas

This will be the main canvas for creating visualizations in Grafana. Let's start by creating the first panel for the **Device Health** dashboard.

State and latency metrics panel

For our first panel, we aim to display the device's status – whether it's **Up** or **Down**. Following our principles for effective data visualization, utilizing colors to indicate the state alongside text provides clear communication. Thus, choosing the **Stat** panel type will suit our needs perfectly.

To find out whether a device is **Up** or **Down**, we'll send a query to our Prometheus database that looks for a specific piece of information, or metric, that shows us just that. For network devices, that key metric is `ping_result_code`, where 0 means everything's good, 1 means the device can't be found, and 2 signals a ping error.

This `ping_result_code` metric comes from our Telegraf setup, which regularly pings the network devices to check on their management interface. If you want to see how all this fits together, take a look at the lab requirements diagram for an overview of the setup and traffic flow. For a deeper dive into how we gather this data, go back to *Chapter 5*.

In the **Query** section of your panel canvas, you'll see Prometheus listed as your data source. Here, you have two options for creating your query. You can use Grafana's **Builder** feature, which helps put together the PromQL query step by step, or you can switch to the **Code** setting to write the full query yourself. We'll use the **Code** setting and the `ping_result_code{device="ceos-02"}` query for our example. We'll also add a title and a description to our panel:

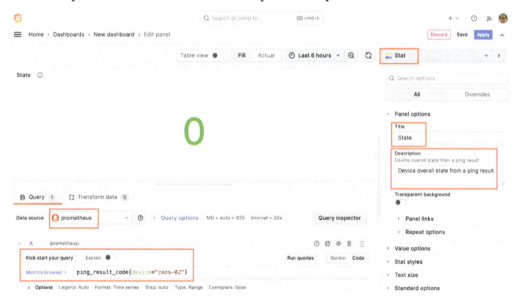

Figure 8.14 – Ping result code

Next, we'll adjust the panel settings to enhance the presentation of this metric. Let's select **Stat styles | Color mode | Background Gradient**, as shown in the following figure:

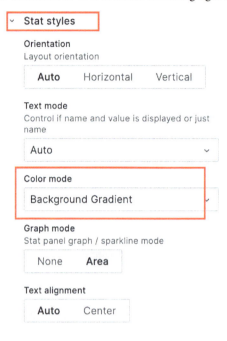

Figure 8.15 – Panel options and Stat Styles

Now, we'll define specific meanings for the `ping_result_code` metric values by updating the **Value Mappings** options. Here's how we'll set up the mappings to communicate the results:

Figure 8.16 – Panel controls – Value mappings

After making these adjustments, you'll notice the updates directly on the panel canvas, which is exactly what we want at this stage. To include this panel in our dashboard, click the **Apply** button in the top-right corner. Remember, you can also resize the panel once it's part of your dashboard so that it fits your layout better:

Figure 8.17 – Device reachability state panel

To construct the **Latency** panel, we'll again opt for a **Stat** type panel. This time, we're going to use the `ping_average_response_ms{device="ceos-02"}` query to gather the device's average response time. Make sure you adjust **Standard Options** | **Units** to **Time / seconds (s)** to accurately represent the metric's scale. The finished panel will look something like this:

Figure 8.18 – Latency metrics panel

Uptime metrics panel

Moving on to the **Uptime** panel, we'll follow the process we used for the previous panels. Start by adding a new visualization in your **Dashboard** canvas. Then, choose Prometheus as the source of our data:

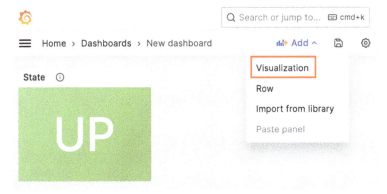

Figure 8.19 – Adding a visualization

For this panel, which will display the device's uptime, a **Stat** type panel is the right choice since the value to display is a metric. We'll focus on the `device_uptime` metric, specifically filtering for the `ceos-02` device, so we'll use `device_uptime{device="ceos-02"}` as our query.

Don't forget to give your panel a title and a short description, as shown here:

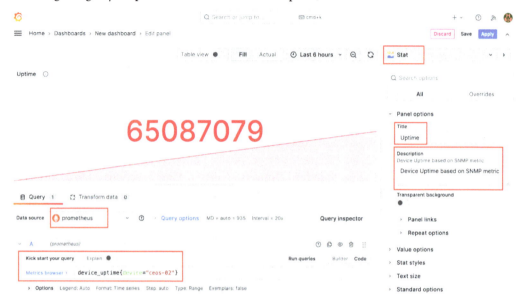

Figure 8.20 – Uptime stat panel

The metric returns uptime in time ticks, which is a non-negative integer representation of time. Let's convert this into a format that's easy to read:

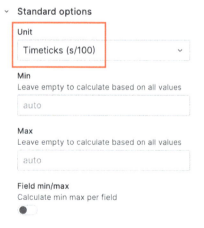

Figure 8.21 – Uptime panel – Standard Options

Additionally, we'll remove the background time series for clarity and choose a neutral color, such as light blue, which is a good color to display contextual information:

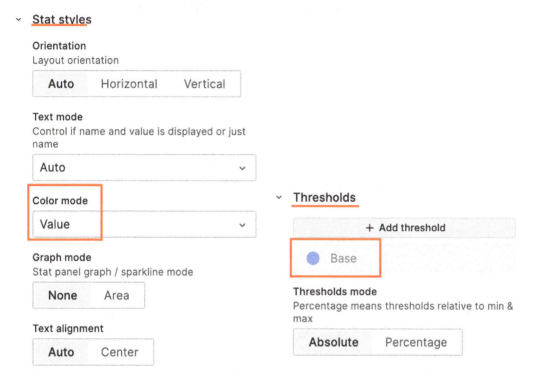

Figure 8.22 – Setting the base threshold color to light blue

At this point, you should have a panel similar to the following:

Figure 8.23 – Uptime metrics panel

Avg CPU and Avg Mem metrics panels

To track the average CPU and memory metrics, we're going to choose a **Gauge** type panel. This type of panel is perfect for these measurements since it visually changes color according to the percentage value – it stays green for values up to 80% and switches to red beyond that. We'll run the `cpu_used{device="ceos-02"}` query to get the CPU usage as a percentage (from 1 to 100). This setup will result in a panel that looks something similar to the following:

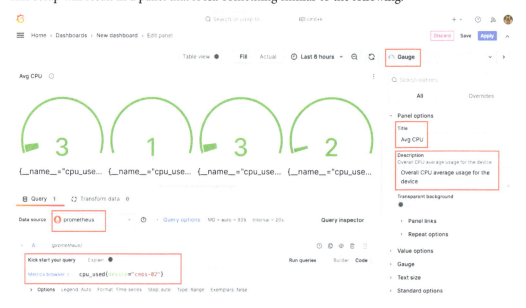

Figure 8.24 – CPU cores usage

The values are shown in percentage format, highlighting data from 4 CPU cores. To find the average CPU usage across all cores, we'll use the `avg(cpu_used{device="ceos-02"})` PromQL expression in our query. By configuring the panel's title and description and setting the **Standard Options | Unit** value to **Percent (0 - 100)**, the resulting panel that's created from this will look similar to the following:

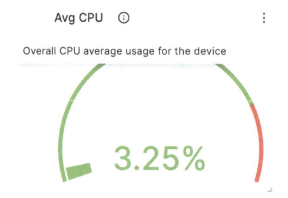

Figure 8.25 – Avg CPU cores of a device

For the memory metrics, we're following a path similar to the CPU metrics but focusing on memory usage. The metrics provide values for both the used (`memory_used`) and available (`memory_available`) memory in bytes for the device. To calculate what percentage of memory is being used, we'll employ the following PromQL expression:

```
(memory_used{device="ceos-02"} * 100)
/
(memory_used{device="ceos-02"} + memory_available{device="ceos-02"})
```

This formula takes the amount of used memory, multiplies it by 100 to get a percentage, and then divides it by the total memory (`used` plus `available`) to calculate the overall memory usage percentage. The outcome will be visualized in a panel that should resemble the following:

Figure 8.26 – Avg Mem used on a device

Bandwidth Usage panel

For this metric panel, we're aiming to calculate the total traffic, both incoming and outgoing, across all interfaces. Reflecting on our lessons from *Chapter 7*, where we learned to calculate the flow of bytes (or octets) through an interface using the rate function, we'll apply a similar approach here. Our expression is as follows:

```
sum(
    rate(interface_in_octets{device="ceos-02"}[2m]) * 8
    +
    rate(interface_out_octets{device="ceos-02"}[2m]) * 8
)
```

Let's dissect this to grasp its workings:

- We use the `rate` function to determine the number of bytes per second passing through an interface, calculating this for both inbound (`interface_in_octets`) and outbound (`interface_out_octets`) traffic.

- Using the `interface_..._octets{device="ceos-02"}` label, we're filtering for interface traffic on the `ceos-02` device only.

- Here, [2m] specifies the period over which we're measuring the rate. In this case, it's the last 2 minutes.

- Multiplying by 8 converts bytes into bits, providing a common measure of network traffic.

- Finally, sum(...) adds the inbound and outbound rates together for a comprehensive total traffic figure.

With our PromQL expression set and the unit specified as bits per second in our standard options, we'll end up with a panel that looks something like this:

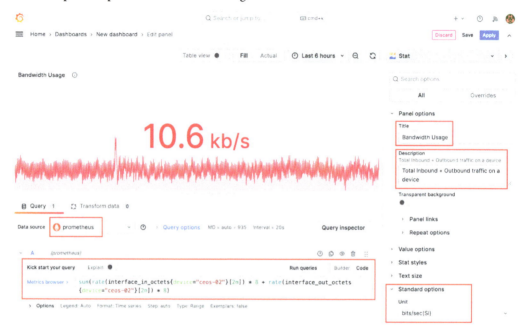

Figure 8.27 – Bandwidth Usage panel

Now that we have a solid understanding of how to create various types of stat panels, this is a good time for us to pause and save our progress.

Saving a dashboard

Let's pause and save our progress on the dashboard so far. First, navigate to **Dashboard settings**:

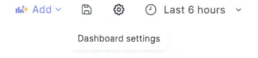

Figure 8.28 – Grafana – Dashboard settings

Here, you have the option to change the dashboard's metadata, such as its name, variables, links, and annotations. For now, let's assign it a name, add a description and tag, and then save our work:

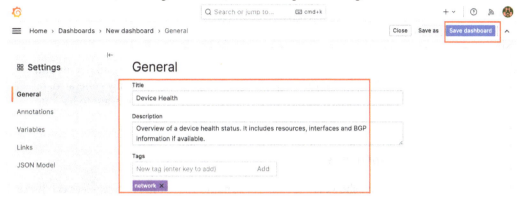

Figure 8.29 – Saving a dashboard

For our lab environment, when you make changes and save them, they're stored in the Docker volume. This setup ensures that your modifications remain intact through stops and starts of the Grafana container. However, should you decide to fully dismantle the lab using commands such as `netobs lab prepare` or `netobs docker destroy --volumes`, you'll find that these saved changes disappear. To avoid losing your work, exporting your dashboard's JSON file and saving it in your repository is a smart move. This way, you can easily reload your Grafana dashboard settings as needed:

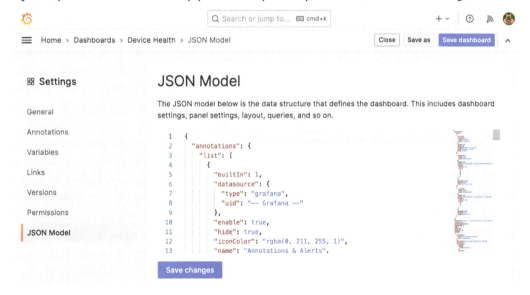

Figure 8.30 – Grafana dashboard settings – JSON Model

To do this, go to **JSON Model**, select the content of the Grafana dashboard settings, and save it under `/network-observability-lab/chapters/ch8/grafana/dashboards/device-health.json`. Remember to `git commit/push` on your forked repository to back up your changes.

> **Note**
>
> This capability of Grafana allows you to treat dashboards as code, meaning all details and settings are saved in a JSON file that can be version-controlled, automatically generated, or modified. This feature is particularly useful for *golden* dashboards that you want to back up and restore quickly.

Table panels with GraphQL queries

Next up, we'll create a table that showcases general information about the device – let's call it the **Device Details** panel.

While telemetry gives us solid data, such as the device's hostname and manufacturer, we're still missing insights into its role or intended state. This is where incorporating data from a SoT or a **configuration management database** (**CMDB**) comes into play, adding depth to our visualization layer – a concept we touched upon back in *Chapter 6*.

For this particular panel, we'll be diving into Nautobot and utilizing a GraphQL query to extract valuable details. Imagine pulling together information such as the device's name, role, manufacturer, model, and intended operational state. To set this up, we'll choose `nautobot` as our data source, apply our GraphQL expression, and opt for **Table** as the panel type to neatly display this enriched data:

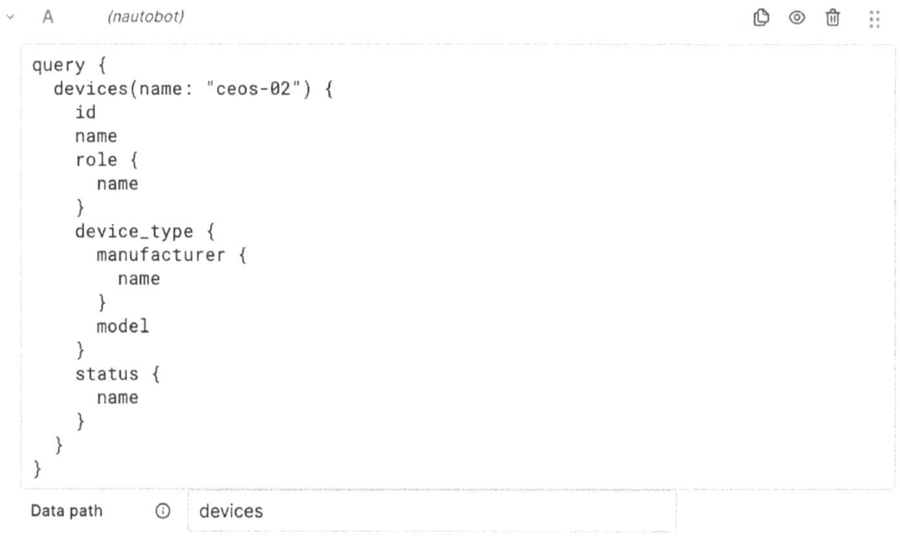

```
 ˅   A      (nautobot)

query {
  devices(name: "ceos-02") {
    id
    name
    role {
      name
    }
    device_type {
      manufacturer {
        name
      }
      model
    }
    status {
      name
    }
  }
}

Data path   ⓘ    devices
```

Figure 8.31 – Nautobot GraphQL query for device information

The finished panel will display a table that neatly organizes all the details that have been fetched from your query's response:

Figure 8.32 – Table panel containing device information

To make the table more user-friendly, we can improve it by renaming the column headers to more intuitive titles. To accomplish this, we'll utilize the **Transformations** feature of a Grafana panel dashboard.

Transforming data

To apply a transformation to your data, simply click on the **Transform data** tab next to the **Query** section. This will present you with various transformation options to choose from:

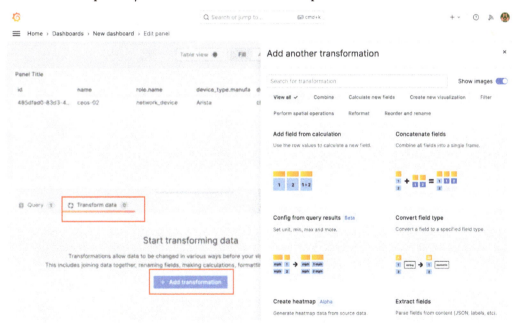

Figure 8.33 – Adding a transformation

Choose the **Organize by Field Name** option. This allows you to reorder and rename the fields that are returned:

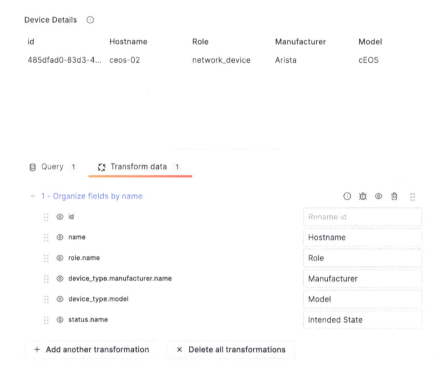

Figure 8.34 – The Organize fields by name transformation

> **Note**
>
> You can add multiple transformations to a panel, and the sequence is important because each transformation affects the next one. For this reason, the panel beside the transformation (as seen in *Figure 8.35*) allows you to re-order, disable, debug, or delete them:

Figure 8.35 – Panel next to the transformation

With these changes, you can now reference fields in your panel options using their new names.

You might be wondering why we include the id value of the device in our GraphQL query. The reason is that we aim to create a clickable link on the device's hostname that directly connects to its detailed information stored in Nautobot. To achieve this, we'll need to use **overrides**.

Overriding data

We aim to remove the id value from the table's display since it doesn't offer much value to an end user, but we plan to use its value to craft a **Data Link** value, which will turn the **Hostname** field into a clickable link that points to the device details in Nautobot. To do this, head over to the **Panel options** section and choose **Overrides**, setting them up as follows:

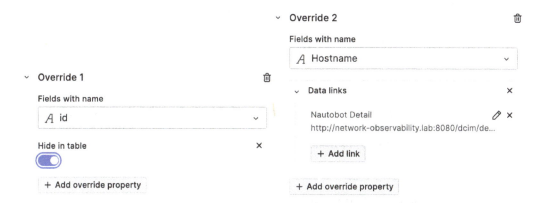

Figure 8.36 – Overriding data

For the second override, we're setting up a **Data Link** value that directs to a specific Nautobot device ID URL (`http://<machine-ip-address>:8080/dcim/devices/<device-uuid>`). This link will use the value derived from Grafana's data processing. So, let's learn how to format this link correctly. To include the device's id value, we must use `${__data.fields.id}` in the URL (notice that there are two underscores before `data`). For detailed guidance on using data variables to configure links in Grafana, check out the official documentation:

Figure 8.37 – Creating a hyperlink with dynamic data

Once you've applied these overrides to the **Device information** table panel, the outcome should look similar to the following:

Figure 8.38 – Device details table with link

> **Note**
>
> As an exercise, try adding a new override for the **Intended State** field. Configure it so that it colors the background of the cell value. Additionally, implement a **Value Mapping** value to convert the Nautobot state – in this instance, **lab-active** – to **Active**, and select a background color of your choice for it.

When you click on the link, it will open a new tab that takes you to the Nautobot details page for the ceos-02 device. After making these adjustments, don't forget to apply the changes and save your dashboard.

Next, we'll explore how to incorporate variables into the dashboard.

Adding variables to your dashboard

At this stage, our dashboard is shaping up, already displaying operational state data in graph form. It's on the right track, so far focusing on just one device, ceos-02. However, Grafana offers a feature that can greatly simplify how we visualize our environment: **dashboard variables**. By introducing variables, we can manage our visuals across multiple devices seamlessly. Let's begin by setting up our first dashboard variable for the device name.

To create a variable, navigate to **Dashboard settings** and proceed to the **Variables** section.

Variables offer flexibility and can be set up in various ways: manually, through a query, or via a text box where users input values. For our purposes, we want to set up a device variable that gets its value from a metric stored in our Prometheus instance. This approach guarantees that the selected device has associated telemetry data:

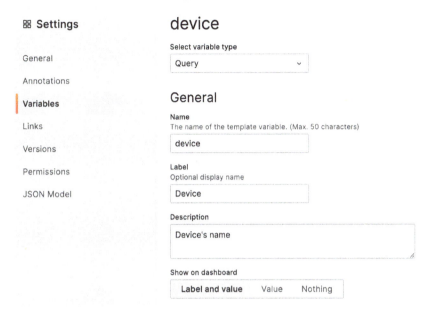

Figure 8.39 – Variable settings – General

Next, we'll specify the query options, which we'll find by scrolling down further:

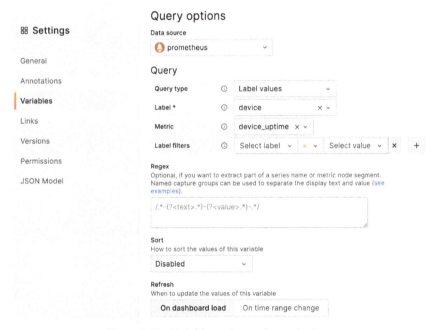

Figure 8.40 – Variable settings – Query Options

Our query is directed at Prometheus and utilizes the **Label Values** query type, a feature of the Prometheus data source. This approach yields a list of possible values for a specified label – in our case, we're interested in the `device` label values. To gather these values, we must select a metric known for its reliability and likelihood of being present, namely `device_uptime`.

Once set up, this query will show the available `device` values associated with the `device_uptime` metric under the **Preview of values** section; it should return both network devices (`ceos-01` and `ceos-02`). The variable will now appear at the top of your dashboard.

Now that the `$device` variable is ready, it's accessible for both existing panels and those you're about to create. Let's put it to use by crafting a new panel that will be customized according to this variable. This one will be a time series chart that tracks the total resource usage percentage (CPU and memory) over time by utilizing our newly established variable.

Instead of guiding you through every step with screenshots, we'll switch gears and present a table that summarizes the query and Grafana panel settings needed to activate this new panel. It's time to dive into setting up our query:

Query Setting	Series A (CPU Usage)	Series B (Memory Usage)
Data Source	Prometheus	Prometheus

Query Expression	`avg(avg_over_time(cpu_used{device="$device"}[2m]))`	`avg(`
		
		`  (memory_used{device="$device"}`
		`* 100)`
		
		`/`
		
		`  (memory_used{device="$device"}`
		`+ memory_available{device="$device"})`
		
		`)`

Options: Legend	CPU Usage	Memory Usage
Options: Format	Time Series	Time Series
Options: Type	Range	Range

Table 8.1 – CPU and memory panel query settings

> **Note**
>
> For CPU usage, we calculate the average usage over time for each core and then combine these to form a single, smoother time series. For memory, we compute an average based on the total memory usage by utilizing the `memory_used` and `memory_available` metrics.

Notice how the **Query Expression** area in the preceding table now incorporates the `$device` variable.

Next, let's adjust the panel controls and settings:

Setting	Value
Title	CPU & Mem Usage (Avg)
Description	CPU and memory usage on average for available cores on the device
Standard Options: Unit	Percent (0-100)

Table 8.2 – CPU and memory panel options

The resulting panel should look similar to the following:

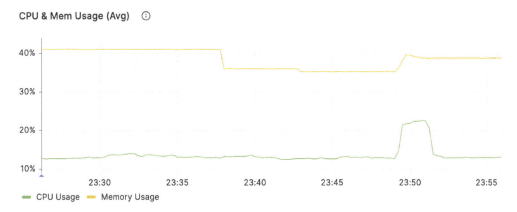

Figure 8.41 – The CPU & Mem Usage (Avg) panel

State timeline panels

Let's move on to adding more panels to our dashboard. This time, we'll focus on the interface state and craft a state timeline panel. This panel aims to visually narrate the status history of the interfaces over time on the dashboard. We'll begin by setting up the query:

Query Setting	Series A (Interface Oper Status)
Data Source	Prometheus
Query Expression	`interface_oper_status{device="$device"}`
Options: Legend	`{{name}}`
Options: Format	Time Series
Options: Type	Range

Table 8.3 – Interfaces State timeline panel query settings

In the legend, the `name` label is used to reference the interface name that's retrieved through the query expression.

Here are the settings and options for the panel control:

Setting	Value
Title	Interfaces State
Description	Interfaces State change over time
Standard Options: Color Scheme	Single Color
Standard Options: Value Mappings	Representation of the Interfaces State code to its text representation and color (see more details ahead)

Table 8.4 – Interfaces State timeline panel options

Here's how we map the numerical codes for interface states to their textual meanings. This helps us showcase the status of interfaces on the dashboard:

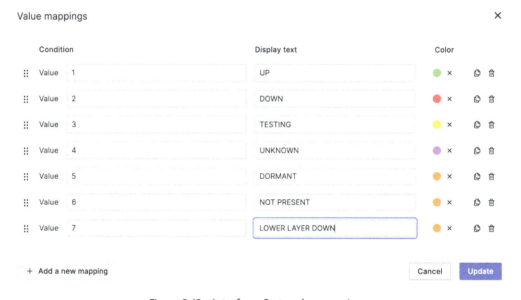

Figure 8.42 – Interfaces State value mappings

The resulting panel should be similar to the following one:

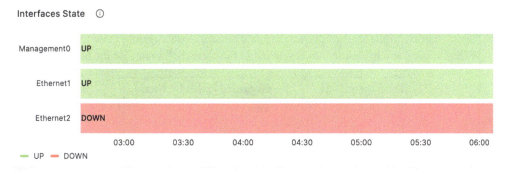

Figure 8.43 – The Interfaces State panel

Metrics inside table panels

One of Grafana's stronger visualizations is its built-in **Table** panel type. We've already used this to display **Device Details** information from Nautobot, tweaking it with transformations and overrides to make the device information clear and concise (check out the *Table panels with GraphQL queries* section for more details).

We're going to use this table panel again, but this time to show details about the device interfaces using the metrics we have collected. We'll see how adding metrics data about the interface traffic usage makes the table more informative.

To do this, we'll start by choosing **Table** as our panel type and then proceed with the specified query information:

Query Setting	Series A (Interface In/Out Octets)
Data Source	Prometheus

Query Expression	`rate(interface_in_octets{device="$device"}[2m]) * 8`
	
	 `+`
	
	 `rate(interface_out_octets{device="$device"}[2m]) * 8`

Options: Format	Table
Options: Type	Instant

Table 8.5 – Interfaces Detail panel query settings

When we set **Options: Type** to **Instant**, it means we're asking for the most recently collected value.

The metrics come with many labels that aren't necessary for our current display, so let's use a transformation to remove some of these labels and enhance the readability of the names of others. Keeping our previous discussion about stacking transformations and how the order they're applied can change the outcome in mind (see the *Transforming data* section), we'll organize these transformations clearly. We'll number them to ensure they're applied in the correct sequence for the best result:

Transform Data	1 – Organize Fields by Name
Field: `name`	Rename: Interface
Field: `intf_role`	Rename: Role
Field: `Value`	Rename: Traffic

Table 8.6 – Interfaces Detail panel transform

Make sure you disable any fields that aren't being displayed by clicking the small eye icon next to the field name. For instance, here's what it looks like when you disable the `site` field:

Figure 8.44 – Disabling the site field

Lastly, we'll adjust the panel controls and settings to introduce more color to the table for a visually appealing display:

Setting	Value
Title	Interfaces Detail
Description	Interfaces Detail table collected from the device
Table: Column alignment	Center
Standard Options: Unit	bits/sec (SI) - Interface traffic value (Inbound + Outbound)
Standard Options: Color scheme	Green-Yellow-Red (by value)
Override 1: Fields with name (Traffic)	Cell Options | Cell type | Gauge

Gauge display mode: Retro LCD

Value display: Text Color |

Table 8.7 – Interfaces Detail panel options

In the settings, we're focusing on customizing the **Traffic** field with a **Retro LCD gauge** visualization. This is based on the combined inbound and outbound traffic flowing through an interface.

The final look of the panel should look as follows:

Figure 8.45 – Interfaces Detail table metrics

From this, you can see how powerful these table visualizations can be when combined with metrics data coming from your observability stack.

Next, we're going to craft a new panel, this time while leveraging our log data.

Device Logs panel

Logs are insightful because they tell us exactly what's happened, such as when a BGP session changes or an interface protocol shifts, and they categorize these events by their severity. When we gather all these logs, we get a full history of everything that's happened on a device. This lets us look back over events and pick out the details we need to make our dashboards even better. For those interested in diving deeper into log data, check out *Chapter 3*. If you want to learn more about how we're collecting and processing logs in our lab environment, be sure to check out *Chapters 5* and *6*.

Thanks to our integration with the Loki data source, we can display syslog information from network devices right alongside the metrics and Nautobot data on this dashboard.

To see logs in our lab environment, we'll need to interact with the devices. Let's dive in and create some activity. To do so, we'll connect to `ceos-02` and cycle the `Ethernet2` interface by shutting and then re-enabling it. This will spur some log generation:

```
> ssh netobs@ceos-02
(netobs@ceos-02) Password: netobs123
ceos-02>enable
ceos-02#
ceos-02# configure terminal
ceos-02(config)# interface eth2
ceos-02(config-if-Et2)#shutdown
ceos-02(config-if-Et2)#no shutdown
ceos-02(config-if-Et2)#exit
ceos-02(config)# exit
ceos-02#
```

This action should have generated several logs, capturing key events such as changes in interface and BGP statuses, as well as when the `netobs` user logged in and made configuration changes to `ceos-02`. Here's a glimpse at what some of those logs might look like:

```
# Interface status change
%LINEPROTO-7-UPDOWN: Line protocol on Interface Ethernet2, changed
state to up

# User enters config mode on a network device
%SYS-4-CONFIG_E: Enter configuration mode from console by netobs on
vty19 (198.51.100.1)
```

After being generated, these logs were picked up and processed by our Logstash instance, and then stored in Loki. This processing step is essential because it helps us pull out key details from the logs, such as where the event came from (the facility) and how serious it is (the severity). These pieces of information are helpful when we're looking to highlight and filter the logs data in Grafana.

Now, it's time to bring those logs into view on our dashboard. We'll proceed by setting up a new visualization, opting for a **Logs** panel type, and applying the necessary query information:

Query Setting	Series A
Data Source	Loki
Query Expression	`{device="$device"}` `|=` `` ``

Table 8.8 – Device Logs panel query settings

This query specifically narrows down the logs to show those related to the `$device` variable exclusively. Now, let's proceed to set up the panel options:

Setting	Value
Title	Device Logs
Description	Device Logs panel
Logs: Wrap lines	True
Logs: Time	True

Table 8.9 – Device Logs panel options

The resulting panel will look something similar to the following:

Device Logs ⓘ

```
> 2024-02-27 10:47:45.000 NGB 10.111.0.1, interface 10.1.7.2 adjacency established
> 2024-02-27 10:47:05.000 Configured from console by netobs on vty18 (198.51.100.1)
> 2024-02-27 10:47:02.000 LLDP neighbor with chassisId 001c.73d0.b1cf and portId "Ethernet2" adde
                          d on interface Ethernet2
> 2024-02-27 10:47:00.000 peer 10.1.7.1 (AS 65111) old state OpenConfirm event RecvKeepAlive new
                          state Established
> 2024-02-27 10:47:00.000 Line protocol on Interface Ethernet2, changed state to up
> 2024-02-27 10:38:43.000 NGB 10.111.0.1, interface 10.1.7.2 adjacency dropped: interface went do
                          wn, state was: FULL
> 2024-02-27 10:38:43.000 sent to neighbor 10.1.7.1 (AS 65111) 6/6 (Cease/other configuration cha
                          nge <Hard Reset>) 0 bytes
```

Figure 8.46 – Device Logs Panel

Incorporating logs into your dashboards elevates your observability capabilities, offering not only a near-real-time view of what's happening with your devices but also enriching your monitoring landscape with valuable context.

As we move forward to explore another standout Grafana feature, known as annotations, we recognize that detailing every panel for the Device Health dashboard could fill many pages of this book with technical specifications and visual examples. To maintain focus and provide a practical learning experience, we've decided to turn the creation of these additional panels into an exercise for you. You'll find the final dashboard, alongside all the panels that have been created, in the completed lab scenario for this chapter.

Annotations

Grafana provides a powerful feature called **annotations** that lets you highlight key moments in your visualizations with detailed event markers. These markers appear as vertical lines and icons on graph panels, offering a visual cue of significant events. Upon hovering over an annotation, you're presented with in-depth details about the event, including a description and any related tags. This feature adds depth to your graphs by offering context, helping you connect the dots between data shifts and actual happenings, such as *changes in interface status* or a user applying a *configuration to a device*. The following figure provides a detailed log view, highlighting just how much valuable context it can offer:

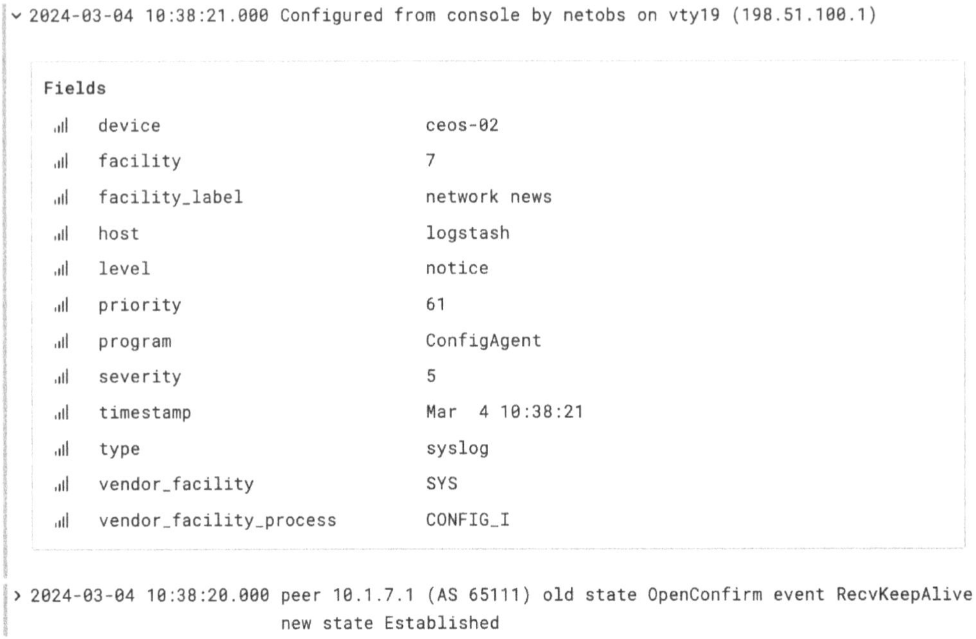

```
⌄ 2024-03-04 10:38:21.000 Configured from console by netobs on vty19 (198.51.100.1)

   Fields
      .ıl  device                    ceos-02
      .ıl  facility                  7
      .ıl  facility_label            network news
      .ıl  host                      logstash
      .ıl  level                     notice
      .ıl  priority                  61
      .ıl  program                   ConfigAgent
      .ıl  severity                  5
      .ıl  timestamp                 Mar  4 10:38:21
      .ıl  type                      syslog
      .ıl  vendor_facility           SYS
      .ıl  vendor_facility_process   CONFIG_I

> 2024-03-04 10:38:20.000 peer 10.1.7.1 (AS 65111) old state OpenConfirm event RecvKeepAlive
                          new state Established
```

Figure 8.47 – A log detailed view of a message and various fields

In Grafana, annotations can be created in several distinct ways to suit different needs and preferences:

- **Manually on a panel**: You can directly add an annotation to a panel. This method is interactive and user-friendly, and it includes the ability to link to external systems for more comprehensive information. It's a quick way to mark events as you analyze them.

- **By configuring annotations in dashboard settings**: This method involves setting up annotations to be automatically generated based on data queries. For example, using the Loki data source, you can construct a query that triggers annotations, linking data patterns or events to annotations on your graphs.

- **Using the HTTP API**: Grafana offers a programmable approach with its API, allowing you to create and manage annotations through code. This is particularly useful for integrating annotations into automated processes or for bulk operations. We'll delve into this method and its possibilities in *Chapter 12*.

Let's put this into practice and create an annotation manually on a state time series panel. Take, for instance, the *interface state panel* we put together earlier. Let's hover our mouse over a point in time and click it. You should see a popup with a section called **Add annotation**:

Figure 8.48 – Interfaces State – creating an annotation

Then, select **Add annotation**; **Description** and **Tags** fields should appear. Let's add some information about the annotation and associate it with a `network` tag:

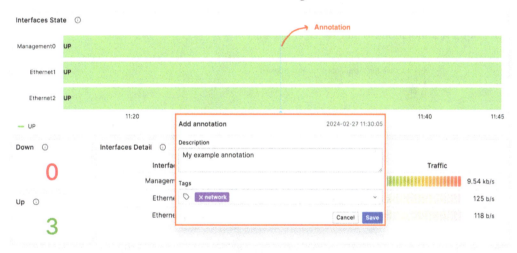

Figure 8.49 – Interfaces State panel – annotation created

Once you create annotations, they're stored in Grafana's internal database, along with their dashboard and panel ID, ensuring they remain as a permanent record. They become valuable tools when you're troubleshooting issues, wanting to share your discoveries with others, or when you need to compile reports.

Now, let's move on and craft annotations using log messages from the devices to give us even more insight into our network's behavior.

Annotating queries from log data

Considering the data from your syslog, certain information could be particularly useful to display alongside other panels. We could focus on two distinct use cases to infer other similar interesting applications:

- **Config mode changes**: A device entering or exiting config mode is a key indicator. Annotating these moments can help us see when device configurations might have changed, which is essential for correlating with the overall network status.

- **Interface flaps**: These are critical when an interface's operational status changes, such as when **Line Protocol** goes up or down. Annotating these events allows us to directly link to BGP state changes and other significant network events.

To begin with annotations from log data, let's tackle the config mode changes first. You'll need to go to the **Annotation** tab in the **Dashboard settings** area to set up the appropriate query.

At this point, you can start creating a new annotation query. Let's name it **Device Config Push** and use **Loki** as the **data** source. We'll choose the panels where we want these annotations to appear. We've created several panels that would benefit from these changes being highlighted. Additionally, let's set the color of these annotations to blue for clear visibility:

Figure 8.50 – Annotation query – Selected panels

Finally, you will have the **Query** section, in which you can use a LogQL expression similar to the following:

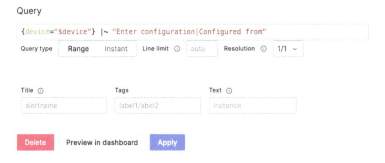

Figure 8.51 – Annotation query LogQL expression

After applying it, you'll see top-level switches in your dashboard that you can use to enable or disable how these annotation queries are visualized in your panels:

Figure 8.52 – Dashboard annotation – control buttons

Now, log in to a device, enter configuration mode, toggle some interfaces (avoiding the **Management** interface), and then exit configuration mode. This should trigger annotations based on the resulting logs. Here's an example of what you might expect these annotations to look like:

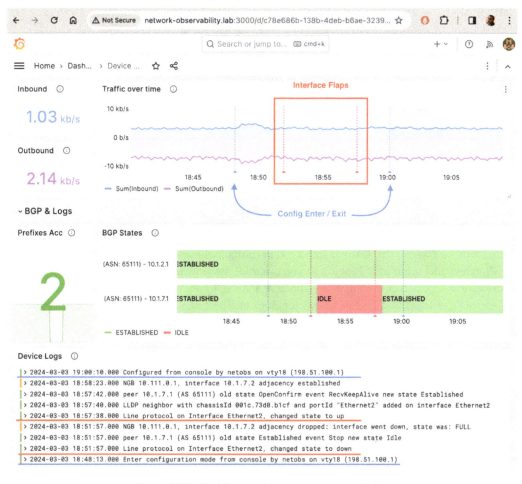

Figure 8.53 – Dashboard with annotation correlation

Here's an example of the details in an annotation when you hover over the annotation with your mouse pointer:

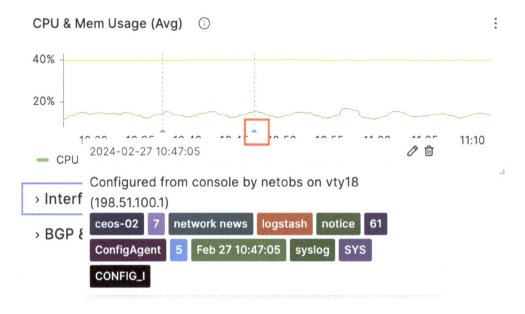

Figure 8.54 – Dashboard annotation details

Let's pause for a moment to understand what's unfolding in our dashboards enhanced with annotations. Imagine that you're a network engineer checking on a network device's health. You'll encounter panels enriched with annotations that clue you into a recent configuration event. For instance, by examining the BGP states panel depicted in *Figure 8.54*, you can trace the sequence of events: a user logging into the device, an interface flap, the impact on the BGP state, followed by another flap that restores the BGP state to established, and finally, the user logging out. These annotations across just a few panels narrate a detailed and meaningful story of what transpired.

Linking metrics with log data like this offers a significant advantage for **root cause analysis** and cuts down the **mean time to resolution** for network disruptions.

Finalizing dashboard details

With the panels and information displayed on this dashboard, you've reached a good stopping point. The key is to avoid visual clutter. To enhance organization, consider using the **Rows** feature in Grafana dashboards. This allows you to group related panels and collapse them, resulting in a more streamlined, tidy dashboard upon loading. You can create these rows easily via the **+Add** button located at the top of your dashboard menu:

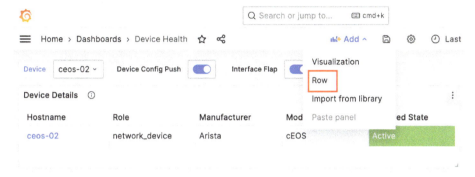

Figure 8.55 – Grafana dashboard – creating a row

Create two rows to organize contextual data, such as interfaces and BGP, alongside device log panels. Place the corresponding panels within these rows, following the layout that was planned for our wireframe visualizations:

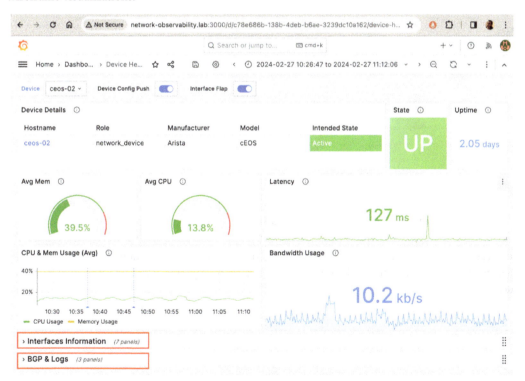

Figure 8.56 – The Device Health dashboard's critical information

The **Interfaces Information** row section is designed to capture key data across several panels, including the count of interfaces that are **Up** or **Down**, detailed information in the **Interfaces Detail** table, average inbound and outbound traffic metrics, and a time series chart showing the trends of overall inbound and outbound traffic over time:

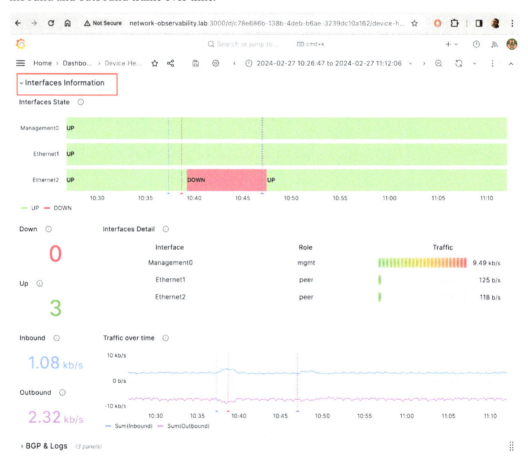

Figure 8.57 – The Device Health dashboard – Interfaces Information

Likewise, the **BGP & Logs** row section aims to consolidate important information through panels that display the number of accepted BGP prefixes, a timeline of BGP session states, and a comprehensive view of device logs:

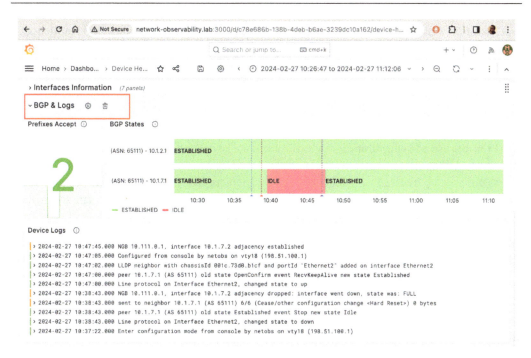

Figure 8.58 – The Device Health dashboard – BGP & Logs information

Note that we haven't explicitly walked through creating the **Down**, **Up**, **Inbound**, **Outbound**, and **Traffic over time** panels or even the BGP panels in this chapter. Instead, we're assigning this as a task for you, encouraging you to apply the skills you've honed from constructing the earlier panels within the **Critical Information** wireframe. To help you in this effort, you'll find the final dashboard with all panels created under the completed lab scenario for this chapter.

Visualization tips and best practices

At the start of this chapter, we examined the principles of data visualization, looking to guide you in crafting dashboards that resonate with your audience. To further refine your visualizations, consider these additional strategies:

- **Conduct interviews**: Displaying data that rarely captures the user's interest offers little benefit. Engaging in interviews and performing an initial discovery phase can clarify which visualizations should take precedence and identify the critical data needed.

- **Sketch your ideas**: As demonstrated with the wireframe example earlier in this chapter, sketching out panel placements and dashboard layouts can provide a solid foundation for both you and your audience to visualize the data more clearly.

- **Maintain a feedback loop**: Regularly seek approval and input from your target audience. Showcasing your progress through demos and check-ins ensures your efforts align with their needs, preventing wasted development on less impactful visualizations.

- **Iterate and refine**: Perfecting dashboards often requires several iterations. This might involve integrating new data sources, adjusting data collection methods, or adding synthetic checks. The aim is always to enhance the dashboard's value to the user.

- **Promote use and adoption**: Since visualizations are key outputs of an observability platform, it's crucial to make them accessible. Incorporate them into company documentation, such as onboarding materials or runbooks, and use them as planning tools for architects. Tailor dashboards to monitor **key performance indicators** (**KPIs**) or provide specific data segments to end users.

- **Leverage automation**: Automate routine tasks, such as dashboard backups, to save time. When possible, utilize API features, such as Grafana's annotations, to streamline processes and enhance functionality.

The ultimate goal is to deliver dashboards that meaningfully present data to their users. So, always be in dialog with your audience to ensure they find value in the current setup and explore ways to optimize it further.

Summary

In this chapter, we explored the visual aspect of network observability, highlighting the significance of how data is showcased. We expanded on the art of crafting intuitive and insightful dashboards in Grafana, building upon the foundational knowledge of databases that was explored in the previous chapters. Through a series of practical steps, we demonstrated how to connect to various data sources within Grafana, construct diverse panels, and seamlessly integrate metrics and logs to present a holistic view of network device health as an example.

Then, we journeyed through the creation of a Grafana dashboard, from the basics of setting up and customizing panels to the more advanced techniques of manipulating data with GraphQL queries and transformations. The power of annotations was also introduced, adding a layer of depth to our visualizations by marking critical events and changes over time.

Next, we'll transition from crafting easy-to-read visualizations to ensuring you're promptly alerted to key issues. The upcoming chapter delves into alerts – those essential signals that draw your attention where it's needed most. We'll take you step-by-step through configuring alerts within your lab environment, equipping you to respond quickly to the insights your data provides. This is about putting your newfound knowledge of PromQL and LogQL to practical use, enabling decisive action when it counts.

9
Alerting – Network Monitoring and Incident Management

Alerting is the last pending component of our recommended observability stack, and we aim to deepen your understanding of how alert mechanisms function in the context of network monitoring and incident management. Having laid the groundwork in previous chapters on data collection, enrichment, and visualization within your network infrastructure, we now turn our focus to the operationalization of this data through alerting strategies.

Alerting serves as an essential bridge between the passive observation of metrics and the active response to anomalies detected within the infrastructure. Most modern monitoring solutions incorporate some form of alerting capability, either natively or through integration with external alerting systems. The alerting component plays an important role in sending the events it captures through various channels, including, but not limited to, email, messaging systems, such as **Slack** and **Microsoft Teams**, and specialized alert and incident management tools, such as **Opsgenie** and **PagerDuty**.

This chapter focuses on the essential aspects of alerting. We will walk you through configuring, customizing, and optimizing an alerting system to improve network monitoring and incident management efforts. Our aim is to share practical insights and best practices to improve the impact of your alerting strategy, making it a valuable tool in your IT operations toolkit.

The main topics covered in this chapter are as follows:

- The art of alerts
- A look into rulers and Alertmanager
- Alerting tips and best practices

The art of alerts

What exactly is an **alert**? You're likely already familiar with the concept, but a quick refresher can be helpful. Consider a car dashboard. It displays various symbols indicating the car's health—such as fuel level, speed, and tire pressure. If something goes wrong, such as an engine issue, a new symbol lights up as an alarm, prompting you to check the problem. This is essentially what an alert is: a notification triggered by an abnormal condition that requires attention. The same principle applies to IT infrastructure, where an alert signals an undesirable state in a resource that needs fixing.

In the networking world, **network management systems** (**NMSs**) are indispensable. They go beyond mere monitoring to provide crucial alerts, serving as a primary output of an observability platform (discussed in *Chapter 4* in *Components of an Observability Platform*). These alerts are vital for upholding the health, security, and performance of IT systems. They act as an early warning system, alerting teams to issues ranging from system overload to potential security breaches. This early detection is key to minimizing downtime, safeguarding data, and ensuring continuous operations.

Alerts are not just about notifying the IT team; they're about communicating with the right people at the right time. This might include network operations teams for network-related issues or product owners for application-specific problems, among others. By targeting these notifications to specific *Personas* with varied roles (more on this in *Chapter 11*), organizations can ensure a swift and appropriate response, thus maintaining service continuity and operational efficiency.

Alert and incident management are key strategies that companies use to address issues found in their systems and applications. One well-known framework for managing these tasks is the **Information Technology Infrastructure Library** (**ITIL**), which provides a comprehensive set of practices for **IT service management** (**ITSM**). This approach is all about ensuring *IT services align with business needs*. Whether or not you're using a framework like ITIL, understanding how your IT services relate to business objectives is vital. It helps you identify which alerts are important and need immediate attention, allowing you to prioritize and categorize them.

Alerts serve as indicators of potential faults within the systems you are observing, and they can escalate into incidents that directly impact your applications or operations. Given their significance, it's important to understand the distinctions between an event, a notification, an alert, and an incident, as well as how metrics and logs interconnect within this framework. The following diagram provides a visual representation of these concepts and their overall relation and creation:

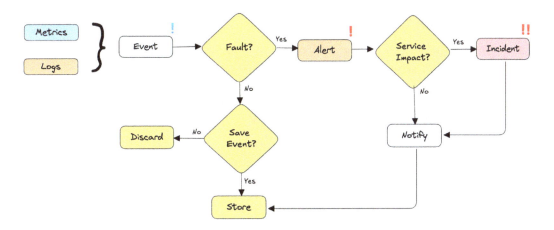

Figure 9.1 – Alerts and incidents generation (Created using the icons from https://icons8.com)

The preceding diagram illustrates the process within an observability system, showing how incoming data can trigger alerts and potentially escalate into incidents. It also highlights the roles of the various components and phases involved in this process. Let's break it down:

- **Metrics and logs**: These are the foundational signals that alert us to potential issues within our infrastructure. It's important to note that we are not only referring to traditional polled or streamed metrics but also to synthetic monitoring, which actively checks the health of services and infrastructure.

- **Events**: Events are the processed data components derived from metrics and logs, providing a standardized representation of significant occurrences within the system. They serve as a universal way to communicate what has happened. When an event indicates a fault or anomaly, it is elevated to an alert, signaling the need for attention.

- **Alerts**: Alerts are generated from events that indicate a fault, typically triggered by an anomaly, unexpected change, or threshold breach in the observed environment. Their primary purpose is to quickly notify the relevant teams about potential issues that require attention.

- **Incidents**: An incident is an event that causes, or has the potential to cause, a disruption or reduction in the quality of a service. It can be triggered by a single alert or a combination of multiple alerts. Incidents are distinguished by their impact on service levels, user experience, and business processes. They are typically identified after analyzing alerts, assessing affected services, or correlating related events within the observed environment.

So, knowing how to act on alerts and when to escalate them to incidents, as well as managing the resolution and aftermath, requires a careful approach. Let's explore how incident management and alert workflow can help put some structure and manage alerts and incidents.

Incident management and alerts

Incident management deals with responding to and managing incidents resulting from alerts. It should not be seen as a single step, but a process that can be summarized as follows:

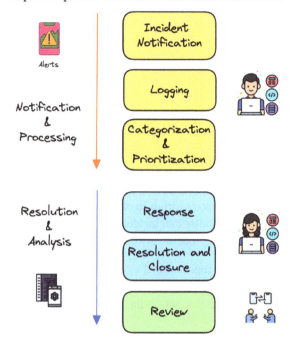

Figure 9.2 – Simple incident management workflow (Created using the icons from https://icons8.com)

The diagram outlines a straightforward incident management workflow, starting from the initial alert and continuing through to the final review. This process is split into two main phases: *notification and processing* and *resolution and analysis*. Each step in the workflow has a distinct role, from identifying and logging the incident to resolving it and conducting a review. Let's walk through each component to understand how they work together in managing incidents:

- **Incident notification**: Alerts signal when an incident has occurred, such as a service disruption, an operational issue, or a decline in quality.

- **Incident logging**: Every alert and event around the incident are logged to keep track and help the analysis process.

- **Incident categorization and prioritization**: The alerts help determine the severity and impact of an incident, guiding the prioritization process.

- **Incident response**: Using the information from alerts, response actions are taken to resolve the incident. These actions are usually guided by runbooks or playbooks, which help operators find the root cause and speed up the resolution.

- **Incident resolution and closure**: After the incident is resolved, the alerts can be closed with documentation of the resolution process for future reference.

- **Incident review/root cause analysis (RCA)/post-mortem**: Post-incident analysis often involves reviewing the alerts and responses to understand the incident's root cause and to improve future alert management.

An incident management workflow helps organizations handle issues step by step, ensuring that each incident is resolved and reviewed for future improvements. From the generation of an alert to the final review of an incident, every part of the process is important for keeping services running smoothly. Simply having an approach like this not only helps resolve issues quickly but also strengthens incident management over time, leading to more reliable systems.

Challenges and considerations on alerting

Alerts play a key role in maintaining the health and performance of IT systems, acting as early warning signs when something goes wrong. However, managing these alerts can be quite challenging, especially given the complexity and rapid changes of today's IT environments. These challenges stem from several factors that can overwhelm even experienced IT teams. Let's dive into these challenges:

- **Alert fatigue**: One of the most significant challenges with alerting is the sheer volume of alerts generated. This can lead to alert fatigue, where IT personnel become desensitized to notifications due to the overwhelming number of alerts, many of which may be false positives or non-critical.

- **Noise reduction**: Distinguishing between critical alerts and noise is a daunting task. Critical alerts can easily get lost in the noise of non-essential alerts, delaying response to actual issues.

- **Varied sources**: Alerts can originate from a multitude of sources, including on-premises data centers, cloud services, and **software-as-a-service** (**SaaS**) applications. Each source may use different formats and protocols for alerts, complicating the aggregation and analysis process.

- **Dynamic environments**: Modern IT environments (including networks) are highly dynamic (think Kubernetes or cloud environments), with constant changes due to deployments, updates, and scaling operations. This dynamism can affect the relevance of alerts over time, making it challenging to maintain accurate alerting thresholds and parameters.

Later in the chapter, we will look at more tips and best practices for dealing with alert and notification issues in modern IT environments. But to really understand these strategies, we first need to dive into the concepts of alert aggregation and correlation. These are important techniques for fighting alert overload, cutting down on unnecessary alerts, and detecting incidents.

Alert aggregation and correlation

Navigating the world of alerts requires more than just recognizing when a problem arises; it involves understanding the connections between multiple alerts and how they collectively point to a larger issue within your system. This is where the concepts of alert aggregation and correlation come into play. **Alert aggregation** helps us group similar alerts together, reducing the noise and making it easier to see the bigger picture. **Alert correlation** takes things a step further by linking related alerts across different services or systems, indicating a potential incident. To fully leverage these strategies, it's important to understand the **common alert data model**—examining the essential data and attributes that define an alert, and **alert enrichment**, a method for adding extra metadata from various sources to provide more context around an alert.

Alert data model

When we talk about an alert data model, we are referring to all the key details that describe an alert event in a system. This includes information such as the name of the component that is experiencing an issue, which tool or agent detected the problem, and a brief description of what is going wrong. These pieces of information are key because they help whoever is responding to the alert—whether it's a person or another system—understand the issue quickly and clearly, and take action promptly. The model is designed to be flexible, allowing for additional data attributes that can enhance understanding of the alert, so it can better support processes, such as alert aggregation and correlation. Typically, this model would include fields, such as the following:

- **Summary/subject/signature**: This is a concise, easy-to-understand description of the alert generated from the system check. It's designed to give a quick snapshot of the issue for those monitoring the system.

- **Description/message**: This field provides a detailed explanation of the alert, allowing for a more in-depth message. Most implementations support rich text formats, such as HTML or Markdown, making it easier to include important details or instructions.

- **Priority/severity**: This indicates how urgent the alert is, ranging from low priority (such as P5/Sev5) to high priority (such as P1/Sev1). The system categorizes alerts based on their severity to help responders prioritize their actions.

- **App/entity**: This identifies the specific component, system, or application that triggered the alert or is experiencing the issue.

- **Type**: This specifies the kind of alert issued, which varies based on the alerting system or engine in use. For example, in Splunk, this category helps distinguish between an *alert* and a Splunk *event* (see Splunk documentation: `https://docs.splunk.com/Documentation/CIM/5.3.1/User/Alerts#Fields_for_the_Alerts_event_dataset`).

- **Status**: This field shows the current status of the alert, indicating whether it is open, has been acknowledged (**ack**), or has been resolved/closed.

- **Tags**: This is a set of values designed to assist in filtering and categorizing alerts. Tags are invaluable for enriching alerts and facilitating their aggregation and correlation, which we'll discuss further in the upcoming section.

In addition to the basic attributes, alert management systems often include features to track the life cycle of an alert, such as the following:

- **Created**: The timestamp when the alert was initially generated.

- **Last updated**: The most recent timestamp when the alert was updated.

- **Activity log**: A record of all events related to the alert, detailing any actions taken or changes made.

Most tools and vendor solutions in the market offer a comparable alert data model.

A topic that deserves highlight and attention is the work of the RFC 8632 A YANG Data Model for Alarm Management (`https://datatracker.ietf.org/doc/html/rfc8632`), which provides a great resource of what an alarm data model should be to help with noise reduction and accommodate support for root-cause analysis, service impact, and alarm correlation. It lays down a solid foundation and provides an extensive data model that aligns well with the needs of modern IT infrastructures. It would be highly beneficial for vendors to consider adopting this standard to optimize and refine alarm management practices.

Next, we'll explore strategies to inject enrichment data into the alerts generated and aggregation and correlation techniques.

Alert enrichment

We have touched on the concept of data enrichment throughout this book, notably in *Chapter 6, Enhancing Insights with Data Enrichment*, where we discussed augmenting our observability data with information from a **source of truth** (**SoT**) system. This process is equally beneficial for alerts, enhancing the criteria for triggering alerts and providing a richer context for the events they signify. By incorporating additional details from a SoT, alert rules become more precise and the information conveyed by alerts becomes more pertinent and actionable.

Typically, the data needed for alerts is included in the tool's data model, often accessible through tags or additional property features, as outlined in our earlier discussion on data models. It's important to note that there are several ways to enrich alerts. For instance, **Keep** (`https://docs.keephq.dev/overview/introduction`), an open source alerting system, allows for alert enrichment by mapping additional data from a CSV file. In the context of Alertmanager, alerts can be enriched with Prometheus info metrics. For a deeper conversation around info metrics and their applications, please check out Info Metrics in *Chapter 6* and *Chapter 7*.

The data that significantly enhances the context and understanding of an alert revolves around the business aspects and the stakeholders of the system or application in question. Incorporating fields such as the following can offer deeper insight and help connect the dots:

- **Tenant/owner:** This identifies the specific stakeholder or business unit responsible for the system or application involved in the alert. This helps pinpoint who needs to be informed or take action.

- **Service**: This represents a broader category that links systems, networks, and applications, offering a holistic view of their status to key business stakeholders. This abstraction level is fundamental for understanding the impact on business operations.

- **Responders/target business unit**: This specifies the teams or stakeholders who should respond to the alert, ensuring that notifications reach the right people equipped to address the issue.

Alerting engines can correlate this data to help reduce noise and alert fatigue. This is typically done by aggregating them based on attributes such as the alert type, priority, tags, and services affected.

Aggregation and correlation techniques

Correlation engines often specify key data attributes required for linking, prioritizing, and analyzing alerts. In the world of **artificial intelligence for IT operations** (**AIOps**), these engines can also be trained to identify patterns and connections across the vast amounts of messages, logs, and event data you manage. This advanced capability enables a deeper understanding of the relationships between different pieces of data, enhancing incident detection and response strategies.

Here are several strategies that will help you aggregate and correlate a vast number of alerts and operational data:

- **Temporal aggregation**: Grouping alerts that occur within a specific period can help identify peaks of activity and potential issues more clearly.

- **Threshold-based aggregation**: Aggregating alerts based on surpassing certain thresholds (e.g., error rates, response times) can highlight issues that need immediate attention. For instance, analyzing network latency in percentiles or categorizing application response times to pinpoint bottlenecks.

- **Tag and category aggregation**: Grouping alerts by tags or categories (e.g., application name, server, error type) enables teams to quickly identify affected areas and prioritize responses.

- **Pattern recognition**: Using machine learning algorithms to identify patterns in the alerts can help in predicting potential issues before they become critical, facilitating proactive management. More on this in *Chapter 13*.

- **Dependency correlation**: Understanding the dependencies between different systems or services allows for correlating alerts in a way that reflects the impact on the overall IT infrastructure, helping to identify root causes with better accuracy.

- **Statistical correlation**: Employing statistical methods to correlate alerts based on similarities in their attributes or historical occurrences can uncover underlying issues not immediately apparent from isolated alerts.

- **Incident matching**: Correlating alerts to existing incidents or known issues can reduce redundancy and ensure consistent handling of similar or related problems.

By implementing these techniques and strategies, we can significantly expedite RCA by uncovering common patterns that point toward the likely source of faults. Additionally, correlating and prioritizing alerts based on service impact enables a clearer understanding of affected services, guiding how to prioritize incident response efforts.

Next, we'll dive into understanding the components that power alerting systems from the inside out.

Alert engine architecture

To better understand some of the concepts discussed around data model, enrichment, and correlation, it might help to present a typical alerting engine architecture. The following figure is a high-level representation of an abstraction of what an alert engine performs:

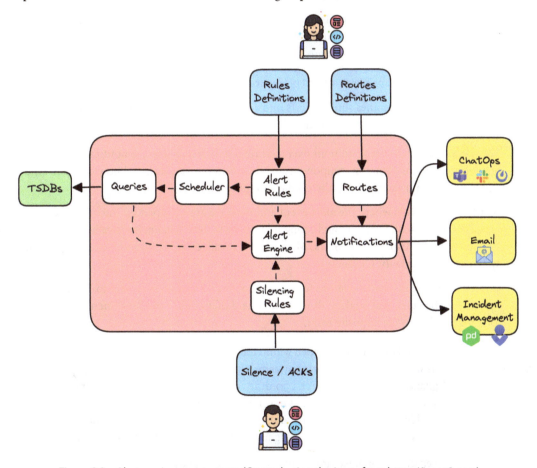

Figure 9.3 – Alert engine components (Created using the icons from https://icons8.com)

As illustrated in *Figure 9.3*, the journey begins with observability data stored in a database, such as a time-series database for metrics or a log storage system. The alerting mechanism builds on this persistence layer, employing a series of components to execute the alerting process:

- **Alert rules**: These are the conditions that evaluate the collected data and create an alert. These definitions are usually what an operator designs to recognize when a target goes into a faulty state. These definitions are mostly static and are in the forms of conditional queries, such as `is_device_down == true`, or evaluating thresholds, such as `cpu_used > 89`.

- **Scheduler**: This component manages the timing for when queries should be executed and actions, such as silencing an alert, are to be carried out.

- **Queries**: These involve the database queries that fetch data for analysis by the main engine.

- **Engine**: The core where data meets alert rule logic to decide whether to trigger an alert. This is also where silencing rules are enforced, preventing further notifications for silenced alerts. Key functions include the following:

  - **Deduplication**: This maintains a count of identical alert events to avoid creating redundant alerts.

  - **Grouping**: This combines similar alerts into a single notification, serving as a basic form of correlation, as seen in systems such as Alertmanager.

- **Silencing rules**: These are conditions over alert rules to basically stop the notification process. Some systems have silencing components, while others rely on alerts, acknowledgment (the famous ACK), and status to stop the notification process.

- **Notification**: This is where the alert event and message are wrapped up into a notification message and sent to the available and configured channels. These channels typically are ChatOps systems (i.e., Microsoft Teams, Slack, Mattermost), email, or alert and incident management systems (i.e., PagerDuty, Opsgenie).

- **Routes**: These are the configurations to which the alert event needs to be sent to. For example, these could be rules that analyze the data or a field of an alert event and then accordingly forward the alert to the requisite team.

Alert enrichment and correlation usually happen within the alert engine of a monitoring tool. However, in large environments with multiple monitoring systems, this process might be handled by specialized platforms such as **BigPanda** (`https://www.bigpanda.io/`). In such cases, the logic for correlation and enrichment is moved to an external layer that integrates with various systems. This setup allows for additional data to be added, enriching the alerts and improving correlation. This enhancement helps in better managing alerts and incidents by providing more context and making alerts more actionable.

You might have noticed that the architecture we discussed doesn't include a specific database for storing alerts and tracking their statuses, notes, and actions—in essence, maintaining a comprehensive history of all generated alerts. Such capabilities are usually found within dedicated alert management systems such as **Alerta** (`https://github.com/alerta/alerta`), **Keep** (`https://github.com/`

keephq/keep), **Opsgenie** (`https://www.atlassian.com/software/opsgenie`), **PagerDuty** (`https://www.pagerduty.com/`), and others. These systems are specifically designed to keep detailed records of alert histories, ensuring that all information is meticulously organized and accessible.

Excellent; we believe this theory provides a good foundation to now dive into a lab environment and witness alert generation firsthand.

A look into rulers and Alertmanager

We chose Alertmanager and Prometheus for our lab and this book because they are key parts of the Prometheus ecosystem. Alertmanager handles alerts for both Prometheus and Loki, centralizing alert rules and testing with collected metrics and logs. This setup lets us directly use our data with PromQL for Prometheus and LogQL for Loki, as discussed in *Chapter 7*. Additionally, their integration with Grafana helps us visualize metrics, logs, and alert events together.

To get familiar with Alertmanager and its rules and alerts, let's take a look at its architecture.

Architecture

The alerting framework within a Prometheus setup includes a `ruler` component for initiating alerts and a `management` component, Alertmanager, responsible for overseeing the alert's life cycle, notifications, and routing. Similarly, the Loki *ruler* component offers comparable functionalities, highlighting the advantages of the composable architecture of our observability stack discussed in *Chapter 4*. The following figure will provide a high-level overview of how these components interact:

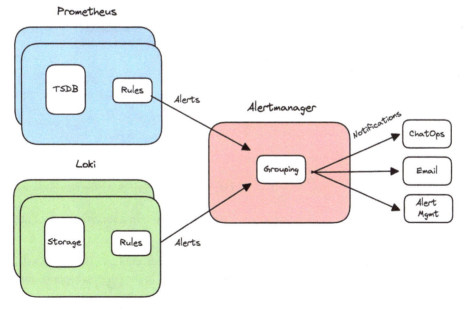

Figure 9.4 – Prometheus, Loki, and Alertmanager architecture
(Created using the icons from https://icons8.com)

As you can see in the previous figure, Prometheus is in charge of using the alerting rules to create alerts from the data it holds. In other solutions, such as Loki or Mimir, this is called the *ruler* component.

Normally these rules are YAML file definitions that Prometheus loads, and based on its schedule, it performs queries to its **time series database** (**TSDB**) to see if the condition of the rule definition is true to then fire up an alert. The process is pretty similar to recording rules, as we have seen in *Chapter 7*.

The alert events created are managed by Alertmanager. This component takes care of the deduplication, grouping, routing, and notification of the alerts, as well as the silencing of alerts.

To fully understand the architecture and the alerting concepts we've discussed, we're going to walk through an example in our lab environment. Let's get everything set up and ready to go.

Setting up the lab environment

The following is a visual representation of the components in the lab environment for this chapter:

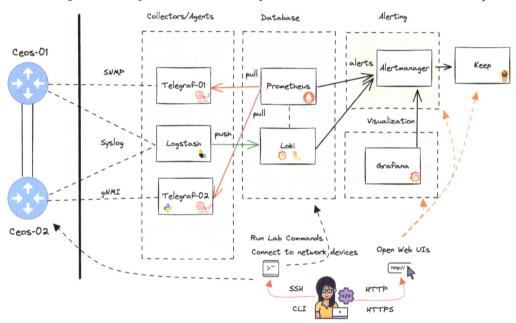

Figure 9.5 – Chapter 9 lab topology (Created using the icons from https://icons8.com)

In this section, we're going to tackle a few key activities to bring our understanding of alerting architecture to life:

- **Configure alert rules**: We'll set up alert rules in both Prometheus and Loki. These rules will assess specific conditions and, when met, trigger alerts that are sent to the Alertmanager component.

- **Observe alerts in Grafana**: We'll observe the configuration and status of our alerts directly within Grafana, providing us with a comprehensive view of our alerting landscape.

- **Setting Keep, and its integration with Alertmanager**: *This is optional.* Configuring Alertmanager to route alerts to an alert management system is an important step, and we'll use Keep for our example. It's important to note that this step is optional for those participating in the lab environment, as it requires creating an account on Keep to proceed with the scenarios. However, Keep's role as an alert management and automation tool exemplifies how we can streamline the notification process and achieve well-coordinated integration with an alert management system. Even if you choose not to set up an account, the insights gained from this section are valuable for understanding the broader alert management process.

Before diving into the lab activities, ensure your lab machine is set up according to the steps outlined in the *Appendix A*, just as we did in previous chapters. Make sure you are in the `~/<path-to-git-project>/network-observability-lab/` directory within the project's repository to execute the necessary commands.

> **Note**
>
> This lab scenario includes a completed version that you can refer to. It is located in the Git repository you cloned, under `network-observability-lab/chapters/<chapter-number>-completed`. This folder contains all the configurations used in the examples throughout that chapter, providing a comprehensive guide to ensure you have everything set up correctly.

Starting the lab environment

Follow the next steps to stand up the lab environment for this chapter:

1. On the machine where you're hosting the lab environment, you'll want to run the `netobs lab prepare --scenario ch9` command to get everything set up.

2. We are using Nautobot as our **source of truth** (**SoT**) and it provides enrichment in our observability data. So, let's verify your connection to Nautobot at `http://<machine-ip-address>:8080` using the default credentials (username: `admin`, password: `nautobot123`). The Nautobot service may take a few minutes to start up completely. You can monitor its progress by checking the logs. To do this, use the `netobs docker logs nautobot --follow --scenario ch9` command. If you want to see only the most recent entries, add `--tail=20` to show just the last 20 lines of the log.

3. Next, to ensure that Nautobot reflects the current state of your network devices in the lab, run the `netobs utils load-nautobot` command. This will import the necessary device data into Nautobot, syncing it with your lab setup. This setup is needed since it enriches the data collected by Telegraf and Logstash, providing simpler ways to query and analyze alerts.

4. Now, check Prometheus to validate the network telemetry data collected. Go to `http://<machine-ip-address>:9090` and run an `interface_admin_status` query in the **Expression** panel:

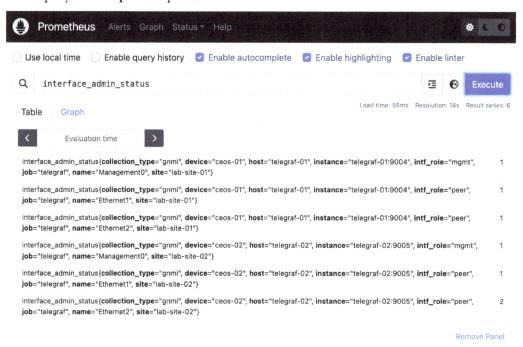

Figure 9.6 – An interface admin status query

OK, now we are ready to start creating some alert rules and generating some events that will trigger them!

Creating your first alerts

Alright, let's start from the beginning: deciding what should trigger our alerts. Crafting alerts is comparable to an art form, much like designing dashboards. It requires a deep understanding of the alert recipients – the personas and stakeholders – and customizing the alerts to meet their specific needs. Building on the approach we took for dashboard creation in *Chapter 8*, let's explore some key considerations for setting up our alerts:

- **What do we want to alert on?**: Answering this question lays the foundation for our alert conditions. Inspired by *Chapter 8*'s examples of creating the *Device Health* dashboard, we aim to monitor specific events: we want to be alerted when a peering interface on our devices goes down, if a BGP session drops, and also if there's an interface flapping on any of our circuits.

- **Who are the intended audiences?**: This normally falls to the persona of your intended audience. In this case, our alerts are primarily meant for the Network Operations Team.

- **What key details should every alert include?**: While alerts typically provide basic information about the system experiencing an issue, in larger networks and infrastructures with alerts spanning hundreds of targets, it can be challenging to identify the specific services or roles involved. Including details such as `circuit_id`, `interface_role`, or `device_role` can immensely enrich alerts, offering the context needed to address the fault indicated.

Keeping these considerations in mind, let's focus on defining the specific alerts requested and the details the network operations team needs to respond:

- **Alert for Peer Interface Down**: This alert is good for monitoring border or edge networks, signaling when a connection to another network, such as a service provider, is lost. It prompts the network operations team to initiate remediation actions swiftly. Critical information for this alert includes `circuit_id` and `intf_role`, providing a clear context for the issue.

- **Alert for Border Gateway Protocol (BGP) Session Down**: These alerts are vital for indicating the loss of a BGP session with ISPs, cloud environments, or any valuable network or service connection. This scenario might result from a misconfiguration or an outage unrelated to physical connectivity issues. Collecting details such as the **autonomous system number (ASN)** and the BGP neighbor's IP address helps in diagnosing and fixing the issue.

- **Alert for interface flapping**: Capturing interface flapping events can be more challenging but is incredibly important. Flapping can cause significant disruptions, especially in environments with dual upstream links without load balancing. These events can lead to network issues, with traffic bouncing between links, which can hurt service quality. It's important to note the occurrence of flapping and gather contextual details, such as the interface's role (`intf_role`), to help understand, prioritize, and eventually fix the problem.

Now that we've detailed our requirements and gained sufficient context, let's move forward with setting up the necessary alert rules.

Prometheus rules

Our alert definitions are closely connected to the operational status of interfaces and BGP sessions. Reflecting on our exercises from previous chapters (5 to 8), we have metrics that monitor the status of interfaces and BGP sessions. Let's work into the *Peer Interface Down* alert definition and craft the necessary rule.

The initial alert we're focusing on is triggered by the transition of an interface state to *down*, specifically for interfaces linked to *peers*. We utilize the `interface_oper_status` metric to monitor the interface states (with 1 indicating *up*), alongside the `intf_role` label that specifies its role, such as whether the interface connects to `peer`.

If you connect to Prometheus on `http://<machine-ip-address>:9090` and run `interface_oper_status{intf_role="peer", device="ceos-01"}` to filter on the `ceos-01` device, you should see results like the following:

```
interface_oper_status{collection_type="gnmi", device="ceos-01",
host="telegraf-01", instance="telegraf-01:9004", intf_role="peer",
job="telegraf", name="Ethernet1", site="lab-site-01"}    1
# Omitted output...
```

From the output shown, a value of 1 indicates that the interfaces are currently reported as *up*.

Now, it's time to get down to the practical part and set up an alert rule in Prometheus. Let's proceed with the following example:

```
groups:
  - name: Peer Interface Down       # Group name
    rules:                          # List of alerting rules
      - alert: PeerInterfaceDown    # Alert name
        expr: interface_oper_status{intf_role="peer"} == 2
        for: 1m                     # Evaluation time
        labels:                     # Alert labels
          severity: warning
          source: stack
          environment: network-observability-lab
          metric_name: interface_oper_status
          device: '{{ $labels.device }}'
          device_role: '{{ $labels.device_role }}'
          site: '{{ $labels.site }}'
          region: '{{ $labels.region }}'
          instance: '{{ $labels.host }}'
          device_platform: '{{ $labels.device_platform }}_{{ $labels.
          net_os }}'
        annotations:
          summary: "[NET] Device {{ $labels.device }}: Interface
          Uplink {{ $labels.name }} is down"
          description: "Interface {{ $labels.name }} on device {{
          $labels.device }} is down!"
```

In this alert rule, our expression (`expr`) is designed to detect when interfaces transition to a down state, represented by the value 2, specifically focusing on interfaces assigned the `peer` role (`intf_role="peer"`). The rule is configured to trigger an alert if this condition persists for one minute. In the `labels` section, we enrich the alert with essential context by incorporating information from the metric's labels, such as the device name, site, and region. The `annotations` section then adds descriptive fields and details that are helpful for crafting the notification message.

> **Note**
>
> Alertmanager utilizes the Go templating system for its notification templates. Thus, `Device {{ $labels.device }}:` will render for `ceos-01` as `Device ceos-01:`.

Now, let's store this configuration in the `network-observability-lab/chapters/ch9/prometheus/rules/alerting_rules.yml` file within the repository of this chapter.

With the Prometheus rule file crafted, our next step is to configure Prometheus to incorporate the `alerting_rules.yml` file. We'll achieve this by updating our Prometheus configuration file (`prometheus.yml`) with the necessary entries:

```
# Output omitted...
rule_files:
  - rules/*.yml
```

This new section tells Prometheus to look for the rule files under `rules/*.yml` and will pick any file terminating in `.yml`.

> **Note**
>
> Given that our lab environment utilizes Docker containers, the rules files must be mounted to a location accessible by the Prometheus configuration. This step has been pre-configured in the lab setup for your convenience. However, you can review the detailed configuration in the Docker Compose file for the completed scenario in this chapter.

Within the same file (`prometheus.yml`), we introduce a new section titled `alerting`. This section delineates the `alertmanagers` that are in use and outlines the method for connecting with them:

```
alerting:
  alertmanagers:
    - static_configs:
      - targets:
          - alertmanager:9093
        timeout: 5s
```

Now, let's implement these modifications in Prometheus by saving the configuration file and applying the changes using the `netobs lab update prometheus --scenario ch9` command. Once the configuration is applied, you can verify it by navigating to `http://<machine-ip-address>:9090/rules` in your browser:

Figure 9.7 – Prometheus rules page for Peer Interface Down

By default, our lab environment sets the `Ethernet2` interface of `ceos-02` to an administrative shutdown state. Unless this setting has been altered, you should start seeing alerts generated because of this condition. To view these alerts, visit `http://<machine-ip-address>:9090/alerts` and check out the alerts triggered by our alerting rule:

Figure 9.8 – Prometheus alerts—PeerInterfaceDown

You'll notice the `PeerInterfaceDown` alert activated by Prometheus. Along with the setup of the alert rules, you can observe the count of active alerts, inspect the labels from the metrics queried from the TSDB, and review historical data about the alert's status and activation time.

The next alert it needs to set up involves BGP neighbor sessions going down. The process for defining this alert rule is similar to what we have done so far. Give it a try! Remember, you can refer to the completed lab for guidance if needed.

Now, for the final alert, we aim to be notified about interface flapping on certain interfaces. Capturing interface flapping events is important because they can cause significant disruptions and are challenging to detect with just polled metrics from network devices. So, we'll use events from our network device logs instead of Prometheus metrics collected to set up this alert rule.

Loki rules

Throughout our labs in *Chapters 5* to *8*, we've not only captured but also visualized logs detailing changes in interface states. Now, as we prepare to define Loki rules, it's important to have a clear understanding of the log's format and content. We will begin by accessing `ceos-02` and activating the `Ethernet2` interface, which we will assume was in a down state:

```
(netobs@ceos-02) Password:
Last login: Sun Mar 24 16:12:22 2024 from 198.51.100.9
ceos-02>en
ceos-02#
ceos-02#conf t
ceos-02(config)#int eth2
ceos-02(config-if-Et2)#no shutdown
ceos-02(config-if-Et2)#exit
ceos-02(config)#exit
ceos-02#exit
```

You should be able to see some state changes in the Grafana **Explore** tab (filtering by Loki source) by using the LogQL query: `{device="ceos-02", vendor_facility_process="UPDOWN"}` `|= ""`. Go to `http://<machine-ip-address>:3000/explore`:

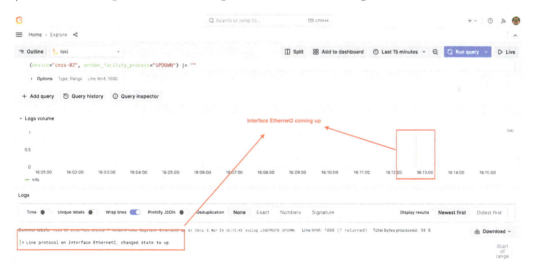

Figure 9.9 – The Loki Logs interface state change

The log messages provide details about interface changes under the `vendor_facility_process` value, which can be found in the query panel situated at the top left of the illustration. In the figure mentioned earlier, we focus on identifying interfaces that are either transitioning to an up or down state.

This visualization enables the network operations team to define an *interface flap event* as occurring when an interface repeatedly goes up or down in quick succession. Such a query helps in filtering the messages that are crucial for identifying these events. To track the frequency of these occurrences, we could create a LogQL metric query (as elaborated in *Chapter 7*), allowing us to count the instances of this event. This count then serves as a basis for comparing and establishing an alert condition.

The following expression calculates the total number of interface state changes for `ceos-02` over the last two minutes:

```
sum by (device) (
  count_over_time (
    {vendor_facility_process="UPDOWN", device="ceos-02"}[2m]
  )
)
```

Before running the expression, you might want to use the `netobs utils device-interface-flap --device ceos-02 --interface Ethernet2 --count 20 --delay 5` command. This command triggers multiple interface shut/no shut operations, making it easier to observe the query results. The command utilizes a pre-written script that connects to the device and performs the specified operations with a set delay between each.

Here is an example run:

```
$ netobs utils device-interface-flap --device ceos-02 --interface
Ethernet2 --count 20 --delay 5
[16:50:08] Flapping interface: Ethernet2 on device: ceos-02
[16:50:09] Bringing interface down...
[16:50:14] Bringing interface up...
[16:50:19] Bringing interface down...
[16:50:25] Bringing interface up...
```

Now, by applying the LogQL metric query expression in the Grafana **Explore** tab, you should see a graph that looks something like this:

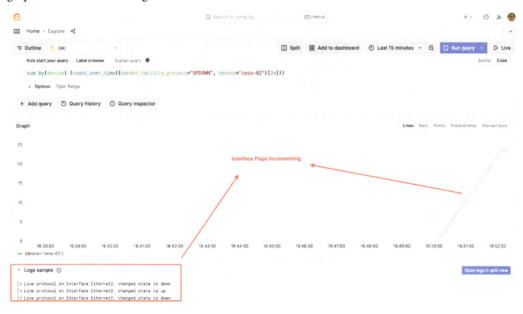

Figure 9.10 – Loki logs metric query for interface flaps

With the LogQL expression in hand, we are now set to establish our Loki alert rule aimed at capturing interface state changes from our log messages, having converted these log events into a metric form. Here's how we configure the alert rule:

```
groups:
  - name: Peer Interface Flapping
    rules:
      - alert: PeerInterfaceFlapping
        expr: sum by(device) (count_over_time({vendor_facility_
        process="UPDOWN"}[2m])) > 3
        for: 1m
         labels:
          severity: critical
          source: loki
          environment: network-observability-lab
          device: '{{ $labels.device }}'
          interface: '{{ $labels.interface }}'
        annotations:
          summary: "[NET] Device {{ $labels.device }}: Interface
          Uplink {{ $labels.name }} is flapping"
          description: "Interface {{ $labels.name }} on device {{
          $labels.device }} is flapping!"
```

This configuration mirrors the approach we have taken with previous Prometheus rules. In this instance, we are tracking interface state changes over a two-minute window. If the evaluation results in a count greater than three, we trigger an alert.

Let's store this configuration in `network-observability-lab/chapters/ch9/loki/rules/alerting_rules.yml`. Just as we did with Prometheus, we need to inform Loki where to locate the rules files. To achieve this, simply add the following configuration to the `ruler` section in the `loki-config.yml`:

```
# Output omitted...
ruler:
  # Output omitted...
  storage:
    type: local
    local:
      directory: /rules
  alertmanager_url: http://alertmanager:9093
  enable_alertmanager_v2: true
```

Now, let's activate the updated Loki configuration by executing `netobs lab update loki --scenario ch9`.

To see the configuration changes and the alerts registered by Loki and Prometheus in action, we'll turn to Grafana. This tool will help us visualize the query results and the alerts they generate.

Grafana for alerts

Grafana, well known for its visualization capabilities, offers various methods to represent telemetry and observability data, as explored in *Chapter 8*. Beyond visualization, Grafana introduces the ability to craft custom alerts directly within its platform, a feature known as **Grafana Alerts**. These alerts can either be managed directly by Grafana or integrated with other systems. To access this feature, navigate through **Home** | **Menu** | **Alerting**:

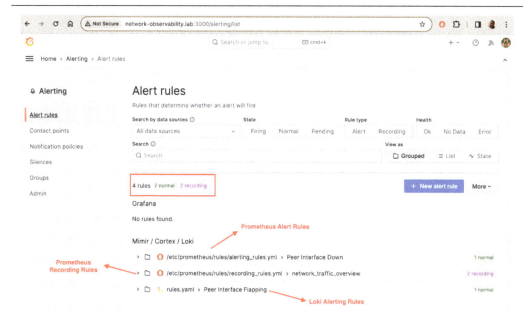

Figure 9.11 – Grafana—Alert rules

Within the **Alert rules** section, you'll notice the integration with both Prometheus and Loki systems. For instance, in this example, a total of four rules have been identified: one alert rule originating from Loki and three from Prometheus, including two recording rules and one alerting rule.

> **Note**
>
> You might have noticed the alert rules are under a section named **Mimir / Cortex / Loki**. This is because these systems use the same alerting architecture derived from the Prometheus ecosystem. This shared architecture allows us to point both Prometheus and Loki rules to Grafana.

You can also view the configurations and the specific alerts they have triggered. If no alerts are visible, don't worry—just rerun the following command to trigger the `PeerInterfaceFlapping` alerts on Loki: `netobs utils device-interface-flap --device ceos-02 --interface Ethernet2 --count 20 --delay 5`. Alternatively, as shown in the next example, you can connect to `ceos-02` and shut down an interface such as `Ethernet2` to trigger a `PeerInterfaceDown` alert. The alert configuration and status are shown in the following figure:

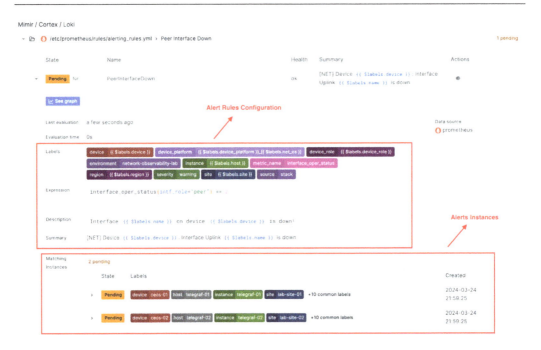

Figure 9.12 – Alert rules and instances in Grafana

When you expand an alert rule configuration, such as the one in Prometheus, you'll see an overview of the setup for the `PeerInterfaceDown` alert rule that we configured. At the bottom, there is a **Matching instances** section, providing a snapshot of the alerts that have been matched, along with general statistics about them.

While Grafana offers strong alert management features, including notifications and silences via Grafana Alerts and Grafana OnCall, we will focus on integrating an external alert management system using Keep. This approach allows us to demonstrate how external integrations work and show the benefits of incorporating external systems into our alert management strategy.

External integrations

Alerting engines gain their true value to organizations through connections to external integrations. These connections can range from simple notifications to complex bi-directional integrations with alert management systems that oversee the life cycle of alerts. Such systems enhance alert capabilities with on-call setups, notification and resolution policies, and even abstract alerts into services and incidents, enabling aggregation and correlation features. In our lab environment, we will demonstrate how to integrate with Keep, an open source alert management system, to manage and display alerts forwarded by Alertmanager.

Keep setup

This step is *optional* for the lab environment, as it requires creating an account with a third-party system. Feel free to bypass this setup and simply engage with the instructions, visuals, and general walk-through provided in this chapter:

1. You can open an account in the Keep platform to send the alert notifications to. For this, go to `https://platform.keephq.dev` and create an account.

2. Next, generate an API key to enable the transmission of alert data from this lab environment into the system. Navigate to your username and then **Settings** | **API Key** | **Create API Key** to generate a webhook API key:

Figure 9.13 – Create an API key for Keep

3. Then, copy and save the API key into `network-observability-lab/chapters/ch9/alertmanager/keep_api_key` file so Alertmanager can utilize it in its webhook connection to Keep. By default, this file is listed in `.gitignore` ensuring it's not tracked by `git`.

4. Finally, we must set up Alertmanager to direct alert notifications to Keep. Use the following configuration and save it under `network-observability-lab/chapters/ch9/alertmanager/alertmanager.yml`:

```
global:
  # Time after which an alert is declared resolved
  resolve_timeout: 30m

# Route tree for Alertmanager
route:
  receiver: "keep"          # Name of alert receiver
  group_by: ['alertname']   # Group alerts by alertname
  group_wait:       15s     # Wait time to group alerts
  group_interval:   15s     # Send interval
  repeat_interval: 1m       # Repeat notifications

# Receivers for Alertmanager
```

```
receivers:
# Receiver for sending alerts to Keep
- name: "keep"
  # Webhook configuration for Keep
  webhook_configs:
  - url: 'https://api.keephq.dev/alerts/event/prometheus'
    send_resolved: true          # Send resolved alerts
    http_config:
      basic_auth:
        username: api_key
        password_file: keep_api_key    # API key file
```

5. Apply the changes to Alertmanager by running the following command:

    netobs lab update alertmanager --scenario ch9

6. Then, generate alerts by either running the interface flap command described in the previous section or by simply shutting down the `Ethernet1` or `Ethernet2` interface on your network devices.

The Alertmanager configuration primarily includes global settings, routing (determining where notifications should be sent), and the configuration of receivers (how to integrate with external systems).

This configuration is merely one example of integration possibilities; there are numerous other options available. For instance, you could direct these notifications to your email.

Managing alerts with Keep

In this part, we're going to explore how connecting Alertmanager with Keep can make handling alerts easier and more effective. Keep helps organize and manage alerts from start to finish.

Connecting Alertmanager with platforms such as Keep simplifies monitoring alerts and enhances how we interact with them. For example, integrating with a ticketing system such as ServiceNow allows us to create notes and updates directly through this connection. This ensures that updates and notifications are also reflected in the organization's **customer relationship management** (**CRM**) tool. This helps keep everyone informed throughout the organization.

For our lab, we started configuring Alertmanager to send alerts to Keep in the previous section. With this setup already in place, we just need to see the alerts reflected on the Keep platform. Now, the alerts generated should appear in Keep's **Feed**. To check the alerts, go to Keep's **ALERTS | Feed** (`https://platform.keephq.dev/alerts/feed`):

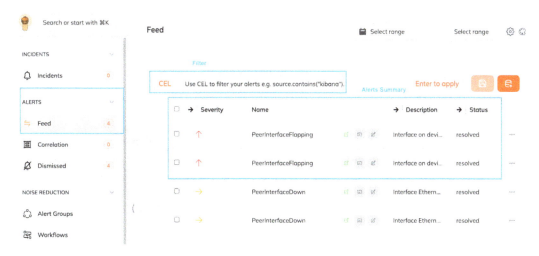

Figure 9.14 – Keep—ALERTS | Feed

This alert management system can help us keep track of the alerts and connect to messaging systems, such as Slack, Mattermost, and MailChimp, or ticketing systems, such as Jira or ServiceNow. It can even help you create automated workflows for alerts for more advanced processing.

Alert and incident management systems are focused on the life cycle of the alerts, what the impact and correlation are, and advanced notification rules and policies for a scalable on-call setup for the teams they are serving. Check out software such as Opsgenie and PagerDuty, as well as other software that falls under this category.

Before closing this chapter, it is time to recap and emphasize tips and best practices when working with alerts in your observability platform.

Alerting tips and best practices

We can group our recommended best practices under four main categories.

Addressing common alert challenges

Let's explore some common challenges faced when dealing with alerts and their ecosystem, particularly the journey from noise to valuable information:

- **Observability stack integration with SoT**: To counter the challenges presented by alert fatigue, noise, and the dynamic nature of modern IT environments, integrating an observability stack with a reliable SoT is key. This approach aims to streamline the alerting process by providing a clearer, more accurate depiction of the IT landscape.

- **Normalization through observability data pipeline**: We're focusing on normalizing data across various sources. This means that regardless of where the data comes from, our observability data pipeline works to standardize it, ensuring consistency and reliability in the alerts generated.

- **Query-driven alerting**: By enabling precise queries on your observability data, we can tailor alerts to be genuinely meaningful. This selective alerting ensures that only relevant changes in the environment trigger notifications, significantly reducing noise.

- **Data enrichment with SoT information**: Enhancing observability data with insights from the SoT allows for a nuanced understanding of which alerts warrant immediate action. For instance, an alert about an interface to an end host going down might not require urgent attention. However, if the alert pertains to a critical inter-site backbone link involved in vital operations, the story changes. Embedding the role of the interface within the data, as informed by the SoT, facilitates the prioritization and categorization of events, helping teams focus on what truly matters.

Build on top of communication and transparency

It's important to keep a clear and updated communication loop with the key stakeholders involved in alert management. Let's discuss some key aspects that can improve the communication process:

- **Building trust**: Open communication builds trust between IT and the rest of the organization. When stakeholders are kept informed about incidents and the steps being taken to resolve them, it creates a sense of confidence in the IT department's ability to manage crises.

- **Minimizing rumors and misinformation**: Transparency in communication helps prevent the spread of rumors and misinformation. By providing accurate and timely information, you can ensure that stakeholders have a clear understanding of the situation, which is particularly important during major incidents.

- **Enhancing collaboration**: Effective communication fosters collaboration across different teams. Transparent sharing of information about incidents and their status encourages teamwork and can lead to faster resolution of issues.

- **Facilitating decision-making**: Keeping all relevant parties informed allows for better decision-making. Transparent communication ensures that decision-makers at all levels have the information they need to prioritize actions and allocate resources.

- **Determining innocence**: In the world of incident management, it's just as important to determine what isn't causing the problem as it is to find the root cause. The popular *It's not the network* saying highlights this concept. By clearly identifying and communicating when the network is not at fault, teams can narrow down the search, focusing efforts on the actual areas of concern.

Healthy incident management process

Getting paged or called when a service goes down can be incredibly stressful, especially when you need to communicate with your team and solve the issue at the same time. This situation can create pressure and tension. Here are some ways to ease the pressure during these moments:

- **Enforce a zero-blame culture**: Promote a zero-blame attitude where the focus is on resolving the issue rather than assigning fault. This encourages team members to report incidents and potential issues without fear of retribution, leading to a more open and proactive environment.

- **Prioritize positive reinforcement**: Use positive reinforcement to acknowledge the efforts and contributions of team members during and after the resolution of incidents. Recognizing achievements, even in challenging situations, can boost morale and motivate the team. This blog post provides greater insight into this topic nicknamed *#HugOps* (`https://www.atlassian.com/blog/statuspage/be-kind-during-downtime-send-hugops-love-today-and-every-day`).

- **Keep time-to-resolve in focus**: While fostering a supportive culture, remember that time-to-resolve or time-to-innocence is critical. Prioritize actions that lead to a swift resolution of incidents to minimize impact on business operations. Implement strategies such as having predefined escalation paths, maintaining an updated knowledge base, and conducting regular training sessions.

- **Implement clear communication protocols**: Develop and adhere to clear communication protocols that outline who should be informed, how updates are to be communicated, and the frequency of these updates. This ensures consistency and clarity in communication during incident management.

- **Utilize a centralized incident management tool**: Employ a centralized incident management tool where updates, documentation, and communication can be centralized. This helps maintain transparency, as stakeholders can view the status of incidents and actions taken in real time.

- **Conduct regular post-incident reviews**: After resolving an incident, conduct a post-incident review to analyze what happened, why it happened, and how similar incidents can be prevented or better managed in the future. These reviews should be constructive, focusing on learning and improvement.

- **Train and prepare**: Regularly train and prepare your teams on incident management processes, tools, and best practices. Simulation exercises can be especially useful in preparing teams for real incidents.

The role of AI in alerting

With the rise of **artificial intelligence** (**AI**) applications, it's becoming clearer how AI can enhance alert and incident management for organizations. Here are some areas where AI can be particularly useful:

- **Intelligent filtering**: AI can significantly enhance alerting by intelligently filtering out noise, identifying patterns that may indicate emerging issues, and prioritizing alerts based on their potential impact on the business.

- **Predictive alerting**: AI and machine learning algorithms can predict potential issues before they occur, allowing for preemptive action based on historical data and trends. This predictive capability can transform reactive alerting processes into proactive measures.

- **Enriching alerts**: AI can add significant value by analyzing the content and source of an alert to automatically categorize it, suggest relevant guides or manuals, and link it to a knowledge base that includes the appropriate runbook or procedure to follow. This enrichment not only speeds up the response process but also ensures that the correct steps are taken, reducing the risk of error.

- **Complexity of implementation**: Integrating AI into alerting systems requires substantial expertise in both IT infrastructure and AI technologies. Organizations must navigate the complexity of implementing, training, and tuning AI models to ensure they accurately reflect operational realities.

- **Trust and transparency**: Relying on AI for alerting introduces questions about trust and transparency. It's fundamental for IT teams to understand how AI models generate alerts and on what basis decisions are made, especially in critical scenarios that may require rapid intervention.

Summary

In this chapter, we took a close look at how alerts work in the realm of IT systems and why they are an important part of an observability platform. Alerts are essentially the signals or messages that pop up to say, *Hey, check this out, something might be off*, much like a notification.

We went over what an alert is, how it differs from incidents, and the key parts that make up an alert's data structure—the standard way these signals are built and used. We also walked through some practical techniques to aggregate and correlate alerts.

We brought some of these concepts to life by setting up a lab environment and following a structured approach that mirrors the real-world needs of IT teams. We discussed how understanding the specific alert requirements of network stakeholders, such as the network operations team, can guide optimal alert rule creation. Using Prometheus and Loki, we demonstrated how to configure these tools to monitor network health and trigger alerts for potential issues. We also explored how Grafana can be used as a powerful visualization tool to display alerts and aid in monitoring efforts. To further enhance alert management, we introduced Keep, an open source alert management system, and showed how it integrates with Alertmanager to streamline the handling of alerts.

Wrapping up the chapter, we shared a collection of best practices and tips, categorized into four key areas: tackling challenges in alert management, fostering communication and trust, nurturing a healthy incident management process, and utilizing AI to enhance alert life cycle management.

As we transition to the next chapter, we'll consolidate the insights gained so far and focus on real-world observability architectures and some of the various tools that are available. This discussion aims to assist organizations in deciding whether to develop their own observability and monitoring solutions or to invest in pre-existing technologies, while also providing a comparative analysis of different approaches.

10
Real-World Observability Architectures

So far, there has been a lot of how-to on individual applications within a modern network observability stack. In this chapter, you will learn about how these tools can be tailored to work within your environment and other tooling that could complement your stack, taking into account the build versus buy trade-offs and these tools' capabilities.

A key early decision point when building your observability architecture (or any other kind of software solutions) is evaluating the components of the solution that you want to build for yourself, as well as deciding which ones you procure or subscribe to. There is a middle ground here as well, with third-party service providers available that can provide solutions as a **software-as-a-service (SaaS)** offering. The capability of systems to interoperate with each other is important in having a flexible and customizable observability stack.

Within the context of this book, the build side is the integration of products or projects together. Primarily, the stitching of products or projects into your observability pipelines will give you a complete pipeline. The strongest part of having an architecture where multiple components can play the roles required of them is that you have flexibility that is extensible as needed. Conversely, the buy side represents observability solutions that come as monolithic solutions (this has been the most common approach in network monitoring so far).

In this chapter, we will cover the following main topics:

- Additional component options to fulfill roles in the observability stack
- An overview of snapshot-based systems
- Comparing build versus buy data points
- Orchestrating an observability platform

Observability stack options

This book uses the following observability pipeline architecture, introduced in *Chapter 4*, as a general reference. These are the components that a modern observability solution should be able to implement, so we will use them to analyze other options rather than the ones already presented in the book:

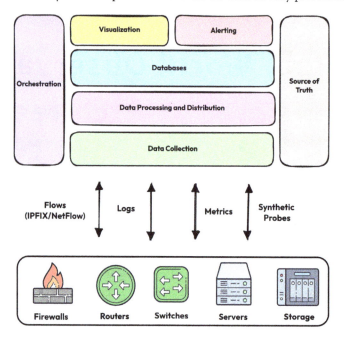

Figure 10.1 – Network observability architecture (Created using the icons from https://icons8.com)

When looking at the options, there are many things to consider. Gaining knowledge and some basic experience with each is beneficial to be able to make an informed decision and path forward for your pipeline. Let's dive into a few of the options within the observability pipeline.

As you can imagine, there are many options when it comes to the applications that fit within the observability stack. For example, for the data processing function, there are (at the time of writing) several message brokers available, including **Kafka**, **NATS.io**, and **MQTT**. However, there will likely be more available in the future. The same applies to the other components of the architecture. Thus, you should be ready to analyze the components analytically to understand what makes more sense in each case.

The next section provides you with several examples of the various types of tool sets available. We will start from the early days and discuss network monitoring with some legacy tools that may bring back some memories. In the same section, we will dive into some of the build versus buying decisions.

All-in-one open source tools

Since the beginning of open source network monitoring systems, the most common (and easiest) way to build a solution was to package together all the key components – data collection, storage, and visualization.

In the early 2000s, the **RRDtool** database was the king database in network observability data. It was used by **Cacti** (`https://www.cacti.net/`), and other similar tools. Cacti is a **Simple Network Management Protocol** (**SNMP**)-only tool that provides graphs of collected data. Here is an example of a Cacti graph from the Midwest Internet Cooperative Exchange (`micemn.net`):

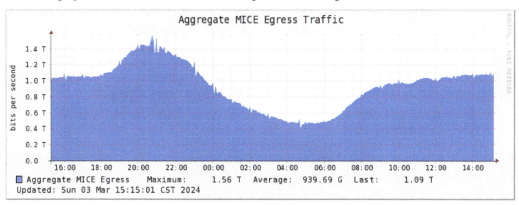

Figure 10.2 – The Midwest Internet Cooperative Exchange Cacti graph

These early monitoring tools shared the same persistence system but also came as a complete solution, including SNMP data collection and visualization.

Technologies have evolved over time, but this approach of packaging all functions together is still very common in solutions such as **Zabbix** (`https://www.zabbix.com/`), **LibreNMS** (`https://www.librenms.org/`), **Observium** (`https://www.observium.org/`), and **Nagios** (`https://www.nagios.org/`). There are likely more available; these are just a few.

Commercial off-the-shelf tools

There are many commercial off-the-shelf tools – so many that we could dedicate multiple chapters to cover each in depth. Each one usually plays most of the roles within the recommended observability pipeline. Let's talk briefly about a few of the most popular tools and how they fit within the reference architecture:

- **Splunk** (`https://splunk.com`), now part of Cisco Systems, made its name collecting logs and is known as a security tool first and foremost, with the ability to fit within the log collection portion of a pipeline, as well as storage and visualization roles.

- **Datadog** (`https://www.datadoghq.com`) is one of the leading DevOps-focused SaaS visibility tools on the market. The product marketing on the Datadog site focuses on cloud and DevOps enablement. This may be a fit for some organizations, especially ones with a large DevOps environment. It also supports an SNMP-based agent for collection.

- **Solarwinds** (`https://www.solarwinds.com/`) is perhaps an early product that many network engineers associate with network monitoring. It is a system that is available, and it integrates many components – whether the base SNMP network monitoring or the collection of logs via the Kiwi Syslog platform – which are then persisted into an internal database, offering a visualization layer integrated into the entire platform. However, it doesn't offer options for architecture customization plugins in other components, such as other collectors, which could extend some missing interfaces such as **gRPC Network Management Interface** (**gNMI**) or API interfaces.

Now that we have reviewed a few of the systems that can work with multiple vendors, let's briefly talk about some of the vendor systems that integrate observability directly into their platforms.

Controller-based systems

One of the earliest decisions is whether you require multi-vendor support in your observability pipeline. If you have a homogeneous environment that consists of a single vendor per domain (routing/switching, wireless, or security), you may want to proceed with looking at a vendor-supplied solution, usually integrated into a network controller. There have been great strides made by each of the vendors in the space for their own, such as Arista's **CloudVision Portal** (**CVP**) or **Juniper MIST**. Tools such as these and other vendors of hardware gear are likely to have a platform with some functionalities related to observability and telemetry. These may meet your needs, especially if you are at an organization that has a single vendor environment. But for multi-vendor environments, there are additional solutions available, as you have seen in this book.

Now that we have covered a few third-party and first-party systems, let's talk about another type of observability data, snapshots. Much of the book so far has worked with metrics and logs. Another type of observability data is the snapshot.

Time series versus snapshot observability

Snapshot observability is where an application takes a look at the current state of a network and then records the state in a database for later retrieval and comparison. This includes routing tables, command output, MAC address tables, and similar data. This is another important observability capability to evaluate to have within your stack. A couple of tools to evaluate with snapshot-based observability are **IP Fabric** and **SuzieQ**.

You may find Snapshot tools valuable when working with network assurance. The tools are typically built to gather data in a text format and to be able to compare it over time, giving additional insights into the state of the environment over time. For example, SuzieQ uses rules called assertions that will run multiple checks on BGP neighbor information, to determine the network assurance of operating as expected.

These tools are independent tools, providing important data components that are valuable to your network environment. They are independent in the sense that they do not fit directly into the pipeline presented in this chapter.

After covering many of the types of systems, now comes the financials. How do you decide whether to purchase an all-in system or build out components to bring your observability stack to life?

Comparing build versus buy decision points

With so many available options, you will face the build versus buy dilemma, and you will need to create a structured methodology that helps you make a decision aligning with your organization's needs, capabilities, and strategic goals.

In this section, we will provide a comprehensive approach to analyzing your decision, starting from what we discussed previously in this chapter.

Unsurprisingly, the first recommended step is to start by collecting and analyzing your network observability requirements, functional and non-functional.

Defining requirements

The *functional requirements* identify *what* an observability system should do for an organization. Some example functional requirements for an observability platform include the following:

- How frequently do we need to collect new data? Maybe real-time monitoring, using model-driven telemetry or SNMP polling every two minutes?

- What requirements are there to transform data in the same fashion via data processing? Do you have to monitor different types of systems, with different key definitions for key/value data that has to be consolidated and normalized?

- What systems need to be integrated with? Perhaps alerts need to integrate into PagerDuty, OpsGenie, GoAlert, OnCall, or other AI/ML tools to ensure that the right people look at events.

- Alerting requirements around how alerts need to be processed, acknowledged, and closed.

- Logging and tracing requirements to collect events from devices or network-related applications

- How many devices/targets will need to be observed?

- What is the retention policy on the data that will be required?

- Visualization capabilities are required, such as being able to create dynamic graphs, annotations, and a combination of data sources or observability data types.

The *non-functional requirements* are considerations that talk about an environment's performance, security, compliance, and so on, such as the following:

- Where can an application be hosted? Does it need to be hosted on-premises or is the cloud the preferred deployment? Can a service/SaaS provider handle this?

- What are the response times for a web page to display the values?

- What is the expected availability of the application?

Next, we will cover evaluating your currently available in-house capabilities before implementing anything new.

Evaluating in-house capabilities and resources

After gathering your requirements, and before starting to evaluate new options, the next step is to review the tool sets that are already available in-house and see whether they fit into any of the roles that are part of the pipeline (*hint – reuse before reinventing*). Afterward, once you have the inventory of the tools, you can start to look at the required capabilities (according to the observability architecture reference from *Figure 10.1*) and compare them with other new solutions that could complement your existing tool set.

As well as looking at the technologies, you need to evaluate the technical expertise to deliver the necessary technical capabilities. This means that you must assess the team's technical expertise, both inside and outside of the networking teams. Network engineers have evolved from CLI ninjas into capable system engineers capable of understanding how to build and evolve a disaggregated observability platform, but they may need some extra training if they're not ready yet. And, last but not least, do your team members have room to take on the related tasks (not only building but also operating and evolving the observability platform to meet your specific needs)?

Moreover, when deploying databases and applications, you need to consider the storage and computing available at the locations where the application will be deployed. For example, if you're looking at an edge deployment strategy, you should ensure that the edge computing and storage have enough resources available to store the observability data collected there (metrics, logs, traces, etc.) and to properly analyze the data for alerting and visualization.

Finally, another internal consideration is the opportunity cost that the observability platform has for other internal initiatives. What are the other opportunities that may be passed up to establish and operate the infrastructure? Is this a required capability to have the flexibility that the platform enables? Would you rather have a service provider set up and maintain the environment to allow your engineers to focus on other opportunities?

These are just a few of the questions that should be looked at internally when evaluating the observability platform. The next dimension to analyze is the cost associated with the platform.

Cost analysis

Often, cost is a decisive factor when deciding between possible solutions, and it impacts both the build and buy decisions.

With the buying decision, it is usually a bit more straightforward in that you get the cost presented to you by the supplier of the system, and you pay the bill. If it's not SaaS, there is still server upkeep, connection upkeep, system management, and yearly maintenance, at a minimum, with any system. In SaaS, you pay as you go, but there may be some additional costs, and you will want to account for that in your decision.

On the build-out-yourself side, there are generally fewer costs for the licensing of the software on the system because you capitalize on using open source solutions. However, you will have costs that will need to be accounted for on the human side (knowledgeable engineers come with a cost) of building out and maintaining an environment, as well as the IT infrastructure components.

In both cases, when looking at these costs, it is important to not only account for the first year, or the build-out of the systems, but also incorporate the recurring costs, looking at a multi-year (likely, a five-year) **total cost of ownership** (**TCO**). The TCO should incorporate each of the items that are associated with the cost of operating a system, not just the upfront cost. You will need to look at what training may be required, the yearly IT infrastructure maintenance and licensing, and the people needed to operate the system. Also, if there is an opportunity to have the system as a revenue-generating system, the projected income each year should be incorporated into the TCO calculation as a positive item.

A starting point for the financial TCO calculation may look like *Table 10.1*. You can simplify it by merging some similar items, such as licensing and maintenance. Make sure to incorporate any recurring costs as a multiplier on the calculation.

	Option 1	Option 2	Option 3
Upfront cost			
Yearly maintenance (4x)			
IT infrastructure maintenance			
Employee training			
Employee headcount			
Total			

Table 10.1 – An example TCO calculation sheet

Next, let's review the risks associated with deploying a new system.

Assessing risks

Time is the most valuable asset. When the time to deliver a solution (or time to market) is a critical factor for the success of observability solutions, you have to define how quickly you need a solution in place and evaluate the trade-offs between building and purchasing a ready-made solution.

Building a custom solution tailored to your specific needs allows for maximum flexibility but can be time-consuming. Design, development, testing, and deploying the software or system are all involved solutions. Conversely, using off-the-shelf services or acquiring commercial software provides a faster route to deployment. However, it may require adjustments to fit your unique requirements. Balancing time constraints with customization needs is crucial in this decision-making process.

Having addressed the pressure to deliver, eventually, the quality and reliability of a solution impact its success. When opting for a built solution, consider risks associated with unknown bugs or performance issues. Testing and quality assurance are critical for reliability. Commercial products have typically undergone extensive testing and are more likely to be reliable. Vendors invest in maintaining and improving their products, which can save you time and effort. However, thoroughly researching a product, reading reviews, and understanding its track record are crucial steps to assess its quality and reliability.

When using a third-party solution, there are a few more risks to account for. These include the following:

- **Vendor lock-in**, with which you lose the ability to adjust for your business requirements due to what a vendor provides. It may be costly to gain business-required capabilities in time to meet business needs if locked into a solution.

- Licensing and support pricing may change over time, causing an unexpected budgetary expense.

- Product discontinuation, or the lack of new patches and feature releases, may risk compliance and pose other business risks.

> Note
>
> An example of the licensing and support pricing change happened in 2023 when Broadcom acquired VMware. As Broadcom completed the acquisition, pricing levels changed. Some of the previous **value-added resellers** (**VARs**) were unable to continue selling VMware licensing, ending relationships that were established, and many terms and agreements were changed.

These are all risks seen with home-grown and third-party solutions, but they have to be evaluated together with what you require.

Comparing features and flexibility

Going in line with the TCO calculation, the features and capabilities of the tooling also fit. A calculation in which you define your must-have requirements, such as whether streaming telemetry is required as a feature, may drive you to require a gNMI collection method. By weighing requirements as more important than just being nice to have, you can start to determine whether products may be a fit for your needs. Depending on how strong the need for a requirement is, a solution can be eliminated if it does not meet those requirements.

It is important to have the capability to interoperate between systems in both scenarios, buy and build. In that, there are likely to be systems that need to integrate into the observability pipeline components as well as receive configuration and updates from other systems. There may be an enterprise-wide alerting platform, such as PagerDuty or OpsGenie. Conversely, an alerting application should be able to receive inputs from multiple different applications, whether through Webhooks or another integration capability.

In terms of features, you should assume that the requirements will evolve over time. How easy will it be to add new features to your observability solution? If you opt for buying, you should consider the vendor's roadmap for innovation and updates (e.g., how long it takes a feature request to be implemented) versus, in the homegrown solution, your ability to keep evolving observability practices in terms of knowledge and engineering cycles.

Now that we have covered factors that may help weigh a purchasing decision, we will examine documenting the decision.

Making a decision

Finally, it's time to make a decision. At this point, you should provide the following information:

- **Pros and cons**: List the advantages and disadvantages of all the options based on the preceding analysis.

- **Recommendation**: Based on the gathered information, make a recommendation that best suits your organization's current needs, resources, and strategic direction.

- **Document the decision**: Everything evolves, and what is the best recommendation today may not be the best in six months or a few years. Writing down the decision process will allow you, or other team members, to understand it in the future and challenge it when circumstances change.

Now that we have covered a few of the general ideas about how to select the tools, let's take a look at how to simplify some of the complexity of managing and operating solutions with many components interconnected, lowering the barrier of adoption.

Orchestrating an observability platform

Building a platform like the one described in this book may seem easy to accomplish on a small scale (for example, the lab scenarios). However, in real environments, scaling it by hand may seem daunting. That is where infrastructure automation and orchestration come to the rescue.

The idea of orchestration within automation is that you define your patterns on a small scale. Then, you scale out those patterns on the whole platform side of things to accomplish the larger goal. Orchestration is what takes these modern telemetry and observability tools, which are great on a small scale, and makes them extremely powerful platforms at scale.

Let's take a look at how this works with one of the tools we have covered, the Telegraf collector. As you learned in *Chapter 5*, Telegraf requires a configuration file that tells an agent which targets (i.e., network devices) and methods must be used to collect data. At scale, this means that you need to provision hundreds or thousands of Telegraf agents, distributed geographically to cover different zones, and each one needs to come with the proper configuration file. Doing this manually (i.e., writing the configuration files) is not going to work because of the amount of changes needed and how likely errors are to occur.

To solve this problem, we need to analyze a workflow as a whole. The first question is, what is the original change that may require a configuration change? It may be a new device added to the network source of truth. While the overall automation workflow manages configuration provisioning and other tasks, from Telegraf's perspective, it is necessary to identify which configuration file needs updating and to which Telegraf agent it should be deployed.

> **Note**
>
> In the chapter's Git repository (`https://github.com/network-observability/network-observability-lab/tree/main/ch10`), you have a simple example of how to implement this approach with **Ansible**, using a simple inventory and **Jinja2** templates to generate the configurations.

In this section, we will answer both questions – how and when.

Deployment methodologies and orchestration

Scaling out an observability platform requires providing a reproducible and consistent way to deploy the components of the architecture and dynamically adapt to contextual changes, such as a new target to observe.

In the previous Telegraf example, we introduced the idea of rendering a new configuration file once a change happens. This is defining the monitoring intent, and then making this configuration active by deploying a new agent or restarting an existing one. Luckily for us, nowadays, we have many different options that allow us to manage the infrastructure dynamically.

Before we get into enumerating a few of the available approaches, it's important to highlight that there shouldn't be limitations around how to build an observability solution. It doesn't matter whether we use bare-metal servers or a virtualized VM in a private or public cloud, run the services as Linux processes or containers, or even implement some component as a SaaS. Ultimately, for each method, there is a programmatic way to interact without human intervention.

In the aforementioned Telegraf config generation example, we explained how to automatically render the configuration file. However, this is only one of the challenges in the automation process. To streamline the process, there are two other crucial functionalities – an orchestration solution to redeploy on changes and scale out the components, and a triggering mechanism to understand when it is necessary to run the process.

Let's start first with a high-level overview of the different viable approaches to automate the deployment of your observability pipeline (one simple example is the lab described in *Appendix A* that you have used in this book).

How to scale out and reconfigure the pipeline

Nowadays, every IT infrastructure should be managed using an **infrastructure as code** (IaC) approach. In IaC, all the IT components are managed by writing code instead of clicking through a UI or typing commands in a CLI. The main goals that IaC pursues are as follows:

- Ensure consistent deployments across environments, handling snowflake or custom configurations with code. If it works well, it should work well everywhere, and if it fails, it can fail in a development environment.

- Increase speed, reduce errors, and allow repeatability by automating all the steps in the process.

- Provide trackability and a rollback of changes while promoting collaboration.

As you may have already deduced from the previous goals, leveraging a version control system (e.g., Git) or similar solution provides the intended reference of the desired configuration state (i.e., the source of truth) and the steps to reach it. Because of the predominance of Git in this area, it is also known as **GitOps** (**Git Operations**).

Building a network observability pipeline is not an exception. All the IaC principles are applicable here. Thus, in this section, we will provide you with a high-level overview of the different approaches (and some tool recommendations) to help you get started with some of the requirements of each one.

> **Note**
>
> There is plenty of literature about the topic, but one of the first books to introduce it was *Infrastructure as Code* by Kief Morris.

The IaC concept can be applied to any type of IT infrastructure that offers a programmatic interface. However, not all of them offer the same capabilities. Since the eruption of dynamic infrastructure (e.g., virtualization and cloud services) or **Infrastructure as a Service** (**IaaS**), the deployment process can be streamlined without human intervention. That's ideal in terms of IaC because it can control everything. However, even in more traditional environments that still use bare-metal servers without an automated bootstrap process, there is still room to automate the provisioning.

Thus, it's the infrastructure itself that defines how IaC can be applied in each case. To provide a general classification, two main approaches have been defined:

- **Imperative**: Specifies the exact steps to move an infrastructure's configuration state from an initial state to a final/desired one. The tools that implement this approach are also known as configuration management tools, and some popular ones are Ansible, **Puppet**, **Chef**, or **Salt**.

- **Declarative**: Defines the intended state of the infrastructure without worrying about what needs to happen to get there. Popular tools in this area are **Terraform**, **Pulumi**, **Flux**, and **Argo CD**.

There is no good or bad approach. Each one comes with its pros and cons. For example, when you need fine-grained control of a process, the imperative one is more capable, but if you have a common setup, then the declarative one is simpler and easier to reproduce.

Also, sometimes, both play well together. For example, you can deploy Linux servers in a public cloud in a declarative approach (e.g., using Terraform) that brings them into an initial state (maybe with some basic hardening), and once deployed, you apply the imperative approach (e.g., using Ansible) to tune some configuration aspect.

Finally, another important related concept is the immutable infrastructure. This approach treats the infrastructure components as unchangeable after deployment. Whenever an update is needed, a new instance needs to be deployed, going through a recommended continuous integration process that validates it.

We strongly recommend using this approach, which can be implemented in many ways. For example, you can use containers (e.g., **Docker** or **Podman**) or base images for different infrastructure providers (e.g., AWS **Amazon Machine Images** (**AMIs**) or Droplet images in **DigitalOcean**), created by Packer. These artifacts are defined imperatively and, once built, deployed in a declarative way to the target infrastructure.

Detecting when it's time to update the pipeline

Once we have everything automated, the only unanswered question is when to run the pipeline. In fact, many events may require a change on the observability stack, which is built on top of many components, as we discovered in *Chapter 4*. This may be an area where you would want to introduce continuous deployment that automatically deploys the changes once they have been merged. For additional material from Packt on CI/CD pipelines, see `https://catalogue.packt.com/skill/ci-cd-pipelines`.

The need for change may come in different forms. The most natural one is a human introducing changes, such as adding a new metric to collect, including more metadata in the persistence layer, or adjusting the dashboards. These are all direct changes in an observability platform, so there is no confusion about the need to update. All these changes follow an IaC approach. This means that the changes are proposed in a version control system, which eventually will be accepted and will trigger the process to generate and distribute the new configuration.

> **Note**
>
> Using an IaC approach doesn't mean that you cannot play directly with a system if it allows you to. You can experiment with new changes in a non-production environment, but to persist and deploy it requires expressing it as code.

However, the need for change often originates within a monitored environment. For instance, the addition of a new network device to the inventory necessitates updates to the collection processes, or perhaps an existing collection process begins to exhibit unexpected behavior.

Thus, the infrastructure management should be ready to react once a signal is triggered, and the following signals are the common ones to consider:

- A change in the intent of the observability stack configuration – for example, when a new visualization dashboard is added, or when a new metric needs to be collected.

- An alert in the behavior of some of the components of the stack. The observability stack has to be observed itself and react accordingly.

- An external event is received (synchronously or asynchronously) that requires a change – for example, a new device is added to the network source of truth. Two common implementations are via Webhooks or a message in a message bus (for example, **Kafka**, which we introduced in *Chapter 6*).

> **Note**
>
> The authors of this book have observed a trend toward adopting distributed event stores or message buses to allow more decoupling between components, and many automation platforms provide frameworks to leverage them natively – for example, **Ansible EDA** (`https://github.com/ansible/event-driven-ansible`).

In this section, we have provided you with an overview of our recommended approach to implementing the observability stack in a healthy manner, independent of which infrastructure you run on. We know it is not detailed enough to be actionable, but it should be the starting point for you to look at the options you have.

Summary

In this chapter, we introduced some considerations to make when looking at what components to include in your observability stack and what needs to be considered in the build versus buy dilemma. Some solutions will provide a good amount of modern capability but may still come up with a few misses along the way, such as how to enrich data to get contextual information. This is where the open source projects provide flexibility and a complete solution to let you see the bigger picture.

Finally, we have provided some hints about how to mitigate the increased complexity that implies managing many different systems, instead of a monolithic one. This is possible via the adoption of an IaC approach where the automation and orchestration of the platform can react to changes, updating and scaling out the observability pipeline as needed (without human intervention).

With this chapter, we come to the end of *Part 2* of the book, where we covered the architectural reference design of the modern observability stack and dived deep into the different components – visualization, data processing, data collection, alerting, databases, source of truth, and orchestration.

We started with data collection with Telegraf and Logstash. You were introduced to the power of transforming data (through an ETL process) into a single format to simplify the otherwise complex areas of visualization, alerting, and data analysis.

Then, we covered the time series databases (with a focus on Prometheus) to work with metrics and the logging solution Loki to work with logs, traces, and flows. Using these data persistence capabilities led to the visualization of data with Grafana and alerting on the conditions of metrics and logs.

The open source tooling introduced during these chapters has proven to be flexible enough to interoperate with many enterprise IT systems and solutions. Yet they are also powerful enough to be at the heart of modern telemetry solutions for providers and enterprises of all sizes, with a little bit of automation and orchestration to help eliminate the challenges of complexity and scaling.

In *Part 3* of the book, we will dive into helping this data and system provide business success. You will learn more about providing business value relating to modern observability data. You will also learn more about using AI and ML in conjunction with data to inform your decisions.

Part 3: Using Your Network Observability Data

The last part of the book will help you understand what you can do with the rich data that is gathered in today's observability platforms, with examples on how to drive network automation from the data to incorporating artificial intelligence and machine learning.

This part contains the following chapters:

- *Chapter 11, Applications of Your Observability Data – Driving Business Success*

- *Chapter 12, Automation Powered by Observability Data – Streamlining Network Operations*

- *Chapter 13, Leveraging Artificial Intelligence for Enhanced Network Observability*

- *Appendix A*

11

Applications of Your Observability Data – Driving Business Success

This is the final part of the book. In this part, we'll be diving into the practical business applications of observability, the utilization of automation, and how to use your observability data. We'll also look at how to get real-world benefits from your investment in the observability platform.

In this chapter, we will dive into a few of the applications of observability data. We will put together a plan of how to design your dashboards, building on what was covered in *Chapter 8* about building dashboards, and we'll see how to build out your service level agreements once you can measure those components.

This chapter covers the following topics:

- The business value of observability data
- Treating your network as a service
- Architecting dashboards

The business value of observability data

In today's modern business, a network is a valuable asset that transports data to a company. You could be operating a retail environment that needs to have connectivity to process transactions on a card network to a software development shop, because of which the network is vital to providing connectivity to the resources that a business relies on. Therefore, knowing the operational state of the network is critical to the day-to-day operations of the business. In short, the network is a critical part of the business.

In *Chapter 10*, we introduced real-world architectures for observability. The collection of network metrics, logs, and events is a key driver of the capability to measure the state of a network. Is the network performing to its capability? Have the network service providers you have contracted met their SLAs? These are questions that you should be able to determine with a modern network observability solution, as well as the following:

- Do you know when you need to add additional capacity to the **WAN** (**Wide Area Network**) environment?

- Does the application that resides in the data center have enough network capacity to talk to the database server?

- How do the connections to the database server hosts look?

These are all questions that can provide very important information to the end customers. Is there enough network capacity up and down the application stack? What about unexpected latency from the web application server to the database server? These are key components that need to have the appropriate metrics in place to be able to confirm that business applications respond as required or expected.

During the early stages of the COVID-19 pandemic, there was a sudden change in where companies did their work. Suddenly, there was a significant push toward having those able to work from home do so. This saw a significant increase in the usage of VPN clients for those employees who previously would only occasionally use the VPN to connect in the evening if they had some extra work to do. The capacity of VPN concentration devices is typically done via license counts on simultaneous user counts. This had a huge change in the day-to-day VPN usage count for organizations to manage. Without having metrics to look at, there are real challenges to understanding how many licenses are in use, how many are needed, and how many to plan for future capacity.

By collecting telemetry metrics, logs, and events, you can measure and observe the performance of a network. This data will also allow you to help plan out the capacity of the network and any upgrades that may be required, and you can create high-level metrics that help you report the status and health of your network. These are valuable data points to provide concrete metrics to help justify projects, keeping the network running optimally.

Let's look at a few of these examples and how you can manage them.

Capacity planning

Capacity planning is managing in advance the demand for a system environment. In manufacturing, the planning of the capacity of a manufacturing plant is done based on the demand for what is being manufactured. The same principle applies to the components within a network and IT infrastructure. Storage space on a **SAN** (**Storage Area Network**) requires there to be enough storage in terms of disk space to meet the demand of storage needs. The server infrastructure needs to have enough CPU/ memory and general compute power to handle computation workloads. In the network space, capacity planning typically revolves around bandwidth on circuits/interfaces.

Some of the reasons for capacity planning for a network include the following:

- **The growth of network services**, which will require additional bandwidth for applications to reside in. Many organizations have an internet edge for their network. Planning for the capacity of the internet is vital, especially if network services to end customers are being provided through the internet edge.

- **Cost savings**, which is the opposite of the growth of network services. Perhaps there was an application that was recently retired that had a significantly large network profile, or perhaps better optimization techniques are being applied that have reduced the need for as much bandwidth on the network.

- **Ensuring application performance** to help determine the impacts on an application, such as network performance, compute performance, and the underlying application components, such as database connections.

So, what is the best way to measure these components? Thankfully, the study of statistics has provided network engineers with a terrific way to measure the general capacity of a network interface.

Percentiles

Percentiles are used in many different scenarios that are part of everyday life. Percentiles are a statistic (single point) that indicates the position of a data point in an overall dataset. Let's look at this in more depth, as it is an important measurement of observability in general.

When a baby is born, there are percentiles applied to the measurement of them. The measurements of the length and weight are taken and then compared to a growth chart, based on the current averages of newborn babies, such as whether the weight is in the 50th percentile. What does that mean? It means right in the middle. Percentiles are made in relation to a percentage. So, the 50th percentile is the measure right in the middle percent-wise. Another way to put it is if there were 100 samples taken (for a single percent for each of the 100), then there would be 49 points that are below the measurement and there would be 49 measures above.

95th percentile

In network capacity, the de facto standard of measurement has been to look at the 95th percentile. This is because, based on all of the data points collected over the period of time on a measurement, *95% of all of the data points will be below this point*, and the remaining data points will be above this. Let's take a look at the 95th percentile with 20 digits. These are randomly generated, but when you look at the bandwidth, it is likely to be a bit more free-flowing. At the start of the data, it does not mean much in this format to the human eye:

Measurement Number	1	2	3	4	5	6	7	8	9	10	11	12	13	14	15	16	17	18	19	20
Unsorted Numbers	594	9	1	121	842	880	340	411	398	901	737	83	661	598	987	423	292	596	197	549

Table 11.1 – An example distribution of numbers for a 95th percentile

So, after the initial dataset, let's, as an exercise, sort these numbers from the smallest to largest. The numbers now look like the following for the ordered 20 data points:

Measurement Number	1	2	3	4	5	6	7	8	9	10	11	12	13	14	15	16	17	18	19	20
Sorted numbers	1	9	83	121	197	292	340	398	411	423	549	594	596	598	661	737	842	880	901	987

Table 11.2 – Ordered numbers for a 95th percentile

With the data points in place, you can now get the measurement of the 95th percentile. That statistical measurement point is 901. This is because the measurement that is at the 19th (i.e., 19/20 = 0.95) position from smallest to largest is 901.

The 95th percentile measure for network interface or circuit capacity accommodates periodic spikes in burstable bandwidth, while still reflecting the majority of traffic.

This is the best practice measurement number to use when determining how much an interface has been used over time.

We will take a deeper look at how to get the 95th percentile from your metric data a bit later in the chapter when we build a dashboard.

Forecasting

In contrast to analyzing historical data, forecasting involves predicting future outcomes based on current and past data. Two common methods for forecasting today include linear regression, a **machine learning** (**ML**) algorithm that projects future values based on past trends, and using tools such as Meta's open source project Prophet to predict future states with greater flexibility and accuracy.

In *Chapter 13*, we will work through a practical example of forecasting to anticipate the utilization of a network interface according to past data, determining whether we will have enough capacity in the future, or predict which should be the present utilization to compare with the current status, determining whether it is performing as expected or there is an anomaly occurring (i.e., forecasting in the past).

Defining health status

When you look at metrics, whether they are network metrics, project health metrics, and so on, these are often summarized in a traffic light system. The traffic light system is a condition where you assign colors to quickly identify the status of a metric. Seeing the possible values from the *Chapter 8* visualizations, you can get a clear understanding of a state based on the following coloring values:

- Green is associated with everything being in a good state

- Yellow is a state of upcoming caution, suggesting that you may need to keep an eye on things

- Red means that things are not in a good state and some action will need to be taken

In general, the color schemes are what you make of them yourself. Just make sure that if you deviate from the generally accepted color schemes, it is documented somewhere in the dashboard so that those looking at the dashboard for the first time are aware of what everything means. Take a look at the following visualization of an interface state, which clearly shows that the interface was down for a period of time:

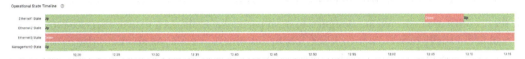

Figure 11.1 – A health status with a change in state

After the arbitrary decision on the colors of their meaning, let's map this to the health of metrics. Each metric definition is different for different areas of an environment. When defining the health status for each metric, it's important to consider both the local health of the component (e.g., device, circuit, or interface) and the overall health of a system.

These values may be different. Let's take a look at a custom value mapping, with colors that correspond to their Prometheus metric value:

Figure 11.2 – Value mappings for the conversion of numbers to words

When examining individual interface utilization, a utilization rate under 50% is often considered a green state. However, from a system-wide perspective, if both of your WAN links are at 49% utilization, this might seem acceptable for each component. But there's a risk – if one circuit fails, the remaining circuits would be pushed to 98% utilization. This could lead to dropped traffic and trigger a broader systemic impact.

When looking at the creation of visualizations and alerting, think about how it relates to providing value for an organization. A few areas that come to mind specifically within the networking space for observability data are providing capacity planning, customized alerting to match business requirements, and dashboards that can be understood at a glance.

Treating your network as a service

Often in IT, the work is related to the manufacturing industry, and certain manufacturing requirements can help us understand the concept of treating your network as a service. One crucial requirement is knowing when to secure funding for your service. The ability to measure service availability by the applications that use it is key. This highlights the importance of observability.

The services that are delivered are designed to meet particular goals. As a network service provider, the goal is to usually provide network services to customers at scale. As an enterprise entity, the goal is to provide network services that help enable applications to serve enterprise goals. The use of observability to provide metrics about the health and state of a network helps to provide a clear understanding of the network services. This can also help to provide data-driven metrics to help get upgrades to the network equipment that you need.

When you start to treat a network as a service, you are acting as a business entity to your organization. To help facilitate running the network as a service, operating like a network service provider is a good start. Needing to know what measures are critical for delivering a service versus the measures that are good to know about the environment guides you on which metrics to gather.

Monitoring SLIs, SLOs, and SLAs for optimal network performance

In *Chapter 2*, you learned about SLIs, SLOs, and SLAs. Let's dive into starting to look at what some of these measurements may look like. The first set of SLIs in the realm of network observability looks at the technical details of transport on a network. Here are a few SLI examples:

- Latency measures
- Availability:

  - Application availability

  - Data center network availability

  - WAN availability

- Error rate

- Interface throughput

- Application response time

- **Mean time to repair** (**MTTR**): The average time that it takes to get something fixed after it has broken.

- **Mean time to change** (**MTTC**): The lead time to get a change into production:

These are the indicators that, as a network engineer, you may be thinking about a lot of the time when trying to place some metrics in a network

The SLOs put measures/goals to the SLI. So, when you say that you have an SLI of latency, that doesn't really mean much without having additional metrics to do so. Let's look at the preceding bullet list and put some SLIs to help illustrate the example:

SLI	SLO
Average latency between Los Angeles, CA, and Chicago, IL	Under 60 ms on average per hour for polls once per minute.
Maximum latency between Los Angeles, CA, and Chicago, IL	Maximum latency observed of 250 ms during the measurement periods.
Standard deviation latency between Los Angeles, CA, and Chicago, IL	The average **mean deviation** (**mdev**) shall be less than 1 ms.
Error rate	There shall be less than 1% errors on an interface.
Throughput	The interface shall support 500 Mbps with a burst rate of 1 Gbps, measured at the 95th percentile for a given calendar month.
MTTR	The average amount of time to restore services is less than four hours. Measured in a calendar year.
Packet loss per queue	The packet loss for a given **QoS** (**Quality of Service**) queue shall be less than 0.1%.
Volume of traffic	The total volume of traffic on a backup link shall support 50 GB. Think of cellular network service plans with a volume of traffic.

Table 11.3 – Example SLI and SLOs

SLAs are contractual obligations, usually within a **SOW** (**statement of work**) or operational contract. SLAs are often directly related to the aforementioned SLOs; however, there is often a financial remedy or penalty associated with a metric not meeting its agreed-upon metric value.

SLAs are often measured internally by the provider providing the services, but for an end customer of the service provider, it may be advantageous to have a mechanism to verify that metrics are being reported appropriately.

How to treat a network as a service

When you treat your network as a service for the organization you are looking at, you should have service availability similar to what you would if you were to get a cloud service. Treating the network as a service for the entire organization in the same way that you would for a cloud service means that you should provide the following:

- Notification of changes
- A dashboard of the current status of key systems
- Fault alerts

Notification of changes

Notification of changes to a network is something that is simply good business practice. The practice of sending out change notifications is typically handled within an **ITSM** (**IT Service Management**) framework of an organization. If your organization is practicing ITSM and has a notification system available, it is best to use it. Changes are one of the number one causes of incidents through normal daily operations. For that reason, the changes must be communicated with the proper awareness (i.e., notification) to help reduce recovery time. Changes are important for an organization to move in strategic directions. Therefore, it is important to help reduce the impact that may be caused by even the most routine change.

The first step in notifying your *customers* of a network service is to know who those customers are. You should look to maintain the primary contacts of each network service in a database that can be referenced to provide notifications. Once you have the contact information, whether a direct individual contact or an email distribution list, you can provide notifications of upcoming changes.

Changes should be communicated in advance, with enough lead time to accommodate business continuity plans being planned and executed. If a change is minor, this may be a shorter lead time than if the change were to have a significant impact, where outages are expected.

Dashboarding network systems

Building dashboards around your network systems provides many benefits, including visualizations that represent the network, provide insight and correlation capabilities, and, if done properly, provide a quick report to the key stakeholders of the network service.

On the dashboard page, you should feature what is important to represent the health of your network service. When looking at a network-specific page, you may want to include things such as the following:

- Current alerts that are firing in the environment
- An overview of how much traffic is flowing through the network
- The current number of interfaces in an unexpected down state
- A general log dashboard that shows the logs that come in
- A feed of changes within a time frame and an upcoming period of time

You may have noticed operational state visualizations missing from a planning dashboard. Keep the visualizations on topic and geared toward the audience of the dashboard. There'll be more on architecting dashboards shortly.

Fault alerts

The point of alerting on faults is to allow the operations team member to be able to react to unexpected states in a network. When working on determining what alerts should be created, the first step is to pinpoint what will be actionable. Having a device offline from an observability point of view is one of the immediate must-have alerts, as it indicates a major fault in the network. Creating an alert that a campus access interface went offline is probably not an alert that is interesting from an operations perspective; however, that data is good to have in an observability platform to help with any investigation that may need to happen later, such as understanding when a laptop may have been connected via its LAN interface.

Remember that when treating your network as a service, a good starting point is to think about what a customer of a network service provider would want to see, which includes proactively responding to alerts, notification of a fault in the system, a restoration plan, and completing the restoration of service during an interruption.

Architecting dashboards

Architecting the appropriate visualizations that were introduced in *Chapter 8* requires a bit of understanding of who the intended audience is for dashboards. Once there is an understanding of who the audience is, then the design and architecture phases of building a dashboard start to come into play.

Network-related personas

Personas are defined as fictional characters that represent the typical user of a product. In this case, the product is the dashboard that is generated. So, it would be a good idea to start to define who the personas are and give some names to the personas, making things more engaging:

Figure 11.3 – Persona names lined up

When working with personas, you give each person a name to associate with them, and then the persona can be referenced by name rather than their role each time. The personas in an organization tend to last longer:

- **Maxine, the network manager**: She is responsible for overseeing the network infrastructure and ensuring that it meets the business needs and SLAs. She wants to have a holistic view of the network's current health and performance, enabling her team to identify and resolve issues quickly. She uses network observability data to monitor key metrics, logs, and troubleshoot problems.

- **Olivia, the organization leader**: She is the executive or senior manager who sets the vision and strategy for an IT organization. She wants to align the IT goals with the business objectives and demonstrate the value of IT investments. She also wants to ensure compliance and security across the network. She uses network observability data to track business outcomes and get a general high-level state of an environment at a glance.

- **Nicholas, the network operations center (NOC) team member**: He is the engineer or technician who implements, maintains, and supports network infrastructure. He wants to have a granular and real-time view of the network components, devices, and traffic. He also wants to automate and streamline the network operations and workflows. He uses network observability data to collect and analyze network data, configure and update network devices, and execute network commands and scripts.

- **Samantha, the security operations**: She is the analyst or specialist who protects the network from cyberattacks and breaches. She wants to have a comprehensive and proactive view of the network threats, vulnerabilities, and incidents. She also wants to respond and remediate the network security issues effectively. She uses network observability data to detect and investigate network anomalies, enforce and audit network policies, and integrate with security tools and platforms.

- **Nate, the network engineering team member**: He is the architect or designer who plans, designs, and tests network infrastructure. He wants to have a deep and historical view of the network behavior, patterns, and trends. He also wants to innovate and improve the network architecture and performance. He uses network observability data to model and simulate network scenarios, validate and optimize network designs, and implement and evaluate network changes.

- **Alex, the application owner**: He wants to ensure that an application meets the needs and expectations of the various personas who use it. He also wants to keep the application updated, secure, and compliant with the best practices and standards. He uses network observability data to manage the application settings, features, and integrations, as well as to monitor the application usage, feedback, and performance.

Now that we have briefly explored the personas, let's build a collection of visualizations into the primary dashboard, relevant for each of the personas as well as anyone else who may want to consume the dashboard types.

Dashboard types

Dashboards range in both detail and audience. When we design summary dashboards, we bring many metrics/logs into a summarized view. Are things in a good state? Then , we start to build out more detailed dashboards that provide the details necessary to support the summary metrics.

Let's take a look at what the organizational structure of the dashboards can look like. We'll start with the least detailed on top, a geography-based map that shows a regionalized summary. This allows a leader to get an idea of whether an environment is in a healthy state or whether there needs to be some additional follow-ups. There may also be some summary dashboards:

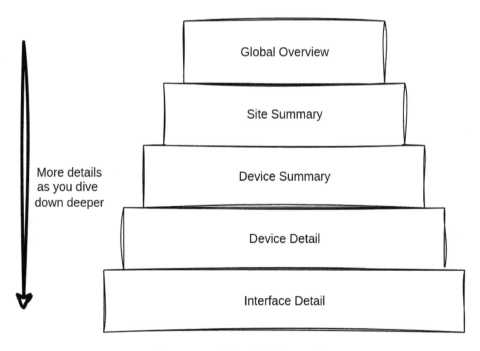

Figure 11.4 – A detail-level pyramid

The following table is a dive into what you will generally look to incorporate into the dashboards. From this data, you can take the dashboards that were developed as part of *Chapter 8* and customize them:

Dashboard type/description	Components	Audience
A global overview: Provides a quick-glance status of the observed metrics and logs.	• Often may include a **geography-based map (GeoMap)**, such as branch locations or POPs • Summary metrics in a **RYG (Red/Yellow/Green)** state	Olivia (organization leadership)
A site summary: Summarizes at a site level, giving the operational status of an individual site.	• A site-level metric summary • Circuit health summary and status • Site logs • A summary of a device status • Upcoming maintenances • Site-level assurance test status	Maxine (leadership)
A device summary: Start of troubleshooting dashboards, to find which components may have issues at a site.	• A device summarized by its RYG status • **IGP (Interior Gateway Protocol)/EGP (Exterior Gateway Protocol**, such as the **Border Gateway Protocol (BGP))** status • A device log summary visualization	Nicholas (NOC) and Samantha (security)

Dashboard type/description	Components	Audience
Device details: Continued troubleshooting, diving into the details of each device. Provides device-level details, while allowing you to dive deeper into component-level details at a lower level.	• Device metric details • CPU • Memory • Uptime • Fan status • Temperatures • Connections/second • Interface status pages • IGP and EGP (BGP) statuses • Neighbor state • Prefixes sent • Prefixes received • Throughput summary • Device logs	Nicholas (NOC) and Samantha (security)
Interface details: This is at the interface-level time series, giving the exact detail of each interface component of a device.	• Interface state over time • Light levels • Error rates • Packet loss • Latency • Throughput	Nicholas (NOC)

Table 11.4 – Dashboard summaries

When you look at the dashboards and the level of detail with an associated dashboard type, you can see the metrics that each individual persona would be interested in. This is not to say that an NOC team member wouldn't be interested in the highest summary dashboard; it is just not where we expect them to spend their time and where their curated set of visualizations would live.

There is still one more persona that has not yet been given a dashboard – Alex, the application owner. Alex is going to be interested in a summarized version of the network as well as other areas. He is interested in getting server logs and metrics summarized, as well as the application logs and metrics:

Dashboard type/description	Components	Audience
An application summary: Provides a quick-glance status of the observed metrics and logs for the application. Primarily, this is something to be glanced at, not focused on for a long time.	RYG summary statistics for the application components: • Application response RYG status based on the metrics/logs • Response time • Error codes • Error logs • Compute resource status • Kubernetes cluster status • Server status • Database status • Network status	Olivia (organization leadership), Maxine (leadership), and Alex (application owner)
Application details: A detailed view of an application over time, using the time series to see trends and events.	• A detailed view of the components over time • The response time of an application over time • The 95th percentile of an application • Error rates • Compute	Alex (application owner)

Table 11.5 – An application dashboard summary

With this in mind, let's go into more detail and look at some dashboard wireframes.

The global overview dashboard

The target audience of the global overview dashboard is anyone who wants to get a quick glance to determine the general health of an environment. Typically, this will be the leadership of an organization.

Here, we will look at displaying organizational SLIs, and perhaps SLOs and SLAs, on the **Global Overview** page:

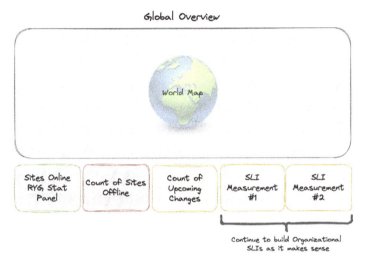

Figure 11.5 – A Global Overview wireframe

The world map dashboard is a great example of a dashboard that shows a geographical region's current status. In this example, there are errors in the New Jersey area. The manager or leader personas will have an immediate understanding of where there may be an impact on the provided services:

Figure 11.6 – A ping dashboard from a world view map

Be careful not to go too deep when building your site dashboards. When building out the dashboards, categorize the dashboards into a summary dashboard and detail dashboard. A summary dashboard is a visualization that provides high-level details in the form of a generalized status quickly, something that may go to the site owner as a quick status of a network and IT systems. Detailed dashboards help provide deeper details and are meant for troubleshooting and understanding all the finer details of a metric.

You can see more about the Grafana Worldmap Panel plugin from earlier on the plugin page at `https://grafana.com/grafana/plugins/grafana-worldmap-panel/`. There is strong development activity around the Worldmap Panel plugin, and you should expect significant changes to and development of the plugin.

A site summary dashboard

The site summary dashboard gives you a glance into what is happening on a site. The NOC and NOC leadership audience would be starting out at this page when looking at the dashboard. It brings together a list of relevant logs for the site, along with a status of key components such as online circuits and devices:

Figure 11.7 – A site summary wireframe

The site summary page shown here brings many pieces together on a page to give an overview of a site. Its intention is to help you understand even further where to look next. It gives information about the site, such as the address, contact information, and other site design information.

The world map shown here is more of a regionalized view, helping you to understand whether a particular site is the only one impacted in a region, or whether there may be other sites nearby that can help diagnose if issues are on a larger scale than a single site.

The alert count and list of alerts help give you an understanding of what active faults are currently present or have occurred recently.

The summary of circuits, including a short status description (up/down), helps to give you an idea of what may be happening from a WAN perspective on a site. The count of devices online and offline does a similar thing but from a device perspective, rather than circuits. The latency table in the middle helps you to understand, from a metric-gathering point of view, what the response time looks like.

The logs for the site help you to understand what events occur from a logging perspective, and the environmental metrics help you to understand the current physical state of an environment. These environmental sensors may include temperature, humidity, and a door/window/lock status, helping you to understand an environment when not there. This may also be a place to put weather forecast information, which is generally available.

The device summary dashboard

The device summary dashboard provides information about a device on a quick-glance page for the operations and engineering teams. The summary is meant to provide confidence that the device is performing as it should. This is where information about routing protocols and individual component statuses are summarized. The uptime measurement is the one measurement that is specific to an individual metric on the device. This is to help determine whether the device has recently been rebooted, either intentionally or unintentionally. In both cases, the uptime is an important summary metric to give an idea of the state of the device:

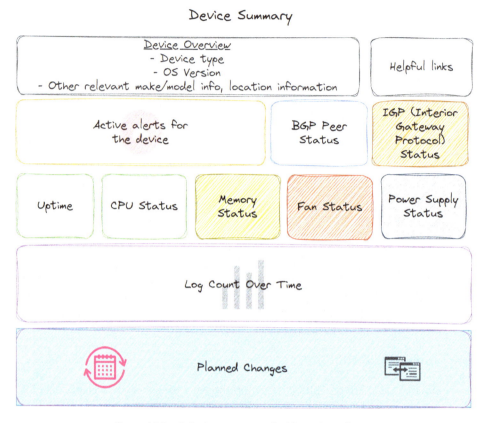

Figure 11.8 – A device summary dashboard wireframe

The **Device Summary** page offers a quick summary of the various areas of a device to help pinpoint where to go further. The first row shows information about the device. This allows for possible identification of issues by just knowing what platform may be affected. This is where useful links, such as a link to the source of truth or other systems, are placed to allow for quick access.

Placing active alerts for a device on the page will help to determine whether there is an ongoing issue with the device that the system has detected. The intention here is to help show what system faults are present, front and center, in a summary view, without going into too much detail.

The BGP peer status and IGP status tables look at a summary of the metrics, bringing an RYG status to the dashboard. Are all the expected neighbors up with the correct number of routes? That should be enough to make the dashboard green. It is up to you to define what is healthy, what a warning level is, and what an error level is.

The middle row, which includes **Uptime**, **CPU Status**, **Memory Status**, **Fan Status**, and **Power Supply Status**, offers a bit more detail than a typical summary dashboard. These metrics are best shown in their current status, with the time series of these being left for the device detail page.

The logs over time help to provide an indicator of whether a particular event happened at a specific time. This is a bar graph over time that shows the counts of the logs. Details of what the logs are can be found on the device detail page.

The last row taps into the ITSM system and retrieves the planned daily changes. Since changes are a major cause of events on a network, having a dashboard visualization that provides information on the daily changes makes sense to help understand what may cause an event.

The device detail dashboard

After having a summary of device health, the deeper troubleshooting dashboard for device details reveals all of the metrics, logs, and observability data about a system. This dashboard is likely going to be one of the largest ones. It targets the operations teams to help them identify what causes issues. This dashboard is deeply rooted in displaying time series information – that is, you see data transitioning a state over time. The summary dashboards are typically a point-in-time snapshot, rather than a historical-looking glass to see what happened previously.

Figure 11.9 – A device detail dashboard wireframe

The **Device Detail** dashboard has a lot more details and a lot more time series. This is the page that will assist with deeper device-level troubleshooting. The top provides similar links and information about a device.

Uptime and alerts provide valuable information about a device's status. If the device just rebooted, it indicates that something occurred, whether due to a power interruption or another reason. The alerts provide details about the faults detected. In this example, the last part of the second row offers an aggregated view of the device's traffic. This time series data helps determine whether the network device recently experienced a surge in traffic.

The **BGP** sections in this dashboard offer a deeper analysis, presenting a table with comprehensive information about BGP neighbors, including their metadata and status. Key metrics, such as the number of prefixes advertised and received, are detailed, with accompanying time series dashboards for both inbound and outbound metrics. This provides valuable insights into any changes that may have occurred during routing.

The **Logs for Device** section on this dashboard is crucial for monitoring and diagnosing the health of your network device. This section provides a detailed record of all events and activities on the device, enabling you to trace issues back to their source. By capturing and displaying these logs within the observability platform, you gain real-time insights into the operational status of the device, helping you to recognize and address potential problems before they escalate.

At the bottom of the dashboard are the time series for the CPU and memory details of the device. It may make sense to make this a higher priority in a system if there tend to be issues in an environment with CPU or memory utilization.

The interface detail dashboard

The interface detail dashboard dives deeper into the individual interfaces of a device. As with the device detail dashboard, it is deeply rooted in time series data, giving a historical viewpoint of observability data. The difference here is that the dashboard leverages rows within Grafana to allow you to select individual interfaces, which helps reduce the amount of data that needs to be displayed, especially if the goal is to look at only a single or handful of interfaces:

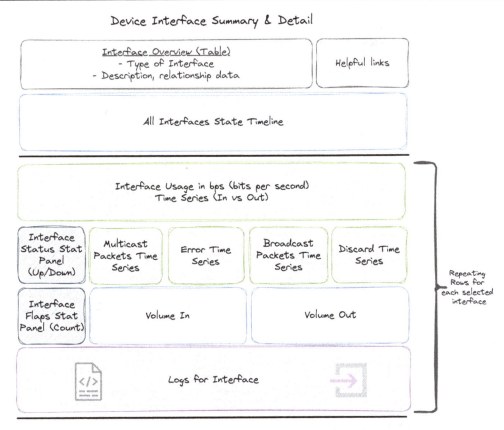

Figure 11.10 – An interface detail dashboard wireframe

The interface detail dashboard dives deeper into the interfaces of a device, making it similar to the device detail dashboard. It treats each interface as its own entity. The top row is familiar throughout the environment in having links to helpful systems.

The *interface state timeline* offers a time series view of all interfaces, providing a visual representation of when an interface changes state. This allows for quick identification of patterns or anomalies, helping you to better understand the operational history and stability of each interface over time.

The unique thing about these dashboards is the leveraging of rows. The rows are for each interface, which is selected as a variable on the dashboard. Each row will hold several time series data visualizations that show various components of data, including the bandwidth time series, interface errors, broadcasts, and discards. Each of these graphs is quite small, but you can use the view method to get a larger view of them.

The logs are then parsed out for each interface to provide the events that occur on the interface itself, giving you a detailed event view of what the logging system sees.

The application summary dashboard

The application summary dashboard is similar to other summary dashboards in that there are point-in-time measures that provide the current status. There may be some minimal time series visualizations to appear on the summary dashboard, but typically, this dashboard gathers information from the components that make up an application:

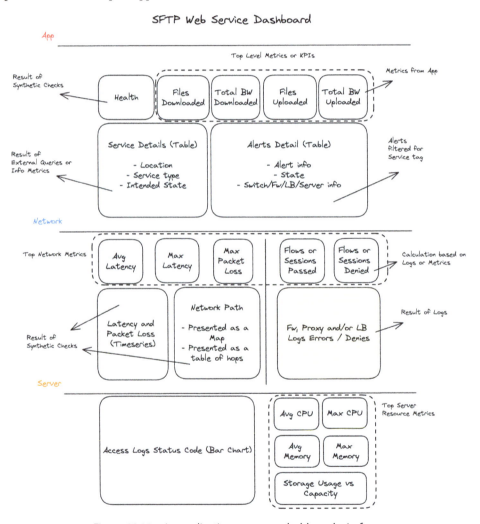

Figure 11.11 – An application summary dashboard wireframe

The preceding example web service dashboard dives into a similar methodology of building a dashboard, this time showing summary data from various domains across an organization. This brings application information, such as design details and application alerts, to the top of the dashboard. Then, summary views of the network and server/compute areas are brought to the forefront, aiding your understanding of where issues may occur within the entire environment.

The web application detail dashboard

The last wireframe example in this chapter is the web application detail dashboard. Similar to the site summary dashboards, this dives into specific details to help the operations team and application teams troubleshoot deeper application issues. This may include database connections over time for an application, the requests served, and response time latency. These measures help to understand where there may be underlying issues relating to an application. The network detail dashboards already provide the information about the network and may be linked to each other:

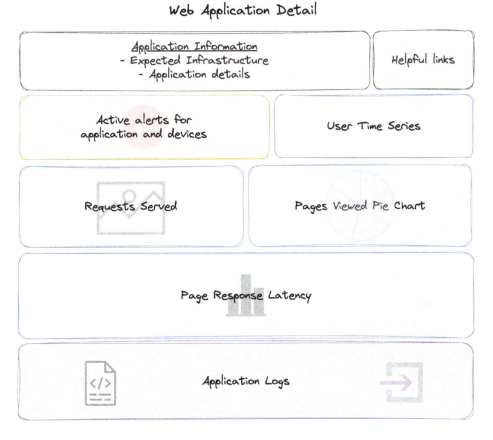

Figure 11.12 – A web application detail dashboard wireframe

The **Web Application Detail** dashboard is similar to other detail dashboards but is specific to web applications. The top section brings application information onto the page and helpful links. This is followed by an alert visualization. In the second row, there is also a time series to show active users in the application over time.

With applications such as Python Django, you can get visibility of the number of requests and the particular pages that are being served. The latency of each of these pages is also given and shown in percentiles. Refer back to the *Percentiles* subsection in this chapter for more on percentiles.

Other detailed dashboards to consider

There are many other detailed dashboards that you can consider to create your own visualization page. Use the general guidance and frameworks covered in this chapter to create a summary view that will then link to a detailed view, allowing you to dive even deeper. Some component considerations include the following:

- BGP
- **OSPF (Open Shortest Path First)**
- **IS-IS (Intermediate System to Intermediate System)**
- **PDU (Power Distribution Unit)** lines and phases
- TCP/IP-level metrics
- Firewall connections
- **VPN (Virtual Private Network)** connections
- **ICMP (Internet Control Message Protocol)** response

As you can see from just this list, you can continue to create many different dashboards that are all relevant to your organization's needs. As you follow the pattern of diving from the highest-level detail (a 50,000-foot view), moving closer and closer to the lowest-level detail (a 10,000-foot view), you will find different visualizations to present data from the proper viewpoint.

In summary, when creating dashboards, be cognizant of what the dashboard is intended to deliver. Ensure that summary dashboards just summarize the viewpoints. Summary dashboards just offer an at-a-glance view to let you know whether everything is in a good state. Then, as you drill deeper into layers, the dashboards and visualizations become more detailed, providing troubleshooting information, not just at-a-glance reporting.

Summary

By gathering your data, you are able to act on and visualize many aspects of your metrics. You have seen just a few examples to *get started*. Dashboard visualizations and the ability to correlate multiple metrics from multiple devices will empower you, allowing you to create great dashboards to represent your network and rich alerting notifications that are actionable by the operations organizations. The collection and enrichments used to color data with metadata allow you to get just as creative as a human looking at the data would be able to do so.

In the next chapter, we will take a look at how you can empower automation from observability data and alerts to accelerate the resolution of network events.

12
Automation Powered by Observability Data – Streamlining Network Operations

Network automation (understood as *automating network operations*) is a powerful tool, but its full potential is unlocked when combined with observability data (the topic of this book). This integration not only improves the implementation of automation in a trusty and reliable manner but also sharpens the decision-making process for subsequent actions.

Consider the evolution of modern vehicles as an analogy. One notable advancement in automotive technology over the past 50 years is the introduction of the **anti-lock braking system (ABS)**. ABS enhances safety by preventing the wheels of a vehicle from locking during emergency braking, thus maintaining vehicular control and reducing stopping distances. This system can be thought of as an early form of automation where the driver initiates braking, and the ABS automatically modulates the brake force to prevent wheel lockup.

Taking this a step further, modern vehicles incorporate advanced sensors that continuously scan the environment. These sensors measure the distance to objects ahead, among other metrics, providing real-time data that is critical for safety. In certain situations, if the vehicle detects an imminent collision, it can autonomously initiate braking faster than a human might react thanks to these observability data points.

Similarly, in network and infrastructure engineering, automation can execute routine tasks efficiently. However, when combined with observability data – such as real-time network performance metrics – it allows for smarter, proactive decisions. This could include dynamically adjusting network bandwidth based on traffic patterns or pre-emptively addressing system vulnerabilities before they impact services.

Just as sensors in vehicles anticipate and respond to road conditions, observability tools in network management detect and address network issues, often before they become apparent to users.

In this chapter, we'll explore how observability data enhances automation workflows for network operations. By understanding how to effectively merge these elements, you will gain knowledge so that you can start engaging with event-driven and closed-loop automation, an approach that helps us automate the initial response to incidents in the infrastructure.

We'll take a deeper look at the following topics in this chapter:

- Interacting with observability data programmatically
- Advanced automation techniques with event-driven automation.

The first step of being able to leverage observability data is learning how to consume it.

Interacting with observability data programmatically

Observability isn't just for creating detailed dashboards or sending alerts when something goes wrong, topics we explored in previous chapters. As automation and **artificial intelligence** (**AI**) become more integrated into network management, it's crucial to understand how to access and use data programmatically – that is, through automated methods, not just manually.

Here are a few key use cases where accessing data programmatically is essential:

- **Capacity planning and anomaly detection**: Using data to predict when more resources are needed or to spot unusual activity that could indicate problems. *Chapter 13* goes deeper into this topic.
- **Automated network assessments**: Regularly checking the health and performance of the network without human intervention.
- **Automated alerts and incident responses**: Setting up systems to automatically react to potential issues quickly, often before anyone even notices there's a problem.

These examples, and others, require interaction, via APIs, to retrieve the data that the observability stack has gathered and processed.

Next, we'll learn how to use this data in real-life scenarios. We'll show you how to retrieve and display this data through some scripts and a command-line app by using the lab setup for this chapter.

Setting up the lab environment

This lab will allow you to see how to work with data from our observability tools firsthand and open the door to many different types of similar integrations.

Figure 12.1 defines a diagram outlining all the components we'll use in our lab:

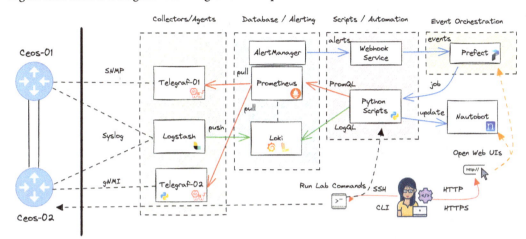

Figure 12.1 – Chapter 12 lab topology (Created using the icons from https://icons8.com)

In this chapter, we'll pay special attention to two key areas in this diagram – the *Scripts/Automation* and *Event Orchestration* sections. These areas are important because they involve writing programs (scripts) that help automate tasks and manage how data-driven events are handled.

While most parts of our lab environment are already set up and ready to go (mainly from the work performed in the previous chapters), we'll focus on writing the Python scripts. These scripts are the tools that will automate our network operation tasks.

> **Note**
> We're using Python for our examples because it's simple and widely used in the network automation community. However, the techniques we'll cover can be applied using any programming language you might be more comfortable with.

First, we'll look at some basic Python programs. These programs will introduce you to the basics of data structure from the network device data collected in the observability platform. Then, we'll advance to a more complex **command-line interface (CLI)** app. This type of app allows for more sophisticated interactions and will focus on representing said data in a user-friendly manner.

Later, in the next section, we'll wrap up with an event-driven automated workflow. This workflow will be triggered by an alert from the lab environment, showing you how to set up systems that react automatically to specific events or conditions.

Before diving into the lab activities, ensure your lab machine is set up according to the steps outlined in the *Appendix A*. Make sure you're in the ~/<path-to-git-project>/network-observability-lab/ directory within this project's repository to execute the necessary commands.

> **Note**
>
> This lab scenario includes a completed version that you can refer to. It is located in the git repository you cloned, under network-observability-lab/chapters/<chapter-number>-completed. This folder contains all the configurations that will be used in the examples throughout this chapter, providing a comprehensive guide to ensure you have everything set up correctly.

Follow these steps to stand up the lab environment for this chapter:

1. On your machine hosting the lab environment, run the netobs lab prepare --scenario ch12 command to set everything up.

2. Ensure that Nautobot is up and running by visiting http://<machine-ip-address>:8080. Log in with the default credentials: a username of admin and a password of nautobot123. Nautobot enriches our observability data and supports the event-orchestration examples that will be covered later in this chapter. The Nautobot service may take a few minutes to start up completely. You can monitor its progress by checking the logs. To do this, use the netobs docker logs nautobot -follow command. If you only want to see the most recent entries, add --tail=20 to show just the last 20 lines of the log.

3. Next, to make sure Nautobot reflects the current state of your network devices in the lab, run the netobs utils load-nautobot command. This will import the necessary device data into Nautobot, syncing it with your lab setup.

4. Finally, connect to a device named ceos-02 and activate the Ethernet2 interface, as shown in the following example. By default, the lab environment comes with all the interfaces in a 'UP' operational state, except ceos-02 Ethernet2. For this lab scenario, it's important that all interfaces in our network lab are turned on and functioning ('UP' status) so that our examples work correctly. Let's run the following commands:

```
> ssh netobs@ceos-02
(netobs@ceos-02) Password:
Last login: Sun Mar 24 16:12:22 2024 from 198.51.100.9
ceos-02>en
ceos-02#
ceos-02#conf t
ceos-02(config)#interface ethernet2
ceos-02(config-if-Et2)#no shutdown
ceos-02(config-if-Et2)#exit
ceos-02(config)#exit
ceos-02#exit
```

Now that the lab has been set up, let's begin by looking at how to get information from Prometheus.

Using the Prometheus API

For this first exercise, we'll retrieve some metric data using the Prometheus API and **Prometheus Query Language (PromQL)**. In *Chapter 7*, we dug into Prometheus's **time series database (TSDB)**, its API, and its query language work. Now, we'll use that knowledge to create scripts that will pull data from the metrics that have been collected from our network devices.

Retrieving BGP metrics

Let's begin by querying Prometheus to see the metrics we have collected and examine the results. We'll be using the `prometheus-api-client` library (`https://github.com/4n4nd/prometheus-api-client-python`) to help us set up a Python client to interact with the Prometheus API. This tool makes it easier to connect to Prometheus and perform queries rather than leveraging the API directly with the `request` library, for example.

First, create a file named `retrieve_bgp_metrics.py` in the `network-observability-lab/chapters/ch12/` directory. Now, take a look at the following code snippet, which will help us start collecting metrics from Prometheus:

```
from prometheus_api_client import PrometheusConnect
from rich import print as rprint

# Initialize Prometheus API client
prom = PrometheusConnect(
    url="http://localhost:9090", disable_ssl=True
)

# Get bgp_neighbor_state metric for ceos-01
data = prom.get_current_metric_value(
    metric_name="bgp_neighbor_state",
    label_config={"device": "ceos-01"},
)

rprint(data)
```

You might have noticed the rich Python library (`https://rich.readthedocs.io/en/latest/`) that was used in this example. We're using it to make the outputs from our examples easier to read and more visually appealing throughout this chapter.

From the `prometheus_api_client` library, we use `PrometheusConnect` to create a client instance to connect and query a Prometheus server. Notice that the URL endpoint is set to `http://localhost:9090` to let the script connect to our local Prometheus server in the lab environment. In a real production environment, this would be the Prometheus server address.

The `prom.get_current_metric_value(metric_name, label_config)` method offers a user-friendly way to query Prometheus for the latest value of a specified metric. This method simplifies the process by eliminating the need to write the raw PromQL query manually. The method takes two arguments:

- The `metric_name` argument defines the metric to retrieve `bgp_neighbor_state`
- The `label_config` argument narrows the query for the `ceos-01` device

The Python method is equivalent to the following PromQL expression:

```
bgp_neighbor_state{device="ceos-01"}
```

Now, to run the script, navigate to the appropriate folder by entering `cd network-observability-lab/chapters/ch12` in your terminal. Then, execute the script with the `python retrieve_bgp_metrics.py` command. You should see an output similar to the following:

```
> python retrieve_bgp_metrics.py
[
    {
        'metric': {
            '__name__': 'bgp_neighbor_state',
            'collection_type': 'exec',
            'device': 'ceos-01',
            'host': 'telegraf-01',
            'instance': 'telegraf-01:9004',
            'job': 'telegraf',
            'neighbor': '10.1.2.2',
            'neighbor_asn': '65222',
            'site': 'lab-site-01',
            'vrf': 'default'
        },
        'value': [1713739469.539, '1']
    },
    # Other metrics omitted...
]
```

The preceding output shows a list of objects, each item with `metric` and `value` details. Let's take a closer look:

- `metric`: This part holds the labels associated with the metric, which are essentially detailed tags that describe the characteristics of the data. These labels mainly come from Telegraf. If you're interested in how we gather and enhance data using Telegraf, please see *Chapters 5* and *6* for a deeper dive.

- value: This consists of two parts. The first part is a timestamp in Unix format, indicating when the data was captured. The second part is the actual numerical value of the metric. For example, in this case, the metric value is 1, representing the state of a **Border Gateway Protocol (BGP)** neighbor, which indicates that the status is Established. To understand how the BGP neighbor state mapping works, please refer to the Telegraf configuration performing the data transformation located at network-observability/chapters/ch12/telegraf/telegraf-01.conf.toml.

This simple script showed how to collect data from Prometheus programmatically and what kind of data structure to expect, which is important for when we build more advanced programs later. For now, let's continue exploring the prometheus-api-client library with another example so that we can become more familiar with its features and capabilities.

Retrieving inbound interface metrics

Now that we've viewed BGP neighbor state data, let's query our Prometheus database to gather data on the total inbound traffic across all interfaces on a specific device. This will help us understand how much data is received by the device, which is useful for troubleshooting traffic issues and planning capacity in general.

For this, we must set up a Python script called retrieve_interface_metrics.py. In this script, we'll develop a function to collect metrics data from Prometheus. It will be designed for easy reuse in future examples:

```
from prometheus_api_client import PrometheusConnect


def retrieve_data_prometheus(query: str) -> list[dict]:
    """Collect metrics from Prometheus."""
    # Initialize Prometheus API client
    prom = PrometheusConnect(
        url="http://localhost:9090", disable_ssl=True
    )
    # Query Prometheus and return results
    return prom.custom_query(query=query)
```

Notice that we're using the prom.custom_query method to directly input our desired PromQL expression rather than get_current_metric_value, which crafts a specific type of query. Now, we must write the main part of our script so that we can retrieve the inbound interface metrics of a device:

```
import time
from rich import print as rprint


def main():
    # Query to sum incoming bits/sec for all interfaces on ceos-01
```

```
    query = """
    sum by (name) (
        rate(interface_in_octets{device="ceos-01"}[1m]) * 8
    )
    """
    # Get the data
    data = retrieve_data_prometheus(query=query)
    rprint(data)

    time.sleep(30)

    # Get the data again
    data = retrieve_data_prometheus(query=query)
    rprint(data)

if __name__ == "__main__":
    main()
```

In this example, we're running a specific PromQL query twice with a 30-second delay between each execution. This will allow us to observe any changes in the traffic calculations.

Next, we must use the code snippet we just discussed and save it in `retrieve_interface_metrics.py`. The output should be similar to the following:

```
> python retrieve_interface_metrics.py
[
    {
        'metric': {'name': 'Management0'},
        'value': [1713739924.774, '4061.8666666666663']
    },
    # Omitted output for other interfaces
]
# Wait for 30 seconds
[
    {
        'metric': {'name': 'Management0'},
        'value': [1713739954.864, '2902.0444444444443']
    },
    # Omitted output for other interfaces
]
```

The results display the interface input metrics output at intervals of 30 seconds. You will notice that the metric portion only includes the name label, which corresponds to the name of the interface. If you're interested in incorporating additional labels as selectors to refine your queries further, feel free to experiment with this. For a more in-depth understanding of how PromQL, metrics, and selectors work, please refer back to *Chapter 7*.

Now that we've laid the foundation for querying Prometheus, let's move on to applying similar techniques to query Loki, our logs database in the lab environment.

Leveraging the Loki API

Now that we've managed to collect and visualize metrics data from the Prometheus database, let's shift to an example with Loki and fetch logs from its API.

Retrieving device logs from the Loki API

For this new example, we'll use the requests library in Python (https://requests.readthedocs.io/en/latest/) to construct and send queries to the Loki API.

First, create a new file named retrieve_device_logs.py in the same directory as the previous scripts. In this file, we'll develop a function to streamline our interaction with Loki's API. This function will handle the complexities of making network requests and parsing the responses received from Loki, making it simpler for us to focus on analyzing the log data:

```python
import requests

def retrieve_data_loki(query, start_time, end_time):
    """Retrieve data from Grafana Loki."""
    # Query Loki and return results
    response = requests.get(
        url="http://localhost:3001/loki/api/v1/query_range",
        params={
            "query": query,
            "start": int(start_time),
            "end": int(end_time),
            "limit": 1000,
        },
    )
    return response.json()["data"]["result"]
```

In this function, we set up parameters to send a query to Loki and wait for the results of the search. Specifically, we're providing a query and the start and end times for the period we're interested in. We use the requests.get() method to reach out to Loki's query_range endpoint. This allows us to retrieve log messages that span a specific time frame. Once the query is executed, the function captures and returns the results.

Next, we need to craft the specific query and the parameters that will fetch the information we need from Loki. For our example, we'll focus on collecting logs from the ceos-02 device, covering the last 30 minutes. The following code snippet shows how to set this up:

```python
from datetime import datetime, timedelta
from rich import print as rprint

def main():
    # Retrieve logs from ceos-02 over the last 30 minutes
    query = '{device="ceos-02"}'

    # Set time range
    now = datetime.now()
    start_time = datetime.timestamp(now - timedelta(minutes=30))
    end_time = datetime.timestamp(now)

    # Retrieve logs
    loki_results = retrieve_data_loki(query, start_time, end_time)

    rprint(loki_results)

if __name__ == "__main__":
    main()
```

At this point, we're ready to test this script. But first, we need to generate some fresh log messages. We'll do this by connecting to and entering the configuration mode on ceos-02. This generates an Entering configuration mode log message:

```
> ssh netobs@ceos-02
(netobs@ceos-02) Password: netobs123
Last login: Sun Mar 24 16:12:22 2024 from 198.51.100.9
ceos-02>en
ceos-02#
ceos-02#conf t
ceos-02(config)#
ceos-02(config)#exit
ceos-02#exit
```

These log messages should now be available in Loki. Let's run our script to collect them by executing python retrieve_device_logs.py:

```
> python retrieve_device_logs.py
[
    {
        'stream': {
```

```
        'device': 'ceos-02',
        'host': 'logstash',
        'level': 'notice',
        'program': 'ConfigAgent',
        'severity': '5',
        'vendor_facility': 'SYS',
        # Omitted labels...
    },
    'values': [
        [
            '1713776282000000000',
            'Configured from console by netobs on vty18'
        ]
    ]
}
]
```

Taking a look at the script's result, we can see that it is composed of a list of objects containing a `stream` object and `values`:

- `stream`: This part contains all the **labels**, which are tags or identifiers, associated with that particular log stream. These labels provide context about the log entries, such as the device they came from or the system that generated them.

- `values`: This is another list within each object that records the log entries. Each entry in this list has two parts: the **timestamp**, which tells us when the log was created, and the **message**, which is the actual content of the log.

If you look back at the output we saw from Prometheus in `retrieve_bgp_metrics.py`, you'll notice that the structure of the results is quite similar. Both systems organize their data in a way that pairs timestamps with values or messages.

Great! We've successfully developed scripts for querying Prometheus and Loki. You might be wondering, what is the significance of these examples? Well, in the next section, we'll explore building a simple CLI app that integrates these components. By bringing everything together, we'll demonstrate how to leverage observability data programmatically in real-world scenarios. So, let's dive into the CLI app and explore the practical applications of these techniques.

A CLI app example

So far, our examples have focused on retrieving and displaying raw data from metrics and logs. These exercises have highlighted the powerful querying and data aggregation capabilities of systems such as Prometheus and Loki. However, for network or infrastructure engineers, this raw output might seem overwhelming due to its verbosity. For this reason, we'll explore how to refine these outputs into more practical, concise forms that directly support engineering tasks.

In this section, we'll focus on constructing a compact CLI app. This tool will allow us to interact directly with systems such as Prometheus, Loki, and Nautobot. Our goal is to create commands that not only retrieve data but also display it so that it's easy to understand and efficient for the user (note we must keep the users top of mind because they will validate the effectiveness of the solutions). This will help make the vast amounts of data these systems generate more accessible and actionable.

First, let's consider what we need from this CLI app. Network engineers are highly skilled in using CLIs to manage and configure network devices. Therefore, creating a tool that can quickly inform them about the **state of the network** using just a few commands could significantly enhance their daily tasks. Let's start by focusing on building a tool that meets these needs.

At a high level, the CLI app will accomplish two key tasks:

- **Identify high bandwidth usage**: The app will show which links in the network have high bandwidth usage. To do this, we'll pull data from our Prometheus instance to see which device interfaces are experiencing heavy traffic.

- **Display the overall device status**: The app will provide a comprehensive status update for all the devices at a specific site. For this task, we need additional details, such as the role or manufacturer of each device, which involves integrating with a **Source of Truth** (**SoT**) app – in our case, Nautobot's API – to get the necessary context.

Given these needs, our CLI app must be capable of communicating with the observability stack and SoT system. This interaction will enable it to retrieve essential metrics, logs, and contextual information. *Figure 12.2* shows the interactions between the CLI app and these systems:

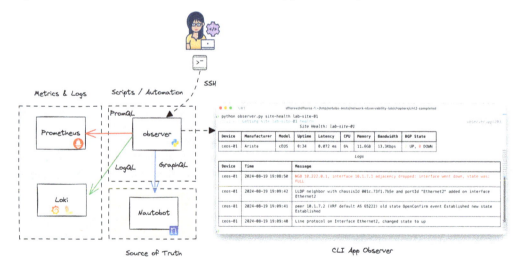

Figure 12.2 – Observer CLI app interactions (Created using the icons from https://icons8.com)

The CLI app will interact with the systems primarily using PromQL for Prometheus metrics, LogQL for Loki logs, and GraphQL for Nautobot's network data. We'll develop the CLI app in a file named observer.py.

CLI app setup

Now, let's begin by setting up the observer.py file. You can create this file in this chapter's lab folder, as you did for the previous examples. Once you have the file ready, follow along with this section to develop its commands.

> **Note**
>
> For reference, a completed version of observer.py is available in the completed lab section at https://github.com/network-observability/network-observability-lab/blob/main/chapters/ch12-completed/observer.py.

Let's begin by setting up the basic (boilerplate) code for our CLI app. This includes the necessary imports and objects we need. We'll be using the typer (https://typer.tiangolo.com/) library, which is a popular Python package for creating CLI apps. It's lightweight and offers a user-friendly framework for developing commands. Additionally, as with our previous examples, we'll use the rich library to enhance the display of our app's outputs, making them visually appealing in the terminal:

```python
import typer
from rich.console import Console
from rich.table import Table
from rich.theme import Theme
from dotenv import load_dotenv

# Load environment variables from our setup
load_dotenv(dotenv_path="./../../.env")

# CLI application definition
app = typer.Typer(name="observer")

# Console object with styled messages
console = Console(
    theme=Theme(
        {"info": "cyan", "warning": "bold magenta", "error": "bold
        red"}
    )
)
```

In the CLI app, we use `@app.command` as a decorator for our functions. This turns them into CLI commands that can be executed from the terminal by running `observer.py`. Additionally, we use the `console` object to display messages in the terminal. You can log these messages by calling `console.log(<message>, style=<info|warning|error>)`, which allows you to format the appearance of messages according to the type of information they convey.

> **Note**
>
> As we go through the following sections, we'll introduce parts of the code alongside the necessary Python package imports to make them functional. Some elements may not be repeated if they have already been defined earlier. We encourage you to build and run the code as you read through this section and then compare it with the completed scenario when necessary.

Next, we'll use the functions we've already developed to pull data from Loki and Prometheus:

1. **Start with Loki**: Open the `retrieve_device_logs.py` script, find the `retrieve_loki_data` function, and copy it. Then, paste this function into `observer.py`.

2. **Repeat the process for Prometheus**: Go to the `retrieve_interface_metrics.py` script, copy the `retrieve_prometheus_data` function and any needed imports, and add them to `observer.py`.

Finally, we'll create a function to fetch device information from Nautobot using GraphQL. We'll start by writing a small function to retrieve NAUTOBOT_SUPERUSER_API_TOKEN from an environment variable to ensure we can make authenticated calls to our Nautobot instance:

```python
import os

# Retrieve the Nautobot API token from environment
def get_nautobot_token():
    """Get the Nautobot API token."""
    token = os.getenv("NAUTOBOT_SUPERUSER_API_TOKEN")

    # Exit with error if token not found
    if not token:
        console.log(
            "NAUTOBOT_SUPERUSER_API_TOKEN env variable not set",
            style="error"
        )
        raise typer.Exit(1)

    return token
```

This code setup means that you need to set the NAUTOBOT_SUPERUSER_API_TOKEN environment variable. This variable is set under your environment variable file network-observability-lab/.env.

Now, we need to create a function that will help us access detailed data about network devices:

```python
import requests

# Retrieve device information from Nautobot
def retrieve_device_info(device):
    # GraphQL query to fetch device model and manufacturer
    gql = """
    query($device: [String]) {
        devices(name: $device) {
            device_type {
                model
                manufacturer {
                    name
                }
            }
        }
    }
    """
    # Retrieve the Nautobot API token
    token = get_nautobot_token()
    # Get the device information from Nautobot using GraphQL
    response = requests.post(
        url="http://localhost:8080/api/graphql/",
        headers={"Authorization": f"Token {token}"},
        json={"query": gql, "variables": {"device": device}},
    )
    response.raise_for_status()

    # Return the parsed device information
    return response.json()["data"]["devices"][0]
```

The preceding code snippet defines the function we'll use to retrieve device information from Nautobot using a GraphQL query. It fetches the Nautobot API token, sends a POST request to the Nautobot API to get the device's model and manufacturer details, and then parses and returns the device information from the response. The data that's retrieved from this call will help enrich the device information we aim to display on the CLI app later on.

Next, let's proceed by writing the code needed to create our first command to *identify high bandwidth usage* on a device's interfaces.

The high-bw-links command

Now, let's focus on developing the first command for our CLI app, which we'll define as `high-bw-links`. This command will identify network device interfaces that have high traffic usage.

To make this command more versatile, we'll include an input parameter for a traffic threshold. This parameter allows users to specify what level of traffic they consider *high*.

Let's design the necessary code and query that can assess traffic on a device interface, based on this threshold, and see what that looks like:

```
@app.command()
def high_bw_links(device: str, threshold: float = 1000.0):
    console.log(
        "Getting links with Bandwidth higher than threshold",
        style="info"
    )

    # Query for interfaces with traffic above threshold
    query = (
        f'rate(interface_in_octets{{device=~"{device}"}}[2m])*8 >
        {threshold}'
    )

    # Perform query
    metrics = retrieve_data_prometheus(query)
```

This snippet adds our first command, `high_bw_links`, to the `typer` app. The command accepts CLI parameters: the first is `device`, which specifies which device you want to query. The second parameter, `threshold`, is optional and is set to `1000.0` bits per second by default. This default value helps determine what constitutes high traffic, based on the metrics we retrieve from the query.

The `retrieve_data_prometheus(query)` function, already shown in a previous snippet, is identical to the one we developed in the `retrieve_interface_metrics.py` script. To reuse this functionality, simply copy the function into `observer.py`. However, the output from this function isn't very readable in its raw form. To improve clarity, we'll format the output into a table using the `Table` feature from the `rich` library. This will make the data much easier to interpret and work with. Here's how to set it up:

```
# Continuing from the high_bw_links command before

# Create table with title and grid format
table = Table(title="High Bandwidth Links", show_lines=True)
table.add_column("Device")
table.add_column("Interface")
table.add_column("Traffic IN", justify="right", style="green")
```

```
    # Build the table with results
    for metric in metrics:
        # Retrieve interface name from labels
        interface = metric["metric"]["name"]

        # Get traffic value, ignoring timestamp at index 0
        traffic = metric["value"][-1]

        # Add new row to the table
        table.add_row(device, interface, traffic)

    # Print the table
    console.print(table)
```

The preceding code snippet continues from where we left off with the high_bw_links command.

First, we create a Table feature with a title and a grid format. We add columns for Device, Interface, and Traffic IN, with the traffic data right-aligned and styled in green.

Next, we build the table by iterating through the metrics values that are returned from our Prometheus query. For each metric, we extract the **interface name** from the labels and the **traffic value**, ignoring the timestamp. Then, we add a new row to the table that contains the device's name, interface, and corresponding traffic data. Finally, we print the table to the console.

As the next step, let's create another command that integrates with Nautobot. This will further demonstrate how to use the CLI app and expand on the capabilities we've developed so far:

```
@app.command()
def device_info(device):
    console.log(
        f"Getting information for device {device}", style="info"
    )

    # Retrieve device information from Nautobot
    device_info = retrieve_device_info(device)

    # Display device information
    console.log(
        f"Device: {device}",
        f"Model: {device_info['device_type']['model']}",
        f"Manufacturer: "
        f"{device_info['device_type']['manufacturer']['name']}",
    )

if __name__ == "__main__":
    app()
```

This `device_info` command connects to Nautobot via GraphQL, retrieves device information, and displays it on the console.

With this, our CLI app now has a couple of commands. Let's test it by running `python observer.py device-info ceos-01` to get information on the `ceos-01` device:

```
·) python observer.py device-info ceos-01
[16:21:08] Getting information for device ceos-01
            Device: ceos-01 Model: cEOS Manufacturer: Arista
```

Figure 12.3 – The device-info command

Here, we can information about the device's manufacturer and model stored in Nautobot, Next, let's identify interfaces with traffic exceeding 1 Kbps:

```
) python observer.py high-bw-links ceos-01 --threshold 1000
[16:22:29] Getting links with Bandwidth higher than threshold
            High Bandwidth Links
```

Device	Interface	Traffic IN
ceos-01	Management0	2827.733333333333

Figure 12.4 – The high-bw-links command

Experiment with the `high-bw-links` command to observe how the app responds under different settings. For example, try lowering the threshold by running the command with `--threshold 100`. This will allow the command to capture more interfaces with lower traffic levels. You can also manually increase traffic to the device by connecting to it and using a `ping` command to send ICMP probes to `ceos-02`. Here's an example of how to do that:

```
> ssh netobs@ceos-01
(netobs@ceos-02) Password:
Last login: Sun Mar 24 17:12:22 2024 from 198.51.100.10
ceos-01>en
ceos-01#ping 10.1.7.2 repeat 500 interval 0.7 df-bit size 1400
PING 10.1.7.2 (10.1.7.2) 1372(1400) bytes of data.
1380 bytes from 10.1.7.2: icmp_seq=1 ttl=64 time=0.050 ms
1380 bytes from 10.1.7.2: icmp_seq=2 ttl=64 time=0.066 ms
1380 bytes from 10.1.7.2: icmp_seq=3 ttl=64 time=0.061 ms
...
```

As an extra exercise, consider formatting the `Traffic IN` variable to make it more readable in the CLI output. For example, change the display from `3390.70` to `3.4Kbps` for clarity. If you need help with this, take a look at the completed example for guidance.

The site-health command

The following command aims to provide a **site health** snapshot. This command will gather comprehensive data (of different types) from various devices to give an overall picture of the site's health. For instance, we collect information on each device's uptime, latency, model, manufacturer, the status of BGP sessions, total bandwidth usage, and even log data from the last 15 minutes.

Let's begin by setting up the entry point for the `site-health` command in our CLI app. This initial step involves creating the boilerplate code for the command. Once we have this, we can pass the collected data to the logic needed for further queries and analysis:

```python
import datetime

@app.command()
def site_health(site: str):
    console.log(f"Getting site {site} health", style="info")

    # Placeholder for the devices information
    devices = {}
```

The `devices` dictionary variable serves as a container for storing all the data we gather from our queries in Prometheus, Loki, and Nautobot. This data is organized by device, with each device name acting as a key.

Next, we'll create a query to retrieve uptime information for the devices. We'll do this by querying the `device_uptime` metric for `site` and then processing the results we get:

```python
# Continuing from the site_health command before

# Collect device uptime and set placeholder for BGP info
query = f"device_uptime{{site=~'{site}'}}"
metrics = retrieve_data_prometheus(query)

for metric in metrics:
    # Convert time ticks to human-readable format
    time_ticks = int(metric["value"][-1]) / 100
    uptime = str(datetime.timedelta(seconds=time_ticks))

    # Store device uptime, removing milliseconds
    device_name = metric["metric"]["device"]
    devices[device_name] = {
        "uptime": ":".join(str(uptime).split(":")[:2])
    }
```

The metrics we query return uptime data in time ticks. To convert these into a more understandable unit, we must divide the numbers by 100 to convert them into seconds. After that, we must use the `datetime.timedelta` method to transform these seconds into a time object. This allows us to format the uptime into a human-friendly format that's easier to read and understand.

Then, an object is created where each device name serves as a key, and its uptime is stored as one of its attributes. This object is then saved in the `devices` container variable.

Since we use `datetime.timedelta` to format the time ticks into a readable duration, remember to include this in your script by adding `import datetime` at the start of the `observer.py` file.

Next, we must gather and analyze the latency data for each device, as recorded by its monitoring agent, Telegraf. We can do this by using the `ping_average_response_ms` metric, which measures the average response time in milliseconds:

```
# Continuing from the site_health command before

# Ping response (latency) for each device
query = f"ping_average_response_ms{{site=~'{site}'}}"
metrics = retrieve_data_prometheus(query)

for metric in metrics:
    # Store device latency
    device_name = metric["metric"]["device"]
    devices[device_name]["latency"] = f"{metric['value'][-1]} ms"
```

The latency data is saved as an attribute for each corresponding device in the `devices` container variable.

Next, we must calculate the total bandwidth usage for each device by summing up all inbound and outbound traffic across their interfaces. The following code snippet demonstrates how to use a query to perform this aggregation:

```
# Continuing from the site_health command before

# Overall BW usage
query = f"""
    sum by (device) (
        rate(interface_in_octets{{site=~'{site}'}}[2m])*8 +
        rate(interface_out_octets{{site=~'{site}'}}[2m])*8
    )
"""
metrics = retrieve_data_prometheus(query)

for metric in metrics:
    # Store device bandwidth usage
```

```
device_name = metric["metric"]["device"]
bw_value = float(metric['value'][-1])
devices[device_name]["bandwidth"] = f"{bw_value:.2f} bps"
```

Just like in previous examples, we store the aggregated traffic usage data for each device in an attribute (`bandwidth`).

Next, we must collect data on the state of BGP sessions and count how many are UP (active) versus DOWN (inactive). To do this, we must query each device for the status of all its BGP neighbor sessions and count them. Here's a snippet showing how to set up this query and process it:

```
# Continuing from the site_health command before

# BGP state for each device in a site
query = f"bgp_neighbor_state{{site=~'{site}'}}"
metrics = retrieve_data_prometheus(query)

for metric in metrics:
    # Create BGP state lists for each device
    device_name = metric["metric"]["device"]
    if "bgp" not in devices[device_name]:
        devices[device_name]["bgp"] = {"up": [], "down": []}

    # Store the device's BGP state
    state = BGP_STATES[int(metric["value"][-1])]
    if state == "Established":
        devices[device_name]["bgp"]["up"].append(state)
    else:
        devices[device_name]["bgp"]["down"].append(state)
```

The preceding code snippet will classify and store the BGP session states in a list under the `devices` container variable. Each session will be marked as up if it is actively established, or down if it is in any other state.

If you've been following along, you may have noticed that we haven't defined the `BGP_STATES` mapping variable yet. This variable converts numeric metric values into understandable BGP state names. You should add this variable to the top of the `observer.py` file as a global variable:

```
BGP_STATES = {
    1: "Established",
    2: "Idle",
    3: "Connect",
    4: "Active",
    5: "OpenSent",
    6: "OpenConfirmed",
}
```

Having completed our queries with Prometheus, let's shift our focus to Nautobot so that we can gather additional details, such as the device's model and manufacturer. The following code snippet demonstrates how to retrieve this information:

```
# Continuing from the site_health command before

# Device Manufacturer and Model from Nautobot
for device_name in devices.keys():
    # Get device manufacturer and model from Nautobot
    device_info = retrieve_device_info(device_name)

    # Store device manufacturer and model
    devices[device_name]["manufacturer"] = (
        device_info["device_type"]["manufacturer"]["name"]
    )
    devices[device_name]["model"] = device_info["device_type"]
    ["model"]
```

Recall that we introduced the `retrieve_device_info` function earlier in this section. It sends a GraphQL query to Nautobot, fetching the model and manufacturer's name for each device. These details are then stored under the `devices` container variable, associating each piece of information with the corresponding device.

Next, to visualize all the information we've gathered so far, we must display it in a table format. Let's check out the following code snippet to see how this is done:

```
# Continuing from the site_health command before
# Create initial table of the site health
table = Table(title=f"Site Health: {site}", show_lines=True)
table.add_column("Device")
table.add_column("Manufacturer")
table.add_column("Model")
table.add_column("Uptime")
table.add_column("Latency")
table.add_column("Bandwidth")
table.add_column("BGP State")
# Build the table with the results
for device, data in devices.items():

    # Count the number of BGP states and color them
    up_bgp = f"[green]{len(data['bgp']['up'])}[/]"
    down_bgp = f"[red]{len(data['bgp']['down'])}[/]"
    # Add the row to the table
    table.add_row(
        device,
        data["manufacturer"],
```

```
            data["model"],
            data["uptime"],
            data["latency"],
            data["bandwidth"],
            f"{up_bgp} UP, {down_bgp} DOWN",
        )

    # Print the table
    console.print(table)
```

Here, we're transferring the attributes stored in the `devices` container variable into table rows for each device. Additionally, we're enhancing the table by adding color highlights to differentiate BGP sessions that are up from those that are down.

Now, let's save the progress we've made. Try running the `python observer.py site-health lab-site-02` command to see the output. It should look something like the following. Before running the command, make sure you set the NAUTOBOT_TOKEN environment variable if you haven't already:

```
> python observer.py site-health lab-site-02
[16:24:30] Getting site lab-site-02 health
                          Site Health: lab-site-02
```

Device	Manufacturer	Model	Uptime	Latency	Bandwidth	BGP State
ceos-02	Arista	cEOS	3:58	0.104 ms	11329.14 bps	2 UP, 0 DOWN

Figure 12.5 – The site-health command

Finally, let's add a section to gather the logs from each device for the past 15 minutes. Take a close look at the following code snippet to see how we can accomplish this:

```
    # Continuing from the device site health table code from before

    # Now let's collect the logs for the site in the last 15 minutes
    log_results = []
    for device_name in devices.keys():
        # Create the Loki query filtering by device
        query = f'{{device=~"{device_name}"}}'

        # Set the start and end time for the query
        now = datetime.datetime.now()
        start_time = datetime.datetime.timestamp(
            now - datetime.timedelta(minutes=15)
        )
        end_time = datetime.datetime.timestamp(now)
```

```
# Retrieve the logs
loki_results = retrieve_data_loki(query, start_time, end_time)

# Add the first 4 logs to the results
log_results.extend(loki_results[:4])
```

First, we must initialize log storage by creating a `log_results` variable to hold the collected log entries. Then, we must perform a LogQL query to retrieve logs from each `device` for the past 15 minutes. To keep the output compact and readable, we'll only store the first four logs of the logs returned by Loki, preventing our terminal from being overwhelmed with too much information. In a production scenario, you could add a parameter to the CLI app to specify the number of log lines to show and their order, making the command more flexible.

Next, we must display the log contents in a human-friendly format. See the following code snippet:

```
# Continuing from the device's log queries

# Create the table that contains the logs
table = Table(title="Logs", show_lines=True)
table.add_column("Device")
table.add_column("Time")
table.add_column("Message")

for log in log_results:
    # Extract the log information
    labels = log["stream"]
    log_time = log["values"][0][0]
    log_message = log["values"][0][1].strip()

    # Extract the log time and make it human readable
    log_time_fmt = datetime.datetime.fromtimestamp(
        float(log_time) / 1000000000
    ).strftime("%Y-%m-%d %H:%M:%S")

    # Add the row to the table
    table.add_row(labels["device"], log_time_fmt, log_message)

# Print the table
console.print(table)
```

Lastly, we must parse the collected logs to extract and separate the timestamp (`log_time_fmt`) and message (`log_message`) from each log line returned. This involves formatting the timestamp so that it's in a readable format and isolating the log message for clarity. Each parsed log entry is then added to a `table` object, which is printed out at the end for a clean and organized display.

To test our setup, we need to generate some log messages. Connect to the `ceos-02` device, enter configuration mode, and then exit. This sequence will create the necessary log activity for our validation. We can use the following commands to generate log messages:

```
> ssh netobs@ceos-02
(netobs@ceos-02) Password: netobs123
Last login: Sun Mar 24 16:12:22 2024 from 198.51.100.9
ceos-02>en
ceos-02#
ceos-02#conf t
ceos-02(config)#
ceos-02(config)#exit
ceos-02#exit
```

Now, we can execute the `python observer.py site-health lab-site-02` command to test the functionality of the command we've just set up, targeting the `lab-site-02` site. You should see an output similar to the following:

```
·) python observer.py site-health lab-site-02
[16:25:36] Getting site lab-site-02 health
```

Site Health: lab-site-02

Device	Manufacturer	Model	Uptime	Latency	Bandwidth	BGP State
ceos-02	Arista	cEOS	3:59	0.077 ms	11262.48 bps	2 UP, 0 DOWN

Logs

Device	Time	Message
ceos-02	2024-08-13 12:32:21	NGB 10.111.0.1, interface 10.1.7.2 adjacency established
ceos-02	2024-08-13 12:27:14	NGB 10.111.0.1, interface 10.1.2.2 adjacency established
ceos-02	2024-08-13 12:27:29	System is initialized
ceos-02	2024-08-13 12:28:40	System restarted

Figure 12.6 – The site-health command with logs

Now, you have a CLI app equipped with two useful commands to monitor the network infrastructure. While the app could benefit from additional error handling and support for more devices (it's not a production-ready example), it serves as a good example of how automation can enhance daily operations for network and infrastructure engineers by providing programmatic access to observability data.

Building on this foundation, the automation techniques we've discussed are essential for developing a robust strategy that supports everyday tasks in IT infrastructure. The sky's the limit for what you can implement with operational data. For example, you can create a CLI app for a full network **change management** (**CM**) procedure. It can perform pre and post-change checks to ensure everything is running smoothly, offering a clear snapshot of the network's overall health.

With this context in mind, let's explore additional use cases for automation. We'll see how these tools can not only simplify daily operations but also significantly improve how we can manage network and systems infrastructure regarding events or failures.

Advanced automation techniques with event-driven automation

Network and infrastructure engineers understand the challenges of managing a robust and highly available IT infrastructure. Automation can significantly ease management and operational tasks. As demonstrated in the previous section, tools and CLI apps that provide instant snapshots of network health are great for managing changes. Additionally, IT automation architectures that trigger automated responses to specific network events can greatly enhance alerting systems or incident management.

In this section, we'll explore two advanced automation strategies: **event-driven automation** and **closed-loop automation**.

Event-driven automation

Consider automatic windshield wipers in modern vehicles. This system is equipped with rain sensors that detect moisture on the windshield, representing a specific event – the start of rain. Once this moisture is detected, the event triggers an immediate response from the vehicle's system: the activation of the windshield wipers. The wipers continue to operate so long as the sensors detect rain, their speed adjusting based on the intensity of the rainfall. This automated response ensures visibility is maintained without any input from the driver, highlighting how the system is directly reacting to changes in its environment.

In the realm of network and infrastructure engineering, event-driven automation plays a similar role. When a specific event or set of conditions is detected by monitoring tools – such as a security breach, a failure in hardware, or unexpected downtime – the system automatically triggers a predefined response. This might involve isolating affected network segments, deploying additional security measures, or initiating a backup process. Just like the automatic wipers that start when rain is detected, these automated responses in IT infrastructure are triggered by specific events to ensure immediate and appropriate action is taken.

So, an event-driven automation platform is a system that's designed to execute automated processes or workflows in response to specific events or conditions within an environment. The following figure demonstrates the concept of an event-driven automation system:

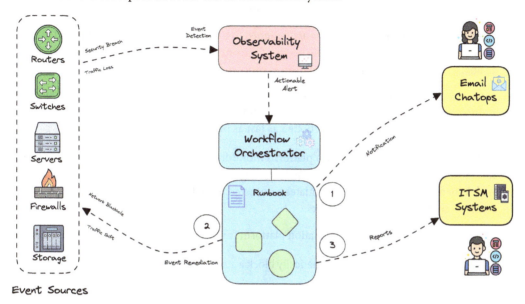

Figure 12.7 – Event-driven automation (Created using the icons from https://icons8.com)

Here are some characteristics and functions of an event-driven platform:

- **Event detection**: The platform keeps a constant watch on the environment for specific events or triggers. In the world of an observability platform, these are often *actionable alerts* and *incidents* that it detects. These events can include anything from a system status change, a breakdown, or unusual network traffic to more complex issues that could suggest security threats or operational problems.

- **Conditional logic**: When the platform detects an event, it checks it against set rules or logic to figure out if action is needed. This process can be straightforward, such as using simple *if this, then that* rules to decide what to do next. Or, it can be more complex, such as in AIOps, where the system gathers additional data from various sources and analyzes it together to decide the best action to take.

- **Automated responses**: If an event fits the rules set by the platform, it automatically starts one or more actions, often referred to as *workflows*. These actions can be as simple as sending out a notification, or as complex as performing a series of tasks across several systems. For instance, in network management, the actions could include rerouting traffic, adjusting how resources are distributed, or starting up security measures.

- **Integration**: Event-driven automation platforms can usually connect with many different tools and systems, such as databases, cloud services, networking hardware, and software management tools. This capability to integrate allows the platform to collect data from various sources and perform actions that impact different parts of the IT infrastructure.

- **Scalability and flexibility**: Platforms are designed to handle large volumes of events and can be scaled up or down depending on the needs of the organization. They are also flexible enough to adapt to different workflows and business processes.

- **Real-time processing**: Event-driven platforms usually handle data immediately or almost immediately. This quick processing ensures that responses to any alerts or incidents detected by the observability platform are actioned as fast as possible.

In summary, an event-driven automation platform monitors an environment for specific events or triggers, such as alerts or incidents, and evaluates them against predefined rules to determine if action is required. When conditions are met, it initiates automated workflows, which can range from simple notifications to complex multi-system tasks.

To understand how an event-driven automation platform functions, consider the following use cases:

- **IT security**: The platform can automatically block a suspected IP address if it detects a potential security threat

- **Network operations**: Automations adjust network configurations automatically in response to changing traffic patterns

- **Business workflows**: The platform triggers steps in a supply chain management system when a new order is placed

An example of such a system is Ansible **Event-Driven Automation** (**EDA**), which operates using the Ansible Automation Platform (you can find more on EDA here: `https://www.redhat.com/en/technologies/management/ansible/event-driven-ansible`). This system is built on three main components:

- **Source**: This is where events are introduced into the system. Event detection is typically handled by an observability platform, which then passes the actionable event to EDA.

- **Rules**: These are defined in Ansible rulebooks, which determine the conditions under which actions should be taken – essentially answering *why* and *when* an action should occur.

- **Actions**: Executed through Ansible playbooks, these define *how* the automation logic is carried out.

Later in this section, we'll demonstrate an automated workflow that's triggered by an alert or incident that's been detected by the observability system.

Closed-loop automation

Continuing with our vehicle analogies, automatic transmission in modern vehicles is a prime example of closed-loop automation. This system automatically adjusts the gear ratio as the vehicle moves, ensuring the engine runs efficiently at varying speeds. Sensors within the transmission system monitor the vehicle's speed, engine load, and throttle position. Based on this data, the transmission control unit makes real-time decisions to shift gears up or down to optimize performance and fuel efficiency.

The *loop* in closed-loop refers to the continuous feedback mechanism where the system constantly receives data, processes it, and adjusts its actions accordingly. This ensures the vehicle operates smoothly under different driving conditions without driver input regarding gear changes. The following figure shows the closed-loop automation cycle:

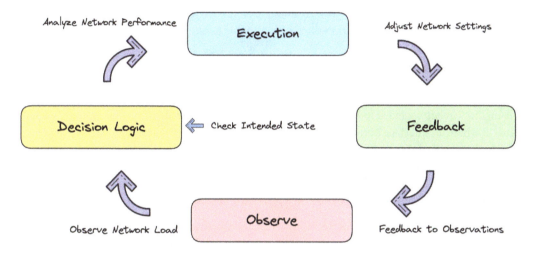

Figure 12.8 – Closed-loop automation (Created using the icons from https://icons8.com)

In the context of network management, closed-loop automation refers to a sophisticated approach where the entire process of monitoring, decision-making, and taking action is automated. This method creates a feedback loop that continuously improves and optimizes network operations without human intervention. Here's how it typically works:

- **Observe**: The system continuously collects and analyzes observability data from the network. This includes metrics such as traffic flow, resource utilization, and performance statistics.

- **Decision logic**: Leveraging reference data (that is, the intent) and based on predefined rules or using artificial intelligence and machine learning, the system evaluates the observed data to detect any anomalies, potential issues, or opportunities for optimization.

- **Execution**: Once an issue or a potential improvement is identified, the system automatically implements changes without waiting for human input. This could include adjusting bandwidth, rerouting traffic, or applying security patches.

- **Feedback**: After taking action, the system monitors the outcomes of its interventions to learn from its decisions. This feedback is then used to refine its future actions, making the automation smarter and more efficient over time.

The *closed loop* in this scenario implies that the system is self-sufficient and capable of operating continuously and autonomously, learning from each loop to enhance its effectiveness. Notice that this is not all or nothing. You can establish a closed-loop system that covers a set of use cases while leaving others outside of the automated process.

At this point, we want to show you a high-level example of how these approaches can be implemented by leveraging the observability platform we have built in this book.

Event-driven automation with Prefect

In this section, we'll apply the concepts we learned about previously by establishing an actionable alert and launching an automated response workflow. Using the lab environment for this book, consider a scenario where an interface between the two devices, `ceos-01` and `ceos-02`, is experiencing instability, known as **flapping**. Our observability system has to identify this as a flapping event. In response, we updated the SoT (Nautobot, in our lab environment) to indicate that the interface is now in alert mode. This ensures that any team member or user consulting the network intent is fully aware of the interface's current status before they proceed with any related work.

Thanks to the work we have done in previous chapters, we already have most of the necessary components in place and configured in our lab. The only component that's missing is an event-driven automation platform that allows us to define the code and steps required to collect information from the intended state of the network device and update the affected interface status. To implement this, we'll build a Python program by using a workflow orchestration system called **Prefect** (`https://www.prefect.io/`), which will help us manage the entire network automation workflow.

Prefect is a workflow orchestrator that is widely used with Python scripts and programs and is especially popular in the data science community for managing data lineage and general data processing tasks. Throughout this chapter, as you have seen, we have developed various Python scripts for data collection, normalization, and even for our CLI app. Continuing with this technology, Prefect integrates seamlessly, making it a good choice (among many available options) for orchestrating our network automation workflows.

The upcoming lab involves setting up a *free* account with **Prefect Cloud** (`https://www.prefect.io/cloud`). Using Prefect Cloud simplifies the process by providing a ready-to-use environment, including a configured Prefect server.

The following figure shows the various components involved in this section of the lab, building on the previous scenarios (see *Figure 12.1* for the entire topology diagram). It illustrates how data flows between these components, highlighting their interactions and how they work together:

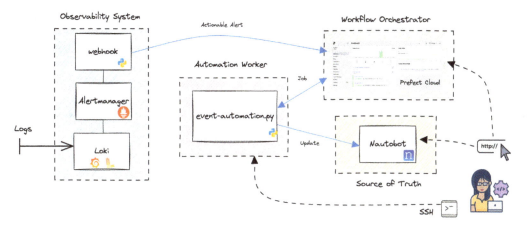

Figure 12.9 – Event-driven automation with Prefect (Created using the icons from https://icons8.com)

The lab's main functions are as follows:

- **Data collection**: Our observability system continuously collects metrics and logs from the network lab. The Loki and Alertmanager components are configured to detect when an interface flap occurs by using setups that we refined in previous chapters.

- **Alert handling**: When an alert is triggered, it is sent to a custom webhook service. This service is pre-configured to receive alerts and format the necessary data to initiate workflows in Prefect Cloud. We'll cover how to set the required environment variables for connecting to Prefect Cloud in the next section.

- **Workflow execution**: Once Prefect Cloud receives the formatted alert, it triggers an automation workflow. A Prefect worker within our lab environment then executes this workflow, which includes tasks such as interacting with and updating Nautobot. We'll also develop the Python code that the worker needs to carry out these actions.

Before we proceed, let's start by setting up a Prefect Cloud account.

Setting up Prefect Cloud

Let's begin by setting up Prefect Cloud so that we can prepare our lab environment:

1. Create a free account on Prefect Cloud at `https://app.prefect.cloud/`. Once your account has been set up, you might be asked if you wish to set up an example workflow tutorial. You can decline this since we'll be doing that shortly. At this point, you should see a dashboard similar to the following:

Figure 12.10 – Prefect dashboard

2. Next, you'll need an API key to interact with your Prefect Cloud instance. Go to your workspace settings and select **API Keys**. From there, create a new API key so that the workflows we develop can access Prefect.

3. Save the API key in your environment file (`network-observability-lab/.env`). Replace `PREFECT_KEY=<your_prefect_cloud_key>` with the actual value of your API key. This key will be used by the webhook service to send actionable alerts to Prefect Cloud.

4. Next, we need to make the Nautobot token accessible to the workflow worker so that it can connect to Nautobot for the final step. In the Prefect server, you can achieve this by creating a **secret block**. Navigate to the **Blocks** section and use the search function to find the **Secret** option:

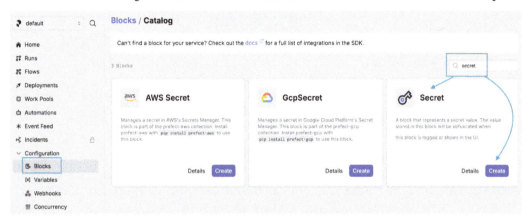

Figure 12.11 – Prefect secret blocks

5. Then, click on **Add Block**, and name it `nautobot-token`. Copy and paste the token value from your environment file (`network-observability-lab/.env`) into the designated field. For instance, the lab provides a default value of `NAUTOBOT_SUPERUSER_API_TOKEN=0123456789abcdef0123456789abcdef01234567` that you can use for this example:

Figure 12.12 – Prefect secrets – Nautobot token

6. Next, connect to the Prefect Cloud server. On your lab environment machine, open the Terminal and run the `prefect cloud login` command, following the prompts. This step ensures that the worker can connect to the Prefect Cloud server. The following is an example of the

command and its output, but for more detailed information on how to connect, please refer to the official documentation (`https://docs.prefect.io/`):

```
> prefect cloud login
? How would you like to authenticate? [Use arrows to move; enter
to select]
   Log in with a web browser
> Paste an API key
Paste your API key:
Authenticated with Prefect Cloud! Using workspace '<user>/
default'.
```

7. Finally, to obtain the PREFECT_URL value (which points to your Prefect workspace), run the `prefect config view` command on your lab machine. This command will display the configuration values that the Prefect CLI uses to connect to Prefect Cloud. Look for the PREFECT_API_URL value in the output, copy it, and then save it in your .env file as PREFECT_URL=<your_prefect_cloud_url>. The following is an example of what the command's output might look like:

```
> prefect config view
PREFECT_PROFILE='default'
PREFECT_API_KEY='********' (from profile)
PREFECT_API_URL='https://api.prefect.cloud/api/accounts/account-
123456/workspaces/workspace-id-123456' (from profile)
```

8. Make sure you save the highlighted PREFECT_API_URL value in your .env file. For example, you might use something like PREFECT_URL=https://api.prefect.cloud/api/accounts/account-123456/workspaces/workspace-id-123456. Remember to replace this example with the actual value you obtain.

Now that everything has been set up, you're ready to proceed with this part of the lab and start developing the workflow.

Developing an automation workflow

Now, let's start coding the steps that need to be executed after an alert is detected. Start by creating a file named event-automation.py in the network-observability-lab/chapters/ch12/ directory. This file will include the following initial boilerplate code. This is required to activate an automation workflow in Prefect:

```python
from prefect import flow

@flow(log_prints=True)
def alert_receiver():
    pass
```

```
if __name__ == "__main__":
    alert_receiver.serve(name="alert-receiver")
```

In the preceding script, the `if __name__ == "__main__":` section identifies the entry point for execution. In this case, it designates the script to `serve` the Prefect flow named `alert_receiver`.

> **Note**
>
> For reference, a completed version of `event-automation.py` is available in the completed lab section at `network-observability-lab/chapters/ch12-completed/event-automation.py`.

In Prefect, `flow` represents a workflow definition, essentially acting as a blueprint for orchestrating tasks. In our lab environment, we use a flow to handle alerts and execute subsequent actions. A flow is defined as a Python function that the Prefect server orchestrates and monitors.

So, in our lab environment, we create Python functions to handle alerts and perform subsequent actions, applying the `@flow` decorator to them to designate them as workflows. The `log_prints=True` argument in the flow decorator ensures that any print statements within the `alert_receiver` function are captured and sent to Prefect Cloud.

Let's begin coding our workflow. First, we need to create a function capable of processing the Alertmanager payload from our alerts. This function will analyze the payload and, based on the alert's name, route it to the appropriate function for further processing.

To proceed, add the following code snippet to your `event-automation.py` file:

```
@flow(log_prints=True)
def alert_receiver(alert_group):
    status = alert_group["status"]
    alertgroup_name = alert_group["groupLabels"]["alertname"]
    alerts = alert_group["alerts"]
    print(f"Received alert group: {alertgroup_name} - {status}")

    # Check alert name and forward to respective workflow
    if alertgroup_name == "PeerInterfaceFlapping":
        for alert in alerts:
            # Run the interface flapping processor
            result = interface_flapping_processor(
                device=alert["labels"]["device"],
                interface=alert["labels"]["interface"],
                status=alert_group["status"],
            )

            # Print the result to Prefect console
```

```
                    print(f"Interface Flapping Processor Result: {result}")

            print("Alertmanager Alert Group status processed, exiting")
```

The alert_receiver workflow starts by receiving alert_group data (a dictionary object) from the webhook service. Then, it extracts key attributes from this payload, such as status, alertgroup_name, and the individual alerts objects. If the alert matches our criteria, specifically PeerInterfaceFlapping, each alert within the group is forwarded to another function for further processing.

Next, we create the interface_flapping_processor function. It handles passing the necessary data to update the interface status of a device in Nautobot. Take a look at the following code snippet to see how we set this up:

```
@flow(log_prints=True)
def interface_flapping_processor(
    device: str, interface: str, status: str
) -> bool:
    # Retrieve Nautobot Interface ID
    intf_id = get_nautobot_intf_id(device=device, interface=interface)
    if intf_id is None:
        raise ValueError("Interface not found in Nautobot")

    # Print the interface ID
    print(f"Interface: {interface} == ID: {intf_id}")

    # Update the device interface status in Nautobot
    is_good = update_nautobot_intf_state(intf_id, status)
    return is_good
```

For the workflow, we need to identify the Nautobot interface ID corresponding to the device involved. Once we have the ID, we can update the interface's status to reflect the current alert status.

Let's develop the functionality to retrieve the interface ID from Nautobot, implemented as a task value (a step in a Prefect flow). Check out the example here to see how we can achieve this:

```
import requests
from prefect import task
from prefect.blocks.system import Secret

@task(retries=3, log_prints=True)
def get_nautobot_intf_id(device: str, interface: str) -> str | None:
    # GraphQL query to retrieve the device interfaces information
    gql = """
    query($device: [String]) {
```

```
        devices(name: $device) {
            interfaces {
                name
                id
            }
        }
    }
}
"""

# Retrieve the Nautobot API token using the Prefect Block Secret
nautobot_token = Secret.load("nautobot-token").get()

# Get the device information from Nautobot using GraphQL
response = requests.post(
    url="http://localhost:8080/api/graphql/",
    headers={"Authorization": f"Token {nautobot_token}"},
    json={"query": gql, "variables": {"device": device}},
)
response.raise_for_status()

# Parse the response and return the interface ID
result = response.json()["data"]["devices"][0]
for intf in result["interfaces"]:
    if intf["name"] == interface:
        return intf["id"]
```

Let's dive into this code and understand its key components:

- **Prefect task:** The @task(retries=3, log_prints=True) decorator marks this function as a Prefect task. Tasks are manageable units of work within Prefect that offer features such as orchestration and observability. They allow for retries on failures and can cache results. Here, the retries parameter has been set so that if an exception occurs, the task will attempt to reconnect to Nautobot up to three times. For more information about Prefect tasks, check out their official documentation (https://docs.prefect.io/latest/concepts/tasks/).

- **GraphQL query:** We construct a GraphQL query to request the interfaces and their IDs from Nautobot based on the device's name.

- **Using Prefect secrets:** Recall the Prefect secret block we set up previously. Here, we retrieve the Nautobot API token stored in it and assign it to the nautobot_token variable.

- **Query execution and parsing:** The GraphQL query is executed, and the results are parsed to search for and extract the interface ID. If no matching interface name is found, the function returns None.

Now, we just need to develop `update_nautobot_intf_state`, the final task in the `interface_flapping_processor` workflow. Take a look at the following code snippet to see how we can implement this:

```python
@task(retries=3, log_prints=True)
def update_nautobot_intf_state(intf_id: str, status: str) -> bool:
    # Retrieve the Nautobot API token from Prefect Block Secret
    nautobot_token = Secret.load("nautobot-token").get()

    # Map Alertmanager status to Nautobot status
    status = "lab-active" if status == "resolved" else "Alerted"

    # Update the interface status
    result = requests.patch(
        url=f"http://localhost:8080/api/dcim/interfaces/{intf_id}/",
        headers={"Authorization": f"Token {nautobot_token}"},
        json={"status": status},
    )
    result.raise_for_status()

    # Print the result to console
    print(f"Interface ID {intf_id} status updated to {status}")
    return True
```

Like the previous function, we start by retrieving the Nautobot token from the Prefect block secret. Then, we align the Alertmanager's alert status with the corresponding statuses in Nautobot and update the interface status accordingly in Nautobot.

Great! Now, gather all the code snippets and save them in the `event-automation.py` file. This completes the development of the automated workflow.

Now, let's test the entire event-driven automated workflow. First, set up the Prefect local worker by registering the job in Prefect Cloud so that it can monitor for incoming alerts. To do this, execute the `python event-automation.py` command, which executes `alert_receiver.serve(name="alert-receiver")`. The output should look like this:

```
〉 python event-automation.py
Your flow 'alert-receiver' is being served and polling for scheduled
runs!
To trigger a run for this flow, use the following command:
        $ prefect deployment run 'alert-receiver/alert-receiver'
You can also run your flow via the Prefect UI: https://app.prefect.
cloud/account/<account-id>/workspace/<workspace-id>/deployments/
deployment/<deployment-id>
```

In another Terminal, log in to the lab machine. Then, execute the `netobs utils device-interface-flap` command, which simulates interface flapping by turning the interface admin status up and down alternately. However, before doing this, make sure the `Ethernet2` interface in `ceos-02` is up. Here's an example of the command being used:

```
> netobs utils device-interface-flap --device ceos-01 --interface
Ethernet2 --count 10
[10:40:47] Flapping interface: Ethernet2 on device: ceos-01
[10:40:48] Bringing interface down...
[10:40:53] Bringing interface up...
[10:40:58] Bringing interface down...
[10:41:03] Bringing interface up...
[10:41:08] Bringing interface down...
```

This action should generate multiple syslog messages indicating UPDOWN events for interfaces in a short period. The `PeerInterfaceFlapping` alert rule in Loki (already predefined) should detect these events and send an alert to Alertmanager. Here's a short snippet for reference, though note you can find this rule in the Loki alert definitions under `network-observability-lab/chapters/ch12/loki/rules/alerting_rules.yml`:

```
groups:
  - name: Peer Interface Flapping
    rules:
      - alert: PeerInterfaceFlapping
        expr: sum by(device, interface) (count_over_time({vendor_
        facility_process="UPDOWN"}[2m])) > 3
        for: 1m
        labels:
          severity: critical
          source: loki
      # Rest of the output omitted...
```

> **Note**
>
> If you're interested in viewing these alerts, you can check them out on Prometheus at `http://<lab-machine>:9090/alerts` or on Grafana at `http://<lab-machine>:3000/alerting/list`. The alerts that are generated by both `ceos-01` and `ceos-02` are grouped into an Alertmanager alert group. For more detailed information about alerting and grouping, please refer to *Chapter 9*.

The alert group data is sent to the webhook service, which prepares the alerts before dispatching them to Prefect Cloud. Once received, Prefect Cloud initiates the `alert-receiver` workflow, passing the alert data to our local lab worker (the Python process we executed in the other Terminal). This triggers the start of the workflow, with Prefect Cloud monitoring all the events that have been registered.

To see this in action, log in to your Prefect Cloud account and navigate to the **Flow** section. Here, you can see the alert-receiver flow, as shown here:

Figure 12.13 – Flows

To check the details of the workflow's execution, click on the last flow run. Here, you should see a flow diagram of the steps that have been executed, including the relevant logs, as shown in the following figure:

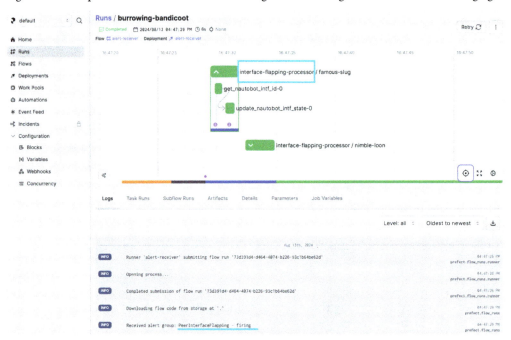

Figure 12.14 – Prefect flow run

The preceding diagram represents a flow run of alerts triggered by Alertmanager and the automation executed by the worker. It displays the entire duration of the `interface-flapping-processor` workflow we recently developed, along with the tasks related to Nautobot.

This indicates that the interface status in `ceos-01` should now show as **Alerted** in Nautobot. You can see this change in the following figure:

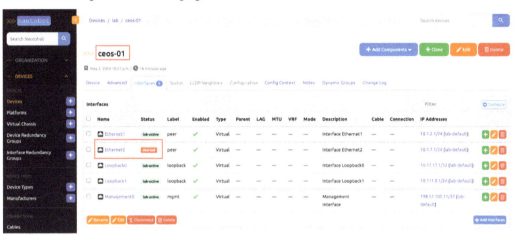

Figure 12.15 – Nautobot Interface Status Change

You can now see that the interface status has been updated to **Alerted**. What makes event-driven automation particularly effective is its *hands-off* capability. In this example, once the script causing the interface flapping finishes, the alert will clear after a few minutes. This will automatically trigger another workflow for a `resolved` alert, which updates Nautobot and returns the interface to its original state.

Here's an example showing the full life cycle of the two workflows activated by this event. One workflow triggers the alerts while the other resolves them:

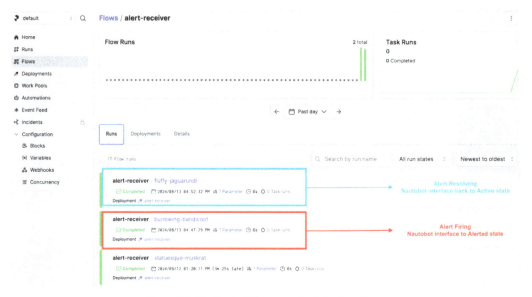

Figure 12.16 – Prefect flow runs for interface flaps

With this, we conclude our exploration of the event-driven automation workflow in this chapter. This example has guided you through the entire process, from the initial detection of an event to the automated resolution, showcasing how such systems can streamline network operations and minimize manual intervention.

Summary

In this chapter, we explored how to use programming tools to access data from observability systems and showed what you can do with that data.

We started with simple Python examples to help us understand how to collect metrics from Prometheus and logs from Loki. This helped us get comfortable with the data structure and develop basic skills for retrieving the information we needed.

Starting from the basics we learned about in previous chapters, we created tools that directly address what network engineers need. We built a simple CLI that uses our earlier data-gathering techniques. This tool can spot which network interfaces are being heavily used or give a quick health check of a network site by collecting data from Prometheus, Loki, and Nautobot. By combining information from these sources, the tool helps network engineers get a clearer picture of network status in one place, saving time and effort.

Then, we tackled more complex scenarios using event-driven and closed-loop automation. These approaches let us automate certain tasks based on specific events detected by our monitoring systems. Essentially, they allow the network to manage itself in certain situations by automatically adjusting or responding based on what's happening in real time.

We discussed the essential ideas behind these frameworks, highlighting the role of a workflow orchestrator. This tool is crucial for managing and monitoring automated workflows. We also covered the important features that these systems need to have to work effectively.

Currently, only a few companies have fully implemented a closed-loop automation framework. The best approach is to start small. Whether you're using a ready-made solution such as Ansible EDA or building your own system with a workflow orchestrator, beginning with manageable steps is crucial. Consider the example we explored using Prefect and the event-driven workflow we developed in this chapter. We started with simple automated actions triggered by an event. Now, you can enhance and tailor this to fit your operations more closely. For instance, if you manage border routers connected to multiple ISPs and have clear policies for shifting traffic between circuits during issues, you can develop this workflow further. By making it more robust, adding detailed logging, and standardizing its processes, you can ensure it runs smoothly without needing middle-of-the-night manual interventions for traffic shifts.

In the next chapter, we'll look at advanced monitoring techniques involving machine learning and artificial intelligence. We'll see how these technologies can improve observability system tools and help you predict patterns or improve root cause analysis in networks and systems, making them even more useful for engineers and developers.

13
Leveraging Artificial Intelligence for Enhanced Network Observability

During the past several years, we have observed the rise of **artificial intelligence (AI)** and **machine learning (ML)** technologies to distill knowledge from data. In this chapter, we'll uncover a few of the network observability use cases that can benefit from them.

With network observability empowered by AI/ML, we can process massive amounts of data and look for patterns that can lead to a human-like analysis, such as finding a common point of failure that explains why many unrelated events relate to the same issue. Even more, with ML, we can learn from these events to predict future similar situations and react to them preventively.

In the first section of this chapter, we'll provide a high-level overview of the most common AI/ML techniques and how we can leverage them to improve our monitoring solutions. We won't dive deep into the internals of these solutions because this topic has been covered in many other books that we'll recommend as we go along.

AI/ML is all about data, and network observability is a massive source of data. Thus, it is a great match! With all the operational data and data from other sources, we'll be able to enhance network operations by providing recommendations (for example, the root cause issue from a set of related events), triggering changes, and digesting many logs to provide more effective summaries.

We'll cover the following topics in this chapter:

- AI and ML fundamentals
- Real-world AIOps

> **Note**
>
> As you may have noticed throughout this book, the scale of network observability goes beyond what humans can digest without extra assistance. AI/ML is the assistant we have been looking for because it imitates human learning at scale, just without our limitations.

Before we dive into concrete network observability examples, it's important to get a high-level overview of the key topics in this area.

AI and ML fundamentals

Like any new technology, there's still a bit of confusion around the scope and meaning of each term. Before diving deep into the challenges that AI/ML can help solve, it's important to demystify some classification terms:

- **AI** is the field of knowledge that tries to make machines reproduce human behavior. Any software that imitates this behavior can be called AI (for example, a simple *if this, then do that* rule).

- **ML** adds the capability to learn/infer patterns from historical data so that it can be applied to new data. It produces new outputs by reusing the learned knowledge.

- **Neural networks** are a subset of ML that emulate how the human brain works while leveraging the concept of neurons and how they're connected. This allows more complex problems to be solved.

- **Deep learning** is a multilayer neural network (more than three layers) that provides more options for building custom neural networks, increasing the capacity to tune its behavior.

- **Generative AI** is the latest stage in AI. It's based on deep learning technologies and allows new contexts to be generated using pre-trained models.

The following figure helps simplify this classification:

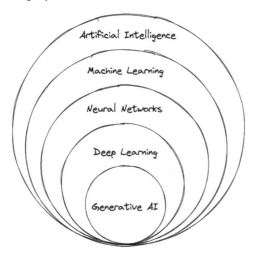

Figure 13.1 – AI and ML classification

You're likely already familiar with using some flavor of AI. For example, AI includes common monitoring tasks such as defining a threshold for triggering an event when the bandwidth exceeds it. Thus, in this chapter, we won't focus on the well-known AI solutions but on leveraging ML techniques.

> **Note**
>
> There are many great book references on AI/ML. We want to highlight *Machine Learning for Network and Cloud Engineers*, by Javier Antich, because it deals with the topic from a network engineering perspective, and *Hands-On Machine Learning with Scikit-Learn, Keras & TensorFlow*, by Aurélien Géron, due to its practical approach.

To give you an idea of potential ML usage in network observability, let's look at a few potential use cases (not exhaustive at all):

- Classify network traffic patterns to determine whether they correspond to a **Distributed Denial of Service** (**DDoS**) attack or legitimate usage, to trigger a DDoS mitigation event

- Forecast the expected number of IP routes in memory to understand if we'll hit some memory limit or to compare it with the actual number to realize that there's less or more than expected

- Understand whether the number of MAC addresses learned in a port matches the normal numbers for this type of port in the network to determine whether there's an anomalous utilization

- Continuous optimization of **Border Gateway Protocol** (**BGP**) peering policies (for example, increasing or decreasing AS Local Preference) by observing the latency of the related network flows to minimize the overall latency

- Generate a **root cause analysis** (**RCA**) by summarizing individual alerts and log messages

All these use cases have something in common: they learn from the data to provide a new output with some degree of confidence. It's important to note that not every recommendation from an ML algorithm is true. It's simply the most likely output considering the context. Sometimes, the odds can be very high (close to 100%), but we must take every recommendation as such.

In the next section, we'll present the most relevant ML techniques at a concept level without even mentioning the specific algorithms used so that we don't go beyond the scope of this book. You can learn more about them using the references provided.

ML algorithms

There are many ML algorithms, and more are appearing as we write this book. We won't analyze them here in detail, but we will explain them in terms of the problem they solve.

Before considering the different types, the first distinction is whether they use unlabeled data (raw data without metadata) or labeled data (that is, data and context metadata to support the learning process). The following figure shows the difference between labeled and unlabeled data:

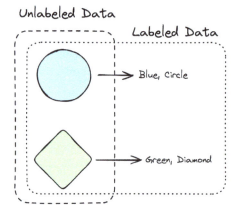

Figure 13.2 – Labeled data

Labeled data requires pre-processing work (that is, associating the input data with the corresponding value), but it isn't always available. An example of labeled data could be a dataset with the **Open Shortest Path First** (**OSPF**) neighbors active in a router (input data) and its associated CPU and memory usage (output or label). Learning from this dataset, the algorithm can infer the most likely numbers for the CPU memory when a different number of OSPF active neighbors is seen in the future.

> **Note**
> Algorithms are the techniques for solving ML problems, and models are specific instances of applying an algorithm to a dataset.

Using labeled data as the classifier, there are three (main) groups of learning: supervised, unsupervised, and reinforcement.

Supervised learning

Supervised learning means that the training process uses a training dataset with labeled data, and these labels are used to train the algorithm (that is, we teach the model with examples). The algorithm learns how to map input features (or variables) to output labels to predict them when they're exposed to new input data. For example, let's say we want to classify the role of the network device into two known categories: border routers and top-of-the-rack switches, according to some parameters such as memory utilization and the number of interfaces up.

> **Note**
>
> In ML, the terms **attributes** and **features** are usually used interchangeably, but they represent one dimension of the data, along with the data type (that is, the attribute) and its value.

The output of supervised learning is a model that *knows* how to imitate the same behavior of the observed dataset with some level of accuracy. Thus, when the observability solution gathers new data from a network device, the ML model can classify it into one of the groups that's been learned during training.

The most common supervised learning types are classification and regression, both of which we'll cover next.

Classification algorithms

Classification algorithms use labels to map the input data to a predefined set of categories known in advance. For instance, continuing with the previous example, we may try to classify input data into one of the possible roles of a network device in our network: border router and top-of-the-rack switch. The learning process uses a dataset with different input data – for instance, the operating system (for example, IOS, JUNOS, EOS, or SROS) and the amount of memory and CPU consumed. From this, with the predefined labels (that is, which role a network device belongs to according to the existing dataset), the algorithm learns how to classify new input data for one of the existing categories (with some probability).

The ML algorithms need to be fed by numbers, so if the input data isn't numeric, it has to be transformed. For example, if the input data is regarding the network operating system, this discrete set of values needs to be transformed into numeric values. This is a simple example, but for complex input variables with many dimensions, you must create a mapping dictionary to encode the data so that it can be fed into the algorithms.

> **Note**
>
> Data representation/visualization is a key component for every ML project to run the algorithms effectively.

The following figure shows how a classification algorithm decides whether the input data belongs to the two roles of circuits you have in the network:

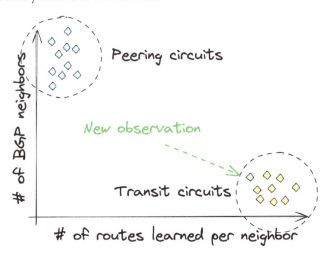

Figure 13.3 – Classification algorithms

Here, we have transit circuits (with a few BGP neighbors but with many learned routes per neighbor) and peering circuits (the opposite). When new input data (the new observation) comes into the classification algorithm, the system decides whether it belongs to one or the other.

Classification algorithms help to categorize new data into predefined classes based on learned patterns.

Regression algorithms

Regression algorithms *predict* a new numeric value based on some input variables. This happens after a model has been trained with labeled data to infer which is the most likely value to be associated with a new feature. The following figure shows how the algorithm provides a new output for a specific sequence from a dataset (that is, the yellow points from where the algorithm has learned):

Figure 13.4 – Regression algorithms

Don't assume that all the prediction algorithms use a linear approach like in this example. There are many types of algorithms; we simply used the simple linear approach to illustrate the concept.

Regression algorithms have many applications in network monitoring to determine which should be the most probable next observation value and then compare it with the actual value to understand the divergence. Likewise, it has many applications in *capacity planning* or any use case that requires the future to be predicted. However, it's important to note that reality has many factors (for example, seasonality or how data is influenced by different events, such as the Super Bowl event) that must be taken into account.

For example, as shown in the following figure, in a time series dataset representing the input packet rate of an interface, we could identify an abnormally high or low data point compared to what's normal in that sequence – with the *normal* sequence being what the regression algorithms may have predicted. This will be covered in one of the real-world examples we'll cover later in this chapter:

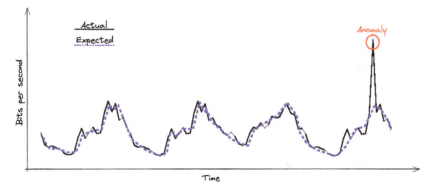

Figure 13.5 – Determining an anomaly comparing actual versus predicted data

Regression algorithms predict continuous numerical values that help with forecasting and trend analysis and can also be combined with others, such as anomaly detection algorithms.

Unsupervised learning

Unsupervised learning doesn't use labels to train algorithms. Instead, the algorithms need to identify the underlying structure of the data without extra context. In contrast to the supervised learning example, where we knew which two roles of network devices we wanted to classify the network devices into, in this case, we don't know the groups in advance.

Unsupervised learning is good for solving different problems such as data analytics, where the amount of data to analyze is beyond human capacity to deal with or some predefined context.

We'll start with one of the most popular types of unsupervised learning: clustering algorithms.

Clustering algorithms

Clustering algorithms try to discover clusters or groups in the data. A cluster contains data that has something in common and is different from other clusters. Thus, the key characteristic of these algorithms is how to measure the similarity of data, or how close the data is.

As an example, let's go over the examples we discussed for classification algorithms. But this time, imagine that we don't know that we have two well-defined role types. We don't know anything about our data a priori. Using only data patterns, the clustering algorithms can determine how many groups (or clusters) exist regarding the two dimensions: the number of BGP neighbors and the number of learned routes per neighbor. These algorithms try to determine how *far* or *close* the data points are between them:

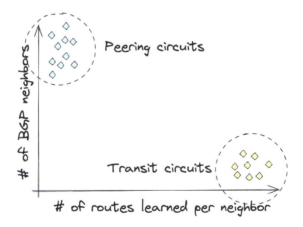

Figure 13.6 – Using clustering to discover existing circuit roles

In the previous example, for humans, it's evident that there are two groups. However, without labeled data and only data observation, it can be a bit more difficult to analyze mathematically. One of the key aspects of tuning the algorithm is how many clusters should be taken into account. To solve this question, the algorithm provides the accumulated error to help you determine which pattern best matches the unlabeled data.

The right amount of clusters comes with a trade-off, with some data points not belonging to one of the clusters. These are outliers, and identifying them helps us understand that this data point may not be a *normal* one.

Anomaly detection

Anomaly detection is a set of techniques (both supervised and unsupervised) intended to identify rare events or observations that do *not fit* with the rest of the observations. Following the previous clustering example, we could identify whether a new network circuit belongs to one type or another.

The following figure shows that the new observation point doesn't fit well with any of the existing groups:

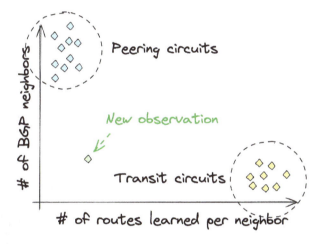

Figure 13.7 – Anomaly detection in clustering algorithms

Indeed, the *distance* (that is, the difference with the observed type representation) is bigger than what makes sense, so the conclusion seems that it doesn't belong to one or another. Thus, instead of doing a poor classification and assigning it to peering circuits (it's a bit *closer*), maybe it's better to raise an anomaly detection event to understand whether the data is correct or whether a new clustering analysis is needed.

In the next section, we'll tackle another important type of algorithm that we can use to prepare data so that it can be used in ML. This algorithm simplifies data by removing data dimensions that aren't relevant enough to solve the problem.

Dimensionality reduction

Often, there's more data than what's needed (that is, an excess amount of data dimensions). Any data that isn't relevant to the problem we're solving is counter-productive: it means more resource consumption and it may *confuse* the algorithms. If a variable doesn't add value, it's better to leave it out.

Humans are used to living in three dimensions. It's very easy for us to imagine the objects in three dimensions. But just adding a new dimension (a fourth dimension) could blow our mind. And, when dealing with data, the number of dimensions can increase as it has many different variables. Having more dimensions increases the complexity of the algorithms, something known as the **curse of dimensionality**. To solve it, **dimensionality reduction algorithms** can be used; they reduce the number of variables in a dataset while preserving as much of the original information as possible. This reduces the complexity of data processing and improves data consumption:

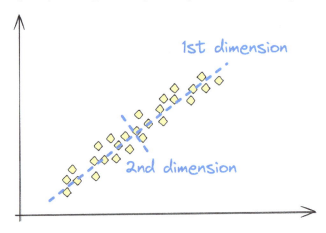

Figure 13.8 – Two dimensions with different amounts of information

In the preceding figure, you can observe how the second dimension has very little variance, adding little value to the information. Thus, we could get rid of it and just work using the first dimension alone. Leveraging dimensionality reduction simplifies ML processing by reducing the number of features while retaining essential information, thereby enhancing computational efficiency and mitigating the curse of dimensionality.

Association rules

Association rules identify frequent patterns and uncover hidden relationships in the data. For instance, when some components appear together in a transaction or event, there's some probability that another component will appear too. In networking, we could apply association rules to alert analysis to determine which components (for example, interface role, device model, temperature, and so on) combination has the highest probability in some alert type.

Using association rules, we can build recommendation systems that understand the probability from the concurrent components and can recommend which is the best action to address some problem. For instance, if we have a high probability of getting a BGP peering issue due to interface issues, the recommendation system will point in this direction to mitigate it.

Reinforcement learning

Reinforcement learning doesn't use input data. It's very similar to how humans learn – that is, through trial and error. We tend to learn by taking action and learning from the positive or negative reward. Depending on that, we know what the most probable output is from our experience, and we can determine our next actions.

In the computer world, reinforcement learning helps us avoid having to provide datasets (labeled or not) to learn from. Learning happens as the agent (the learning process) interacts with the environment, and the outcome (that is, the actions) depends on the reward (shown in the following figure). This trial-and-error learning is a cumulative process that should maximize the reward over time:

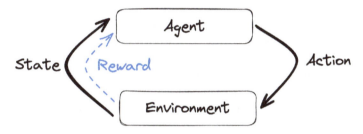

Figure 13.9 – Reinforcement learning loop

The learning process must learn both from mistakes and successes. So, after exploring the various options, it must apply the changes that maximize the reward. This is more complex than it seems because before determining the *best* option, it needs to ensure a certain amount of exploration occurs so that it doesn't drop an option that may have a lower reward in the short term, but a higher reward in the long term.

In networking, there are many examples where this dynamic learning process can help to optimize network operational status – for instance, wireless signal optimization or automatically adjusting AS-PATH in BGP neighbors to improve the latency and load of circuits.

Now that we've provided a high-level overview of the different types of ML that are available, we'll focus on a few terms that have broad recognition and deserve a better explanation.

Neural networks and language models

Neural networks are an approach to ML that mimics the behavior of the human brain. A key concept in neural networks is combining *simple* ML algorithms (for example, the ones described earlier) to capture inner relationships that produce more complex models.

By adjusting the parameters (that is, weights) of the relationships between the ML components of the network, the neural network can improve its behavior with a proper training process that minimizes the difference between the predictions and the actual training dataset.

When we combine several layers (more than three), the neural networks become **deep neural networks** (also known as **deep learning**), which can lead to surprising results with more complex and sophisticated tuning of parameters/features (obviously, with more computing resources needed to train them).

> **Note**
>
> Like everything in ML, neural networks and deep learning by themselves would need several books to explain. For example, there's *Hands-On Machine Learning with Scikit-Learn, Keras, and TensorFlow*, by Aurélien Géron, which offers a hands-on approach, and *Deep Learning* (https://www.deeplearningbook.org/), by Ian Goodfellow, Yoshua Bengio, and Aaron Courville, which offers a more theoretical approach.

Neural networks and deep learning implement models that are trained to capture knowledge that can be applied to new input data to generate similar outputs. As you can imagine, how a model behaves is mostly defined by the neural network architecture (that is, layers, connections, and so on) and the training data. Therefore, if we focus on the second part, every model is biased by the data that was used to train it (no different than humans), so there would be more general-purpose models (that is, foundation models) and more specific ones (that is, vertical models). Depending on what you wish to solve, you would prefer one over the other.

Again, focusing on the data, depending on the type of data (e.g., text, images, or video), each model could be used for different purposes. In networking, we deal with a large amount of text data: from configuration files to different observability data such as log messages, CLI output, or alerts generated by systems. Thus, the models for text processing are very convenient, and these models enter the realm of **natural language processing (NLP)**.

NLP is a multi-faceted topic around how computers can process and analyze large amounts of natural language data. Its scope goes beyond ML, but it's a key part of it. NLP has many applications, such as natural language translation and generation, text summarization and classification, fake content detection, and many more.

> **Note**
>
> NLP can be used in many networking use cases (that is, not only for observability), but we will focus on applying it to network observability use cases in this book.

The ML models that are used to support NLP are called **language models**. Like any other AI/ML mechanism, they try to reproduce a human-like language processing. These models replicate the understanding of human languages and their knowledge (patterns).

Depending on how complex these language models are, we can group them as follows:

- **Large language models** (**LLMs**) are language models with a lot of parameters that have been trained with a huge amount of language data. LLMs are deep neural networks, so they require considerable brute-force resource consumption to train because of the number of parameters that need to be adjusted (for example, rumors say that **GPT-4** has 1.76 trillion parameters, but this hasn't been confirmed).

- Similarly, **small language models** (**SLMs**) are language models but in this case, the number of parameters and trained data is lower (for example, **DistiBERT** has 66 million parameters). Therefore, they have less general knowledge but can be more specific to a use case.

In both cases, a language model outputs the most likely next word given a set of words (that is, the context). And, one word after another, it creates text responses that look reasonable for a human.

Since 2022, with the launch of applications that leverage LLMs (also named chatbots), such as **ChatGPT** by OpenAI, LLMs have become very popular because of the outstanding results they can achieve when interacting with humans processing text. These services have democratized access to these models whose training is only accessible to a few actors (because of the necessary training costs).

> **Note**
>
> In the hands-on section of this chapter, we'll use OpenAI's ChatGPT to provide automated root cause analysis to assist in network operations.

In recent years, these ML models and their related services have been constantly evolving. Due to this, we can see how many players have developed new and more specific models (that is, a model to help with coding tasks such as GitHub Copilot), some of them kept private (as a competence differentiator) and others publicly available. You can find a huge collection of models that can be downloaded at `https://huggingface.co/models`.

Similar to third-party services (e.g., OpenAI ChatGPT), which offer interfaces to interact with the models, you can build and run your own models. Tools such as **Open WebUI** (`https://github.com/open-webui/open-webui`), **Langchain** (`https://www.langchain.com/`), and **Huggingface** (`https://huggingface.co/`) provide frameworks for that.

> **Note**
>
> These models are one-time representations of training, so they aren't trained while in use. For example, the data you exchange with a chatbot (for example, ChatGPT) isn't used to keep training the model (for example, GPT-4). If this was the case, the models could be influenced by users.

Later in this section, you'll learn about the different options you have to build or customize ML models.

AIOps

Artificial Intelligence for IT Operations (AIOps) is defined by Gartner as the process of applying AI to IT operations. AIOps include techniques that can be used to monitor technical infrastructure using AI/ML.

> **Note**
>
> Don't confuse AIOps with MLOps, which refers to the continuous process of managing an ML model's life cycle.

Until a few years ago, most AIOps implementations used unsupervised learning with some basic supervised techniques. These algorithms provided operational insights such as anomaly detection, event correlation, forecasting, and trending analysis.

Nowadays, using ML models, such as LLMs, provides an extra knowledge capability that leads to assisted decisions based on recommendations. These entities that bring knowledge on top of insights are commonly known as autonomous agents.

This knowledge, when reliable enough, can be combined with network automation so that actions can be performed on the network according to the decisions that have been suggested.

As Javier Antich explains in his book (referenced earlier), the following *knowledge creation map* shows that as more domain-specific data becomes available (for example, the relationship of the healthy state of a circuit related to the amount of packet loss observed) without human intervention (via more effective data collection), data-driven knowledge grows exponentially and it can feed deep learning algorithms:

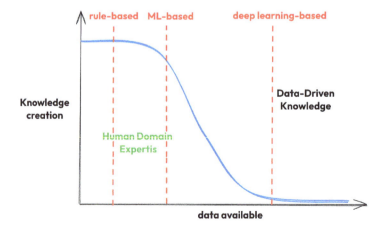

Figure 13.10 – Knowledge creation map by Javier Antich

This increase in available data will accelerate the adoption of more autonomous decision processes that will answer the requirements of higher complexity networks with stricter SLA demands. Without ML, the resourcing challenges of addressing the requirements at scale will make most operation teams struggle.

These algorithms and models have an important role in ML projects, but the most important aspect is the data in use. Depending on how the data represents reality, the learning may be biased by it and impact how much it takes to get proper training. Poor data (for example, noisy data) will need more complex models to filter the information. As you'll see, data management (both for training and to question the model) is going to take the most time in every project as it involves cleaning, removing outliers, filling data gaps, normalizing, and more.

How to adopt ML

Independent of the problem you want to solve with an ML solution (for example, predict the required capacity of a circuit in 6 months for capacity planning or determining whether the current memory levels of a border router are *normal*), there are three basic options:

- Build and train your model from scratch.
- Fine-tune an already existing model, which implies modifying the model according to a new dataset to tailor the model knowledge so that it can solve a more specific use case.
- Combine already existing models, without modifying existing ones, to provide different results and apply techniques such as **retrieval-augmented generation** (**RAG**), which improves the accuracy and reliability of LLMs by including information from other sources in the *context* before getting the answer from the LLM.

In all cases, the first step is to *describe the problem within the big picture*. Even though it may seem obvious, the first step is to define the problem that needs to be solved in terms of business needs. Gathering the context of the problem, including the requirements and assumptions, will help you to consider similar problems that some experience might have already been built around.

Then, you must *prepare the data* needed to train and validate your approach, from getting the data and its implications (for example how much space, legal requirements, authorization, and so on) to exploring the data to be used to understand whether it can be *improved* to train the solution or validate it.

Next, you'll need to *select (and train, if needed) a model* to start validating how it behaves, and then fine-tune the model or the solution's architecture to improve the outcomes. When evaluating the results, you must share the solution with others to get the buy-in from the rest of the business.

Finally, when in production, you must continuously ensure that the model/solutions behave as expected to decide whether any adjustments are needed.

At this point, you may have already noticed that aside from the techniques, the data plays an important role. If the data is too small, not representative enough, or noisy, the model won't learn as well as it could. This is obvious: reading a book about how to repair a truck's engine won't help you write a poem better. Thus, the data is key. Let's look at the common problems you may find with training data:

- **Not enough data**: ML algorithms need a huge amount of training data to learn from. For simple problems, it may be in the order of thousands of examples, while for complex problems (for example, image recognition), there will be millions.

- **Nonrepresentative data**: If your training set is missing some representative data, your model won't learn about it, so it will be biased by the other data, which doesn't fully represent the problem you want to solve.

- **Poor-quality data**: The quality of your training data will influence the quality of the training. If you have training data that contains noisy data (that is, irrelevant features), gaps, outliers, or errors, the learning process won't be able to converge into a satisfying stage.

- **Overfitting or underfitting the data**: If the model is too complex or too simple, respectively, it may lead to inaccurate outcomes.

Now that we've provided a brief introduction to AI/ML, we'll consider some real use cases where you can apply ML to empower your network observability solutions.

Real-world AIOps

As you may have already noticed, the AI/ML realm is extremely broad and deep. We won't pretend to go down this path in this book and instead offer you some clear and effective use cases where applying a few of these techniques can save the day without you having to go into low-level details.

Following the recommended steps outlined in the previous sections, we need to start describing and identifying the problems we want to solve. This first step is crucial so that we can channel our efforts toward effective approaches to solve the challenges. Notice the usage of plural in *approaches* – we always have many options to consider, as well as their pros and cons. These are the two challenges we are going to solve in this chapter:

- **Validating operational changes by observing the network traffic**: In network operations, multiple events involve changes – for example, an operating software upgrade or a configuration change. After the change is executed, we always have to ask ourselves whether the network is behaving as expected. One way to do this is by checking whether the amount of interface traffic usage after the change is in line with the previous historical data.

- **Providing an enriched RCA when an incident is detected**: When a network engineer faces a network issue, the first steps are always about gathering information to understand what the potential causes (not the effects) of the issue are so that they can be mitigated. This exercise takes a lot of time and experience to do it right, so if we could leverage an automated process to collect the information and look for recommendations to guide the analysis, it could lower the time to recover.

To simplify how to create and run the examples in this section, we'll set up a lab (similar to what you already did for the previous chapters) with a new service named `machine-learning-webhook`. Let's start by introducing this scenario and how to get started with it.

Lab requirements

In this lab, you'll be reusing most of the tools from the previous chapter to offer a holistic view of how all these components interact with the new ML service. This service is a simple application that offers a few endpoints that will run the logic to solve the challenges explained in this chapter's introduction:

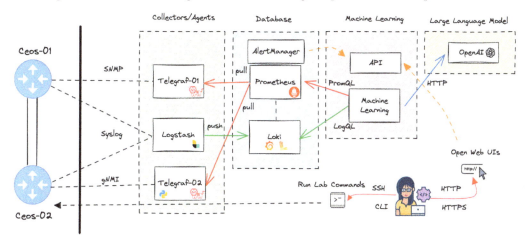

Figure 13.11 – Chapter 13 lab scenario

Thus, most of the components have already been pre-configured for you because the focus will be on creating the two ML services that will help solve the aforementioned challenges.

Similar to what we've done previously, you can get detailed instructions on how to set up this scenario in the *Appendix A*, but in short, you can get started by running the following command:

```
root@netobs-droplet:~/network-observability-lab# netobs lab prepare
--scenario ch13
```

> **Note**
>
> Like the previous chapters, you also have the `ch13-completed` scenario, which has all the examples incorporated and ready to use. The `ch13` scenario needs to be completed while following the steps outlined in the following sections.

The new container (`machine-learning-webhook`) that contains the ML services uses a FastAPI (`https://fastapi.tiangolo.com/`) app with some predefined endpoints that we will enrich as we walk through this chapter.

Once the lab is started, to make sure that everything is ready to go, you can test it in your browser by accessing port `9997` in your lab server (`http://<your-server-ip>:9997`). You'll receive a welcome message in the root path (`/`):

```
←   →   C     ⚠ Not Secure   138.197.95.155:9997

1    // 20240225112443
2    // http://138.197.95.155:9997/
3
4  ▾  {
5        "message": "The Machine Learning service is waiting for your requests!"
6     }
```

Figure 13.12 – Validating that the ML service is working

With the lab up and running, it's time to start applying ML to solve the network observability use cases. We'll start by learning how to validate operational changes.

Validating operational changes

There are many different ways to validate the operational state of a network. We've already presented how – by leveraging automation and the definition of the intended state, you can compare it and determine whether it matches the expectations or not. However, there are other approaches to operational validation where it's almost impossible to predefine the right reference to use. You have to observe it and compare it.

ML allows operational validation to be implemented, which learns from the past to determine whether the present is in line with the projected evolution. For example, if your interface traffic is growing linearly over time and suddenly starts to grow exponentially, something is behaving abnormally. Is this good or bad? Well, it depends. If the operation that triggered the event didn't expect a change, this is bad. But, if the operation was looking to attract more traffic over the interface, it's good.

Thus, you'll have to determine what you're looking for and process the data accordingly.

In this example, you'll use the input packets from the `Ethernet1` interface in the `ceos-01` device to determine whether the interface traffic numbers after a *change* are within the projected ones or not.

You'll take the metrics from Prometheus programmatically (that is, via an API) and define the times when the change started and ended, comparing the actual data with the projected data:

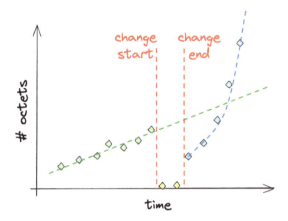

Figure 13.13 – Interface input octets before and after a change

In *Figure 13.13*, you can observe what we'll validate in this example. Before the change started, there was a linear pattern that was transformed into an exponential pattern after the change. This means that the change had an impact that could have been expected. Notice that we define the change as a window that may include a few metrics that we don't want to consider in our analysis because it may be impacted by some transient factors, such as the convergence of routing protocols.

Retrieving Prometheus metrics programmatically

There are many tools we can use to leverage the Prometheus API. In this case, we'll be using the `prometheus_api_client` library (already used in *Chapter 12*), which is already installed in the ML container. It provides an object (`PrometheusConnect`) so that we can interact with the API with ease:

```
from prometheus_api_client import PrometheusConnect
def get_interface_usage(device: str, interface: str):
    time_to_query = "30m"
    prom = PrometheusConnect(url="http://prometheus:9090")
    query = (
        f'interface_in_octets{{device="{device}",',
        f'name="{interface}"}}["{time_to_query}"]',
    )
    response = prom.custom_query(query)
    return response[0]["values"]
```

In the `anomaly.py` file, `get_interface_usage` retrieves the input octets of an interface in a specific device for the last 30 minutes. This is the period we'll use in this example, but you could customize it to your needs.

You can test this in a Python interactive session to facilitate interacting with the code by attaching it to the container shell:

```
root@netobs-droplet:~/network-observability-lab# netobs docker exec
--scenario ch13-completed machine-learning-webhook
... omitted output ...
root@e00b072dbe3e:/#
root@e00b072dbe3e:/# python
>>> from app.anomaly import get_interface_usage
```

Continuing in the Python interactive session, the outcome of this function, by calling it with `get_interface_usage("ceos-01", "Ethernet1")`, is a list of items with the timestamps and their values as a string:

```
>>> get_interface_usage("ceos-01", "Ethernet1")
[[1708842626.917, '12054743'], [1708842641.917, '12054743'],
[1708842656.917, '12054743'], [1708842671.917, '12055911'], ... ]
```

Now that the data has been retrieved from Prometheus directly, we can start preparing the data with a popular Python library to handle time series data: the pandas (`https://pandas.pydata.org/`) library.

Leveraging the pandas library to prepare the data

pandas is the most popular Python library for managing time series data. Moreover, it's the library that's used by the `Prophet` library (`https://facebook.github.io/prophet/`), which we'll use next to forecast what the values should be according to the previous data.

> **Note**
>
> You don't need to worry about installing Pandas as it's already included as a dependency by the `Prophet` library, and already installed in the container. However, if you do encounter errors in terms of the pandas library being missing, it can be completed by executing `pip install pandas`.

The `Prophet` library expects the data to have two predefined columns: `ds` and `y`. This isn't what we get directly from Prometheus, so we have to transform it:

- `ds`: The time in datetime format
- `y`: The value

With the following code snippet, you can retrieve the interface usage (using the function defined previously) and transform it into the DataFrame format for pandas:

```python
import pandas as pd
interface_usage = get_interface_usage("ceos-01", "Ethernet1")
df = pd.DataFrame()
metric_list = []
for data in interface_usage:
    data_dict={}
    # Convert the `seconds` representation into datetime object
    data_dict['ds'] = pd.to_datetime(data[0], unit='s')
    # Save the interface counters value
    # retrieved in a `float` format
    data_dict['y'] = float(data[1])
    metric_list.append(data_dict)
df_metric = pd.DataFrame(metric_list)
```

The content of df_metric is 120 rows (120 is the number of metrics that were collected every 15 seconds during the last 30 minutes) and a table with two columns:

```
>>> df_metric
                          ds            y
0    2024-02-25 06:30:26.917000055  12054743.0
1    2024-02-25 06:30:41.917000055  12054743.0
..                          ...          ...
118  2024-02-25 06:59:56.917000055  12086829.0
119  2024-02-25 07:00:11.917000055  12087908.0
[120 rows x 2 columns]
```

At this point, you have all the data in df_metric. However, to represent the pre and post-change, we want to split the data into data *before* the event (let's name this t1), which represents the normal traffic of the interface, and data *after* the event (let's name this t2), when the data is ready to be comparable, after discarding the data points between t1 and t2, which may be impacted due to the change operations (remember that networking is a distributed system that requires many entities to converge into a stable state).

In this example, you could define t1 and t2 (these values are related to the time when the example was run), and then split the data into two DataFrames – historic and current:

```python
# These two values are taken for the data set in the example
t1 = 1708843856.917
t2 = 1708844051.917
# Create t1 and t2 references in datetime object to compare
reference_time_t1 = pd.to_datetime(t1, unit='s')
reference_time_t2 = pd.to_datetime(t2, unit='s')
# Fabricate historic and current metrics
```

```
historic_metric = df_metric[df_metric['ds'] < reference_time_t1]
current_metric = df_metric[df_metric['ds'] > reference_time_t2]
```

If you check the content in `historic_metric` and `current_metric`, you'll notice how the rows have been split (excluding the data from the in-between window change, which has been excluded from the analysis):

```
>>> historic_metric
                                ds           y
0   2024-02-25 06:30:26.917000055   12054743.0
1   2024-02-25 06:30:41.917000055   12054743.0
..                            ...          ...
80  2024-02-25 06:50:26.917000055   12076889.0
81  2024-02-25 06:50:41.917000055   12076889.0
[82 rows x 2 columns]
>>>
>>> current_metric
                                ds           y
96   2024-02-25 06:54:26.917000055  12081364.0
97   2024-02-25 06:54:41.917000055  12081364.0
..                            ...          ...
118  2024-02-25 06:59:56.917000055  12086829.0
119  2024-02-25 07:00:11.917000055  12087908.0
[24 rows x 2 columns]
```

So, the reference data in `historic_metric` has 82 data points that will be used to calculate the forecast, while `current_metric` contains 24 data points to compare with the forecasted data.

Forecasting values using the Prophet library

The `Prophet` library is an open source Python library sponsored by Facebook (now Meta) to provide forecasting of time series data. It takes into account non-linear trends that seasonal data can incorporate. You can leverage this library for providing capacity recommendations or anomaly detection (as we're doing in this example).

To use `Prophet`, we'll fit a class instance with the historic metric, and then ask it to create the future predicted data:

```
from prophet import Prophet
m = Prophet().fit(historic_metric)
future = m.make_future_dataframe(periods=150, freq="15s")
forecast = m.predict(future)
```

Then, you must merge the metrics in `current_metric` with the forecasted ones (only taking some of the columns):

- `ds`: Timestamp

- `yhat`: Predicted value

- `yhat_lower`: Lower range for the predicted value

- `yhat_upper`: Upper range for the predicted value

This is how to do it:

```
combined_results = pd.merge(
    current_metric,
    forecast[['ds', 'yhat', 'yhat_lower', 'yhat_upper']],
    on='ds'
)
```

The resulting variable, `combined_results`, has the actual data value (`y`) coming from the **current_metric** and the forecasted values, aligned by the timestamp (`ds`).

Then, you must inject a new column (`anomaly`) that, for each row, identifies whether the value is within the forecasted limits (that is, `yhat_lower` and `yhat_upper`):

```
combined_results["anomaly"] = combined_results.apply(
  lambda rows: (
      1
      if ((float(rows.y) < rows.yhat_lower) | (float(rows.y) > rows.
      yhat_upper))
      else 0
  ),
  axis=1,
)
```

Without any real change in the interface, we don't expect anomalies in any of the 24 data points (remember that these are the 24 values from `current_metric` when merged with the forecasted value, `yhat`). As shown in the previous code snippet, the `anomaly` column gets a value of 1 if it's outside of the `yhat` boundary or a value of 0 otherwise:

```
>>> combined_results['anomaly'].value_counts()
anomaly
0    24
Name: count, dtype: int64
```

Now, because we want to get some anomalies to showcase what the validation will look like in the final scenario, we're going to fabricate them by multiplying the data (that is, they column) in all the rows (using :) by two to enforce an anomalous behavior:

```
>>> current_metric.loc[:, "y"] = current_metric["y"] * 2
```

Now, if you recalculate `combined_results` with the new `current_metric` value, you'll notice that among all those data points, one isn't *normal*:

```
>>> combined_results['anomaly'].value_counts()
anomaly
1    24
Name: count, dtype: int64
```

In both `anomaly value_counts()`, we got the 24 values as normal or abnormal. Note that you may have a few of them in each category, depending on your real data.

Checking the `combined_results` data, you can see how the value (y) is above the acceptable limits forecasted by `Prophet` and therefore is an anomaly:

```
>>> combined_results[combined_results['anomaly']==1].sort_values(
by='ds')
                            ds           y           yhat
     yhat_lower    yhat_upper   anomaly
0  2024-02-25 06:54:26.917000055   24162728.0   1.208529e+07
     1.178485e+07   1.235362e+07      1
1  2024-02-25 06:54:41.917000055   24162728.0   1.208562e+07
     1.177738e+07   1.235852e+07      1
. . .
```

In the first row of the previous output, you can see these values (ordered by value), and why the y value is outside of the forecasted boundaries:

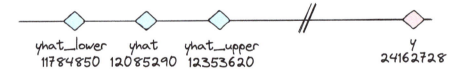

Figure 13.14 – Values outside the boundaries of forecasted data

Being able to compare the actual value (y) against the expected values (yhat) allows us to conclude whether the actual and expected state match or not.

Putting it all together into a real-time service

After analyzing the different parts, it's time to bring them into a combined service that we'll trigger via an API call. We'll leverage the FastAPI service we set up in the lab and tested earlier.

For this example, we'll leverage a skeleton FastAPI endpoint that will take a few query parameters to craft a response. Simply put, we want to understand whether the performance of a metric is behaving as expected after a change, as introduced in *Figure 13.3*.

> **Note**
>
> Don't worry too much about FastAPI (`https://fastapi.tiangolo.com/`). All the boilerplate code is already in place; you only need to focus on the inner logic.

The API endpoint has already been defined for you, so you don't need to worry about it. It leverages the `look_for_anomalies` function, which is defined in `anomaly.py`. This is where you'll implement the ML logic:

```python
@router.get("/v1/api/anomalies")
def check_for_anomalies(
    device: str = fastapi.Query(default=None, description="..."),
    interface: str = fastapi.Query(default=None, description="..."),
    t1: float = fastapi.Query(default=None, description="..."),
    t2: float = fastapi.Query(default=None, description="..."),
):
    log.info(
        f"Checking if the interface {interface} in device {device}",
        "is having a normal activity"
    )
    anomalies = look_for_anomalies(device, interface, t1, t2)
    log.info(f"Detected anomalies: {anomalies}")
    status_normal = True if anomalies.empty else False
    message = (
        f"Interface {interface} in {device} traffic for {t2} was ",
        "normal" if status_normal else "anormal"
    )
    return {"message": message}
```

> **Note**
>
> In the initial setup (that is, ch13, not ch13-completed), the function returns a mock class object with the empty attribute set to True.

This endpoint takes four query parameters:

- `device`: The name of the network device to target
- `interface`: The interface name to get metrics from
- `t1`: The time in UNIX timestamp format to define when the change started
- `t2`: The time in UNIX timestamp format to define when the change finished

> **Note**
>
> To convert the timestamp into the UNIX format, you can go to `https://www.unixtimestamp.com/` or use many other similar websites.

Then, you can craft a GET HTTP query with the current format:

```
http://<your-server-IP>:9997/v1/api/anomalies?device=ceos-01&interface
=Ethernet1&t1=1709398800&t2=1709399100
```

The answer will look something like this:

```
{"message":"Interface Ethernet1 in ceos-01 traffic for 1709399100.0
was normal"}
```

As mentioned previously, this answer (that is, `normal`) has been provided because the `look_for_anomalies` function simply returns that there are no anomalies.

Now, it's time to create the logic that will retrieve the metric data points and check whether the data after the change is behaving as expected by the forecasting.

In the `anomaly.py` file, you already have a `get_interface_usage` function, similar to the one we created earlier.

Now, by bringing all the code snippets from earlier together (which we run in Python interactive mode), we can create the `look_for_anomalies` function, which will return the data points classified as anomalies:

```python
def look_for_anomalies(device: str, interface: str, t1: int, t2: int):
    interface_usage = get_interface_usage(device, interface)
    df = pd.DataFrame()
    metric_list = []
    for data in interface_usage:
        data_dict={}
```

```
        data_dict['ds'] = pd.to_datetime(data[0], unit='s')
        data_dict['y'] = float(data[1])
        metric_list.append(data_dict)
    df_metric = pd.DataFrame(metric_list)
    reference_time_t1 = pd.to_datetime(t1, unit='s')
    reference_time_t2 = pd.to_datetime(t2, unit='s')
    historic_metric = df_metric[df_metric['ds'] < reference_time_t1]
    current_metric = df_metric[df_metric['ds'] > reference_time_t2]
    m = Prophet().fit(historic_metric)
    future = m.make_future_dataframe(periods=150, freq="15s")
    forecast = m.predict(future)
    combined_results = pd.merge(
        current_metric,
        forecast[['ds', 'yhat', 'yhat_lower', 'yhat_upper']],
        on='ds'
    )
    combined_results["anomaly"] = combined_results.apply(
        lambda rows: (
            1
            if ((float(rows.y) < rows.yhat_lower) | (float(rows.y) >
            rows.yhat_upper))
            else 0
        ),
        axis=1,
    )
    return combined_results[combined_results['anomaly']==1].sort_
values(by='ds')
```

There's nothing new in this function; it contains the same code that we explored interactively earlier. However, this time, it's using data retrieved from Prometheus in real time.

To demonstrate this, check out how the `Ethernet1` interface behaves in the `ceos-01` device by default by querying `interface_in_octets` in Grafana (`http://<your-server-IP>:3000/explore`) with `interface_in_octets{device="ceos-01", name="Ethernet1"}`:

Figure 13.15 – Normal interface input octets

With this data, we should expect that if we define a change time between 18:00 (the `1709398800` timestamp) and 18:05 (the `1709399100` timestamp), the interface is behaving normally:

Figure 13.16 – Validated normal behavior without alterations

> **Note**
> My browser timezone is UTC+1, but the timestamps are converted into UTC.

Now, it's time to change this pattern and create an anomaly in the traffic pattern to be detected. We have many options to change it. First, you could shut down the interface and bring it to a zero increase rate. You could also inject some synthetic network traffic to increase the rate of input packets. An easy way to do this is by running a ping that targets the `Ethernet1` interface IP address from the other network device:

```
ceos-02>enable
ceos-02#ping ip 10.1.2.1 size 1500 interval 0.1 repeat 1500
PING 10.1.2.1 (10.1.2.1) 1472(1500) bytes of data.
```

```
1480 bytes from 10.1.2.1: icmp_seq=1 ttl=64 time=0.142 ms
...
1480 bytes from 10.1.2.1: icmp_seq=1500 ttl=64 time=0.106 ms
--- 10.1.2.1 ping statistics ---
1500 packets transmitted, 1500 received, 0% packet loss, time 156332ms
rtt min/avg/max/mdev = 0.067/0.126/0.686/0.027 ms
```

After the ping completes, you can verify that the input rate changed in Grafana (using the same query):

Figure 13.17 – Abnormal interface input octets due to synthetic traffic

Now, if we move the observation window to the new window that contains the anomaly (*19:30* to *19:35* in the preceding figure), we can rerun the anomaly detection service to check that – in this case – something isn't going as expected:

Figure 13.18 – Validating abnormal behavior with synthetic traffic

If you check the logs of the ML container, you'll get some logs about the detected anomalies because the y value is outside of the forecasted values:

```
INFO:       | 2024-03-02 18:34:47 | Detected
anomalies:                             ds
             y          yhat      yhat_lower   \
8   2024-03-02
```

```
18:32:11.916999936   30635716.0   3.012164e+07   3.010473e+07
9  2024-03-02
18:32:26.916999936   30635716.0   3.012191e+07   3.010388e+07
10 2024-03-02
18:32:41.916999936   30635716.0   3.012218e+07   3.010328e+07
... omitted output ...
```

This example, which has been extremely simplified, shows how leveraging data and ML techniques can provide knowledge about what's going on with your network.

> **Note**
>
> You will likely be able to save time leveraging functions that have already been implemented by tools that include this capacity (for example, Elastic provides a sample data forecast function (`https://www.elastic.co/guide/en/machine-learning/current/ml-getting-started.html#sample-data-forecasts`), but it shows you how you can adapt it to your custom needs).

Next, we'll consider another interesting case for ML that brings together an automated process around the observability stack and the potential of LLM models to provide more comprehensive information.

Assisted root cause analysis

In this second example, we're going to reuse the same API-based approach from the previous one (that is, exposing an API endpoint) so that we can use ML to assist us in getting the best from our observability data.

The network observability data is strongly text-based, so digesting it to extract summarized information could be very helpful to speed up the diagnostics process and the time to recover when something goes wrong. As you've learned in this book, we need automation around observability to handle the massive amount of data we have to manage to understand what's going on in the network. LLMs are very helpful in extracting information/knowledge from the data.

In this case, we'll use the following approach:

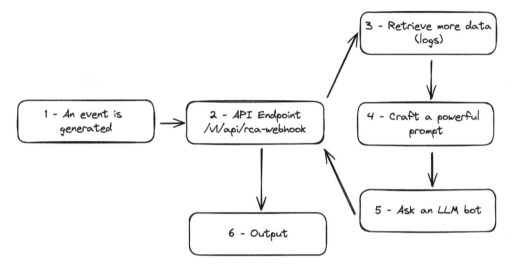

Figure 13.19 – RCA example

The following steps will compose the RCA example:

1. An event (that is, alert) will be generated by Prometheus and routed to the API endpoint by Alertmanager.

2. When the API endpoint receives the request, processing starts.

3. First, more information is collected related to the problem to enrich the question that will be sent to the LLM.

4. Afterward, we craft a concise text that will be used as the prompt for the LLM.

5. Then, we ask an LLM bot, via an API, to obtain the processed text.

6. Finally, we act upon the response. In this case, we will simply leverage the logs of the API service. Note that it could be directed to any type of notification system, such as email or an instant messaging system.

We'll apply this logic so that we can analyze a concrete problem: detecting a BGP session going down. The triggering event will be generated when Prometheus detects that the status of the BGP session is different than Established; this will request an API endpoint to take care of all the aforementioned steps. Let's start by defining the logic that will run within the API endpoint.

The Alertmanager webhook endpoint

Similar to the previous example, we have a FastAPI app with the skeleton already created in `api.py` (that is, `/v1/api/rca-webhook`) that will be triggered from Alertmanager to handle the alert:

```
@router.post("/v1/api/rca-webhook", status_code=204)
def process_webhook(alertmanager_webhook: AlertmanagerWebhook):
    log.info(«Alertmanager webhook status is firing...»)
    log.info(f»Received webhook: {alertmanager_webhook.json()}»)
    for alert in alertmanager_webhook.alerts:
        log.info(alert)
        # here comes your logic
    log.info(f"Alert is {alertmanager_webhook.status}, exiting")
    return {"message": "Processed webhook"}
```

Note that we're performing a POST request here. This is according to the webhook Alertmanager sends. Also, we're using a class, `AlertmanagerWebhook`, that's been already defined to deserialize the request data sent by Alertmanager.

At this point, you have the pipeline that triggers an API request that you have to handle. Next, in the `rca.py` file, you'll define some functions to complete the process that's required for root cause analysis processing.

Retrieving logs from Grafana Loki

According to the plan, once we get a BGP alert, we'll collect the related logs from Grafana Loki. The following code snippet takes a query (that you will define) and the reference time, and then uses `requests.get` to retrieve the logs:

```
def retrieve_data_loki(query: str, start_timestamp: int, end_time: int) -> dict:
    response = requests.get(
        url=f"http://loki:3001/loki/api/v1/query_range",
        params={
            "query": query,
            "start": int(start_timestamp),
            "end": int(end_time),
            "limit": 1000,
        },
    )
    return response.json()["data"]["result"]
```

This is simply a helper function that collects logs from Loki between two timestamps.

Crafting the LLM prompt

To interact with an LLM, you must create a question that contains enough context and the right instructions to provide a useful answer. The following is an example (the one included in ch13-completed) that combines some information from the alert (for example, neighbor ID and ASN) with other data, such as the logs, so that it can ask the *right* question. If you wish, you may play around and craft a different question that provides a better result:

```
def generate_rca_prompt(loki_results, device_name, neighbor_id,
neighbor_asn):
    return f"""
    RCA for a BGP neighbor issue
    - A BGP session in {device_name} has changed from Established to
another non-desired state
    - BGP neighbor IP: {neighbor_id}
    - BGP neighbor ASN: {neighbor_asn}
    - Associated Logs: {loki_results}
    - Based on the available data, what are the potential causes for
the loss of the BGP Established state?
    - Given the findings, what immediate actions should be taken to
mitigate the current issue?
    - What additional data would be helpful to further investigate this
issue?
    ... omitted some content ...
    """
```

The preceding prompt provides the context for the LLM to craft the automated response. It's wise to be as specific as possible – for example, notice how we ask for actionable insights.

Asking the LLM API

Finally, you'll use a popular LLM (GPT-3.5-turbo) that's offered as a service (by OpenAI) to ask the question we crafted previously (using the data from the alert). The following code leverages the OpenAI SDK's chat functionality:

```
def ask_openai(prompt: str, model: str = "gpt-3.5-turbo") -> str:
    client = OpenAI(api_key=os.getenv("OPENAI_API_KEY"))
    response = client.chat.completions.create(
        model=model,
        messages=[
            {
                «role»: «user»,
                «content»: prompt,
            },
        ],
    )
    return response.choices[0].message.content
```

As you can see, the preceding code uses an OpenAI token key because it's a paid service (it isn't expensive – you will only spend a few credits that cost less than a dollar). To test this, you must inject it into the variables of the lab setup (OPENAI_API_KEY=<your_token>, which can be found in the .venv file). There are options to run a local LLM assistant, but we tried to keep it simple by using a third-party one.

Fully assisted RCA

Finally, you'll create the logic that takes the logs for the last 10 minutes in the API endpoint (these logs will contain not only the BGP ones but also the interface status change and OSPF adjacency lost). You'll create the prompt by combining the logs with the alert data so that we can query the LLM. As mentioned previously, in this example, the response is outputted in the logs so that no further complexity is added:

```
@router.post(«/v1/api/rca-webhook», status_code=204)
def process_webhook(alertmanager_webhook: AlertmanagerWebhook):
    for alert in alertmanager_webhook.alerts:
        device_name = alert.labels[«device»]
        now = datetime.now()
        loki_logs = retrieve_data_loki(
            query=f'{{device="{device_name}"}}',
            start_timestamp=datetime.timestamp(
                now - timedelta(hours=0, minutes=10)),
            end_time=datetime.timestamp(now)
        )
        prompt = generate_rca_prompt(
            loki_logs,
            device_name,
            alert.labels["neighbor"],
            alert.labels["neighbor_asn"]
        )
        rca_response = ask_openai(prompt)
        log.info(f"RCA Analysis: {rca_response}")
    return {"message": "Processed webhook"}
```

Once the webhook endpoint is ready to handle the RCA requests, it's time to configure how the events that will connect to the API endpoint will be triggered.

Configuring Prometheus and Alertmanager to trigger an RCA

Reusing the setup from previous chapters and the knowledge you've learned about how to configure alerts in Prometheus, you're going to define a Prometheus recording rule that triggers an alert. Then, in Alertmanager, it'll call the ML service behind the API endpoint.

Prometheus recording rules

Prometheus recording rules were introduced in *Chapter 7* and used in *Chapter 9*. To recap, we create an alert when detecting that a specific BGP `neighbor` (`10.1.2.2`) is down in the `ceos-01` device (we narrowed it down to a very specific use case to limit the number of alerts) for at least 1 minute. This alert will be enriched with labels, some static (for example, `severity` or `metric_name`) and others dynamically taken from the metric labels (for example, `device` or `neighbor`). These labels will provide the necessary context to handle the alert later:

```
---
groups:
 - name: BGP Neighbor is not established
   rules:
     - alert: BGPNeighborNotEstablished
       expr: bgp_neighbor_state{device="ceos-01",neighbor="10.1.2.2"}
       != 1
       for: 1m
       labels:
         metric_name: bgp_neighbor_state
         device: "{{ $labels.device }}"
         neighbor: "{{ $labels.neighbor }}"
         asn: "{{ $labels.neighbor_asn }}"
         ... some labels omitted ...
       annotations:
         summary: "BGP neighbor status flap"
         description: "Device {{ $labels.device }}: BGP neighbor
         {{ $labels.neighbor }} for ASN {{ $labels.neighbor_asn }} is
         not established!"
```

With this alert rule configured in Prometheus, and the BGP session established, you should get the following status in Prometheus Alerts at `http://<your-server-ip>:9090/alerts`:

Figure 13.20 – BGP Prometheus alert rule not triggered

In this example, to introduce a BGP flap, bring the interface down in `Ethernet1` in the `ceos-01` device:

```
$ ssh netobs@ceos-02
(netobs@ceos-02) Password:
ceos-02>enable
ceos-02#config terminal
ceos-02(config)#interface Ethernet1
ceos-02(config-if-Et1)#shutdown
ceos-02(config-if-Et1)#
```

After bringing the interface down, after a minute or so, you should see that the alert has been triggered:

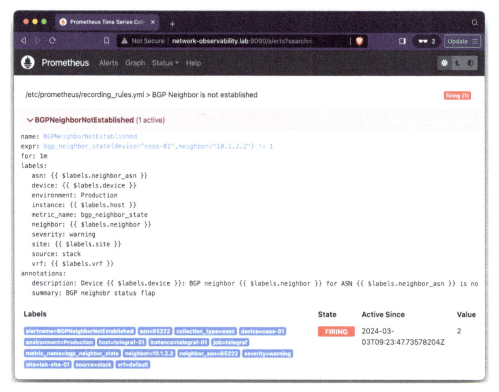

Figure 13.21 – BGP Prometheus alert rule being triggered

Now, you have to *do* something with this alert. This is the role of Alertmanager. In terms of Prometheus, you can define it as follows:

```
alerting:
 alertmanagers:
   - static_configs:
     - targets:
         - alertmanager:9093
     timeout: 5s
```

Once Prometheus has been configured to send the alerts to Alertmanager, it's time to define what to do with them.

Alertmanager configuration

The Alertmanager configuration, as explained in *Chapter 9*, uses routes to define where to send the webhook. In our case, we will send it to http://<your-server-ip>:9997/v1/api/rca-webhook by using the following configuration:

```
---
global:
 resolve_timeout: 30m
route:
 receiver: machine-learning-webhook
 routes:
   - group_by:
       - alertname
     match:
       source: stack
     receiver: machine-learning-webhook
receivers:
 - name: machine-learning-webhook
   webhook_configs:
     - send_resolved: true
       url: http://machine-learning-webhook:9997/v1/api/rca-webhook
```

In the Alertmanager UI (your server, in port 9093), you'll see how the alert that was triggered from Prometheus is being handled:

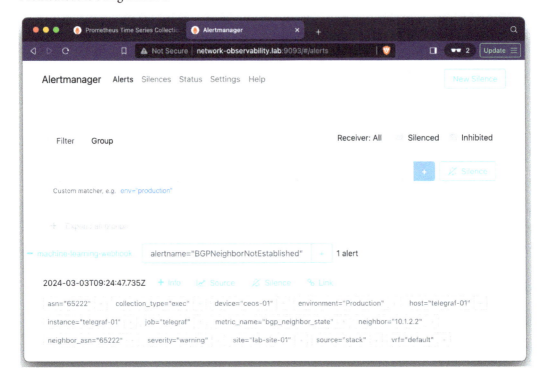

Figure 13.22 – BGP alert in Alertmanager

At this point, we have event generation in place: an alert that's generated in Prometheus is raised in Alertmanager and forwarded to the FastAPI endpoint that will execute the logic defined earlier to collect data and ask an LLM bot for an assisted RCA response.

A few minutes after the interface shuts down, you can check the logs from the FastAPI ML service to see what's happening:

```
root@netobs-droplet:~/network-observability-lab# netobs docker logs
machine-learning-webhook --scenario ch13
```

First, you'll see how the Alertmanager webhook is received and the data it contains. This is where you can see how all the labels are defined:

```
INFO:      | 2024-03-03 09:25:17 | Received
alertmanager webhook: {"version": "4", "groupKey": "{}/
{source=\"stack\"}:{alertname=\"BGPNeighborNotEstablished\"}",
"status": "resolved", "receiver": "machine-learning-webhook", "asn":
"65222", "collection_type": "exec", "device": "ceos-01", "instance":
```

```
"telegraf-01", "job": "telegraf", "metric_name": "bgp_neighbor_state",
"neighbor": "10.1.2.2", "neighbor_asn": "65222", "site": "lab-
site-01", "vrf": "default"}, ... omitted output ...}
```

Then, you'll see the logs that have been retrieved from Loki (in the last 10 minutes), including the interface down, OSFP adjacency dropped, and BGP session change:

```
INFO:      | 2024-03-03 09:25:17 | Loki logs: [{'stream': {'device':
'ceos-01', 'event_type': 'interface_status', 'facility': '7',
'facility_label': 'network news', 'host': 'logstash', 'interface':
'Ethernet1', 'interface_status': 'down', 'priority': '61',
'program': 'Ebra', 'severity': '5', 'severity_label': 'Notice',
'timestamp': 'Mar  3 09:22:30', 'type': 'syslog', 'vendor_
facility': 'LINEPROTO', 'vendor_facility_process': 'UPDOWN'},
'values': [['1709457750000000000', 'Line protocol on Interface
Ethernet1, changed state to down\n']]}, {'stream': {'device': 'ceos-
01', 'facility': '7', 'facility_label': 'network news', 'host':
'logstash', 'priority': '60', 'program': 'Ospf', 'severity': '4',
'severity_label': 'Warning', 'timestamp': 'Mar  3 09:22:30', 'type':
'syslog', 'vendor_facility': 'OSPF', 'vendor_facility_process':
'OSPF_ADJACENCY_TEARDOWN'}, 'values': [['1709457750000000000',
'NGB 10.222.0.1, interface 10.1.2.1 adjacency dropped: interface
went down, state was: FULL\n']]}, {'stream': {'device': 'ceos-01',
'facility': '7', 'facility_label': 'network news', 'host': 'logstash',
'priority': '61', 'program': 'Bgp', 'severity': '5', 'severity_label':
'Notice', 'timestamp': 'Mar  3 09:22:30', 'type': 'syslog', 'vendor_
facility': 'BGP', 'vendor_facility_process': 'ADJCHANGE'}, 'values':
[['1709457750000000000', 'peer 10.1.2.2 (VRF default AS 65222) old
state Established event Stop new state Idle\n']]}]
```

The RCA analysis points out the actual root cause – that is, the interface went down – and then everything else. After, it provides the most common causes, such as physical problems with the connection or configuration changes:

```
INFO:      | 2024-03-03 09:25:24 | RCA Analysis: 1. The potential
causes for the loss of BGP Established state could be related to the
interface status change. The log indicates that the Line protocol on
Interface Ethernet1 changed state to down. This interface is likely
the one connecting to BGP neighbor with IP 10.1.2.2. The adjacency was
dropped because the interface went down, causing the BGP session to
transition from Established to Idle state.
```

System interactions or external factors that could have influenced this event include network equipment failure, configuration issues, or external environmental factors causing the interface to go down:

```
Actionable Insights:
- Immediate actions to mitigate the current issue would be to
investigate and address the root cause of the interface going down.
This may involve checking the physical connection, configuration
settings, and troubleshooting any hardware or software issues related
to Interface Ethernet1.
... omitted some output ...
```

Remember that this output is a one-time response. You'll get a different result, but with similar content, every time you run it.

These are just two examples to showcase how you can leverage AI/ML technology to enhance your network observability. These examples haven't been created so that they can be used in production – they've been created to inspire you to define solutions based on AI/ML once you identify a challenge with your own domain knowledge. Moreover, AI/ML can also help you with many other network automation use cases, such as asking for advice for reported **Common Vulnerabilities and Exposures** (**CVE**) impacting your network operating systems or translating network CLI configurations from one vendor syntax into another.

So, the first step is to enunciate the problem to solve. Our recommended rule of thumb is that *AI/ML can do whatever a human can do* (actually, emulating human learning behavior is the final goal of AI) *but without human limitations*, such as massive data parallel processing or working with more than three dimensions.

However, like humans, learning the patterns requires data and a training process. This means that, like humans, AI/ML can't help us solve problems whose patterns haven't been learned in advance. Indeed, an AI/ML system won't be able to predict any error in the network infrastructure that hasn't been trained for. It can detect anomalies from normal behavior (once the normal state has been identified), but without context (for example, historical data), it won't have the necessary reference to produce knowledge. Thus, collecting and preparing the data to train the ML/AI system is a fundamental part. Throughout this book, we've explained the different types of data a network observability system can collect from the network and how it can be enriched to increase its effectiveness.

Summary

In this chapter, we presented the foundations of ML techniques and explained how they can allow us to scale the observability data analysis to assist and provide educated recommendations. As we noted, this is just a drop in the ocean in the complex and evolving landscape of ML, but we hope it will help you get started and motivate you to dive deeper into the topics you find more interesting.

The use of ML to support network operations – and in concrete, network observability – is just the start of a big revolution that we will see materializing in the next few years, and it will transform how we interact with networks. We'll see many new solutions ready to use coming from the industry, but we wanted to show you two simple examples to illustrate the basics of how these solutions will work.

The usage of forecasted data to validate reality versus expectations that we used to validate operational challenges has many other applications. You can use a similar approach to predict when the capacity of a circuit will hit a level that will require upgrading it or to suggest (or execute) traffic engineering changes under certain conditions. Note that LLMs open the door for digesting text at scale (and networks use a lot of text for configuration and events). Similar to real life, asking good questions (with proper context) usually leads to interesting answers that you may wish to take into account.

Don't forget that an ML system reproduces what it has learned and there will always be some grade of confidence in the response (like what humans provide).

We encourage you to continue following the evolution of ML and start leveraging it to keep your network observability up to date while following the increasing scale of networks, with more devices and more data to analyze.

With that, you've come to the end of this book. By following along, you've gained a comprehensive understanding of modern open source network observability. From the basics to advanced AI and ML applications, each chapter has provided you with the tools, insights, and practical experience to enhance your network's performance and reliability. However, the field of network observability is always evolving, and the knowledge you've acquired here is just the beginning. We encourage you to continue exploring and applying these concepts – the challenges and opportunities in this space are ever-growing. With the foundation you've built, you're well-prepared to navigate and shape the future of network operations. We look forward to seeing where you take this new knowledge and the innovations you'll bring to the field.

Appendix A

This appendix serves as a comprehensive, hands-on guide to setting up a lab environment capable of running the code and configuration examples provided in the book. We provide detailed instructions on the deployment and configuration of an open source observability stack, allowing you to configure, experiment, and validate your network observability efforts in a controlled environment. These steps will pave the way for successful implementation in production settings.

The observability stack comprises several key components that may or may not be used in a given lab scenario (there'll be more on this later).

The following list provides a brief overview of each component and its role in the stack:

- **Telegraf**: An agent for data collection, normalization, and enrichment, well known for the variety of plugins it offers to collect metrics and data from systems. In *Chapter 5* and *6*, we explored Telegraf in detail, highlighting its capabilities and integration methods. For further information, you can refer to `https://www.influxdata.com/time-series-platform/telegraf/`.

- **Logstash**: Logstash is an open source data collection engine with real-time pipeline capabilities, widely recognized for ingesting and parsing syslog information from various systems. We explored the usage of Logstash in *Chapters 5* and *6*, focusing on its powerful data processing features. For more comprehensive information, refer to the official Logstash documentation (`https://www.elastic.co/guide/en/logstash/current/introduction.html`).

- **Prometheus**: A robust tool for storing time-series metrics. We explored the Prometheus architecture and its querying language in depth in *Chapter 7*. For additional details, visit the official Prometheus documentation (`https://prometheus.io/docs/introduction/overview/`).

- **Grafana Loki**: A horizontally scalable, multi-tenant log aggregation system inspired by Prometheus. We explored Loki's components and its querying language in depth in *Chapter 7*. For further details, visit the official Grafana Loki documentation (`https://grafana.com/oss/loki/`).

- **Nautobot**: A **Source of Truth** (**SoT**) that holds our intended network infrastructure data and is used to enrich observability data. Although not the primary focus of the book, Nautobot is featured as an SoT in multiple chapters. For more insights into its role and capabilities, refer to the official Nautobot documentation (`https://nautobot.readthedocs.io/en/latest/`).

- **Grafana**: A powerful platform for data visualization. While it is referenced throughout the book, we take an in-depth look at its features and capabilities in *Chapter 8*. For more detailed information and guidance on using Grafana, visit the official Grafana documentation (`https://grafana.com/docs/grafana/latest/`).

- **Alertmanager**: Alertmanager is a tool that manages alerts sent by applications such as Prometheus and Loki. It's used to handle notifications based on the metrics and logs collected from these systems. We explored its configuration and capabilities in detail in *Chapter 9*. For more information, refer to the official Alertmanager documentation (`https://prometheus.io/docs/alerting/latest/alertmanager/`).

Additionally, this *Appendix A* provides detailed steps to execute the code and lists all the necessary software dependencies for installation.

A lab environment

A lab environment replicates real-world scenarios in a risk-free setup, allowing you to experiment with software and configurations without disrupting a live production system. Establishing this lab is essential for understanding the concepts discussed in this book and testing the included code examples. Preferably, the lab environment should consist of a single Linux **VM** (**virtual machine**), utilizing `containerlab` (`https://containerlab.dev/`) to create container-based network topologies and `docker` (`https://www.docker.com/`) to set up the observability stack. The following diagram offers a visual representation of the proposed lab environment:

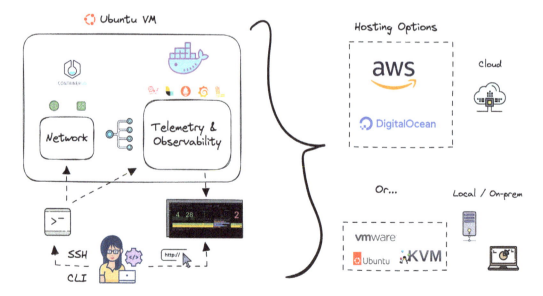

Figure A.1 – A lab environment setup (Created using the icons from https://icons8.com)

As the authors of this book, our goal is to provide a straightforward, reliable method to set up the lab environment. To ensure that the setup process is repeatable and tested, we've opted to use a cloud provider – specifically, **DigitalOcean**. We've prepared two distinct guides to help you provision your VM in this context:

- A step-by-step guide that walks you through building the lab environment manually. You can find this guide in the *Option 1 – manual* section.

- An automated version of the setup process that utilizes Ansible (`https://www.ansible.com/`). This can be found in the *Option 2 – automated* subsection of this *Appendix*.

We'd like to emphasize that these guides aren't restrictive. You're more than welcome to run the lab environment on your own machine, be it Windows with **Windows Subsystem Linux 2 (WSL2)**, macOS, or a Linux-based system such as Ubuntu. Just make sure to not have any constraints with the software requirements. Our main goal is to provide a flexible, effective environment for your learning and exploration, regardless of the platform you choose.

Hardware requirements

For a smooth running of the lab environment, we recommend setting up your VM on a 64-bit x86 CPU architecture. Ideally, the VM should have 4 vCPUs and 8 GB of RAM. However, increasing these resources will significantly enhance the performance and speed of operations.

If you're planning on leveraging a cloud-based VM, such as those offered by Amazon AWS or DigitalOcean, or using a self-hosted solution, such as VMware, make sure your VM specifications align with these recommendations.

Software requirements

The lab environment VM will run `containerlab` to create container-based network topologies and use `docker` natively to construct and manage the observability stack.

The necessary configurations for this lab environment, along with detailed deployment steps for the network topology and observability platform, are all stored in the GitHub repository (`https://github.com/network-observability/network-observability-lab`). This repository also houses the `netobs` utility, which simplifies the deployment and management of the lab environment.

The following is a list of the core dependencies required by the VM to operate the lab environment:

Package	Version
`containerlab`	0.54.2
`docker`	26.1.1
Python (for the `netobs` utility)	3.10

Table A.1 – The main software dependency versions

> **Note**
>
> The preceding versions are accurate at the time of writing. For the most up-to-date information on dependencies, refer to the GitHub repository.

Additionally, the lab environment requires the container images for the components used in the network and the observability stack. Here is the reference for the core components found in the lab scenarios:

Service	Image	Version
Arista cEOS image	`ceos`	4.31.2F
Telegraf	`docker.io/telegraf`	1.31
Logstash	`docker.io/grafana/logstash-output-loki`	main
Prometheus	`docker.io/prom/prometheus`	2.52.0
Alertmanager	`docker.io/prom/alertmanager`	0.26.0
Grafana	`docker.io/grafana/grafana`	10.4.4
Loki	`docker.io/grafana/loki`	2.9.8
Nautobot	`docker.io/networktocode/nautobot`	2.2-py3.10

Table A.2 – The main container dependency versions

The table lists the main components used in the scenarios presented in various chapters of the book. Each component is critical for setting up the observability stack and running the network lab environment. In addition to the primary components, there are other elements such as **Kafka**, which plays a role in *Chapter 6*. To see the exact version and setup of Kafka used in this chapter, refer to the respective Docker Compose file available on the GitHub repository (`https://github.com/network-observability/network-observability-lab/blob/main/chapters/ch6/docker-compose.yml`). This file contains detailed information on how Kafka and other services are set up, ensuring that you have all the necessary details to run the lab scenario.

> **Note**
>
> Aside from the `ceos` network container image, all other images are pulled from Docker Hub and are automatically downloaded when setting up the environment. The `ceos` image is downloaded from the Arista website and imported into Docker (more on this later in the *Downloading and importing Arista cEOS images* subsection).

In the following sections, we will provide a detailed description of the steps to deploy and manage an environment.

Step 0 – Git repository setup

The lab environment tools, configurations, and code examples are all stored in the **Network Observability Lab** GitHub repository (`https://github.com/network-observability/network-observability-lab.git`). To get started, you'll need to fork the repository to your GitHub account.

> **Why fork?**
>
> When you fork a repository, you can freely experiment with changes without affecting the original project. This is especially useful when you need to practice with the code and make changes without affecting the main branch. Forking creates a separate copy of the repository under each reader's account, allowing them to have their own independent version to work with.
>
> With your own fork of the project, you can experiment with the code examples, make changes, and push them to your fork, thus saving any changes. You can also use your fork to submit pull requests to the original repository, allowing you to contribute to the project.

To fork the repository, you'll need to have a GitHub account. If you don't have one already, you can create one for free at (`https://github.com/join`).

To fork the repository, navigate to `https://github.com/network-observability/network-observability-lab` and click the **Fork** button located in the top-right corner. Then, select your GitHub account as the destination for the fork. For more detailed information on working with forks in a GitHub repository, refer to the official GitHub documentation (`https://docs.github.com/en/pull-requests/collaborating-with-pull-requests/working-with-forks/fork-a-repo`).

After forking the repository, you can clone it to your local or lab machine. To keep your fork up to date with the latest changes from the main repository, simply visit your fork's GitHub page and click the **Sync** button. For detailed instructions on syncing your fork, refer to the GitHub official documentation (`https://docs.github.com/en/pull-requests/collaborating-with-pull-requests/working-with-forks/syncing-a-fork`).

The next steps focus on building a lab machine and deploying a lab environment.

> **Note**
>
> If you don't have a GitHub account, you can clone the repository directly from the original repository. However, any changes you make to the repository will not be saved.

Step 1 – VM provisioning

In the upcoming section, we build the lab environment using a cloud provider. This method requires minimal external dependencies, making it a streamlined and hassle-free option. As the authors of this book, we recommend this approach because it has been tested and guarantees a reliable and consistent experience for you when running the examples provided by the book.

It is important to highlight that there is no need to pay to play around with the lab environment. You can use the free trial offered by cloud providers such as DigitalOcean or Amazon AWS. Alternatively, you can use a self-hosted solution such as VMware or VirtualBox, or host natively in your system (a Windows WSL2 or Linux system, but keep in mind that the `docker` and `containerlab` installations may vary for them).

If you are self-hosting the lab environment, here is a table with links to the `docker` and `containerlab` installation for the different operating systems:

Operating system	docker install	containerlab install
Windows WSL2	Docker Desktop on Windows (`https://docs.docker.com/desktop/windows/install/`)	Containerlab on Windows WSL (`https://containerlab.dev/install/#windows-subsystem-linux-wsl`)
MacOS	Docker Desktop on Mac (`https://docs.docker.com/desktop/mac/install/`)	Containerlab on macOS* (`https://containerlab.dev/install/#apple-macos`)
Ubuntu	Docker Engine on Ubuntu (`https://docs.docker.com/engine/install/ubuntu/`)	Containerlab with apt package manager (`https://containerlab.dev/install/#package-managers`)

Table A.3 – The operating system software installation guide

> **Note**
>
> Note that, at the time of writing, there are limitations to running `containerlab` natively on macOS. For more information, refer to the official documentation. It is important to mention that all development and testing of the lab environment has been conducted using Ubuntu.

This guide uses DigitalOcean as the cloud provider, and it is important to remember that charges may incur. You can use other cloud providers (for example, AWS), but the steps to provision the VM may be different.

DigitalOcean account setup

To set up an Ubuntu VM (or **droplet** as it's called in DigitalOcean), you must first meet certain requirements and follow specific steps:

1. Create a DigitalOcean account.

    Go to their website and create an account if you don't have one already. You can use this link (`https://www.digitalocean.com/try/free-trial-offer`) to sign up, with $200 in credit over 60 days for new users at the time of writing.

2. Create an SSH key pair and upload SSH public keys. For example, on Linux or macOS, you can use the following command to create an SSH key pair:

    ```
    $ ssh-keygen -t rsa -b 4096 -C "network-observability-lab" -f
    ~/.ssh/id_rsa_do
    ```

    This will create an SSH key pair in the `~/.ssh` folder, with the public keys named `id_rsa_do` and `id_rsa_do.pub`. After it is created, you can copy and upload the public keys to your DigitalOcean account.

    Next, go to your DigitalOcean account control panel, click the **Settings** menu, and then click **Security**. Then, click **Add SSH Key** and paste the content of the public key file. The **name** field is optional, but you can use `network-observability-lab` as the name. Then, click **Add SSH Key**.

You can use this link (`https://docs.digitalocean.com/products/droplets/how-to/add-ssh-keys/to-account/`) for more instructions and other available alternatives.

The next sections will guide you through building the lab environment step by step. You can choose between a manual installation or the automated version of the deployment.

Option 1 – manual

To manually provision a VM to host the environment, you'll begin by creating a droplet. You can do this through the DigitalOcean web interface or via the command line. For step-by-step instructions on using the web interface, refer to this guide: `https://docs.digitalocean.com/products/droplets/how-to/create/`.

When setting up your droplet, be sure to select the following options:

*   **Distribution**: Ubuntu 22.04 (suggested)
*   **Plan**: Standard, 4 vCPUs, 8 GB RAM, 80 GB SSD (at a minimum)
*   **Datacenter region**: New York 1 (or the closest to your location)

- **Authentication**: SSH keys

- **SSH keys**: Select the SSH key you created in the previous step

- **Create**: Give the droplet a name and create it

You can monitor the progress of your droplet being built through the DigitalOcean interface. Once the process is complete, it's time to log in and verify that everything is set up correctly.

To do this, SSH into the droplet using its IP address and the SSH key you created earlier:

```
$ ssh root@<droplet-ip-address> -i ~/.ssh/id_rsa_do
```

For more detailed instructions on connecting to your droplet via SSH, refer to the official documentation (https://docs.digitalocean.com/products/droplets/how-to/connect-with-ssh/).

Installing docker and containerlab

Now, we are ready to install the main dependencies for this project: docker and containerlab.

1. First, install docker using the following commands:

   ```
   $ sudo apt-get update -y
   $ curl -fsSL https://get.docker.com -o get-docker.sh
   $ sudo sh get-docker.sh
   ```

2. Then, install containerlab using the following commands:

   ```
   $ bash -c "$(curl -sL https://get.containerlab.dev)"
   ```

   For more detailed instructions, visit the official Containerlab installation guide (https://containerlab.dev/install/).

Next, we'll clone our forked lab environment repository. This repository serves as the primary reference for all the lab exercises presented in the book. By cloning it, you'll gain access to all the necessary scripts, configuration files, and documentation needed to set up and operate the various observability tools within your lab environment.

Cloning the lab environment repository

To set up the lab environment, clone the repository from your fork and navigate to the directory:

```
# Clone the repository from your fork
git clone https://github.com/<your-github-username>/network-
observability-lab.git
```

```
# Replace <your-github-username> with your GitHub username.
# After cloning the repository, navigate to the directory
cd network-observability-lab
```

You are now in the root directory of the cloned repository. This will serve as the main reference point for files and configurations throughout the lab exercises presented in the book.

Next, copy and modify the necessary environment file (example.env to .env) for the lab scenarios to run. This file contains variables used by the components deployed in each lab scenario, as well as the settings for the netobs utility that helps configure its operation. Here's an example:

```
$ cp example.env .env
$ vim .env
```

The .env file holds variables as follows:

```
#######################################
# NetObs utility
#######################################
LAB_SCENARIO=batteries-included
LAB_SUDO=false


#######################################
# Telegraf
#######################################
# Creds
NETWORK_AGENT_USER=netobs
NETWORK_AGENT_PASSWORD=netobs123
SNMP_COMMUNITY=public
...
```

You are encouraged to modify the contents of the copied .env file to fit your setup, but it is not mandatory. Next, let's download the network device images and prepare them for Containerlab's usage.

Downloading and importing Arista cEOS images

The lab environment uses Arista cEOS container images. You need to register and download these images from the Arista website. Once registered, go to https://www.arista.com/en/support/software-download to search and download the images under **cEOS-lab** | **EOS-<version>** | **cEOS64-lab-<version>.tar.xz**.

The website has a limited selection of images available for cEOS. You can download the latest version, which should be compatible. We have tested versions from 4.28.5.1M to 4.31.2F at the time of writing.

After downloading the images, use the following commands to import them as Docker images:

```
# Import cEOS images as Docker images
docker import <path-to-image>/cEOS64-lab-<version>.tar.xz ceos:image
```

Replace `<path-to-image>` with the path to your downloaded image and `<version>` with the appropriate version number. The `ceos:image docker` tag is used by the lab's default configuration.

> **Note**
>
> If you are running the automated VM provisioning method covered in the *Option 2 – Automated* subsection that follows, you don't need to import the image. The upload and import of the image to the DigitalOcean droplet will be taken care of automatically by `netobs`.

These images are used by Containerlab for the network topology setup. Your forked repository's `containerlab/` folder includes a configuration file named `lab.yml` and default startup configurations for the network devices (`ceos-01` and `ceos-02`). These configurations provide the details of the network topology setup. If you want to understand the topology better or make adjustments, you can review these configuration files.

Next, let's install the utility tool to manage the lab environment.

Installing the netobs utility

The `netobs` utility is a Python **Command-Line Interface** (**CLI**) tool that allows you to manage the observability and network stack.

> **Note**
>
> It is recommended to install the `netobs` utility in a Python virtual environment, especially if you are running it on your local computer. Refer to the official Python documentation for creating and managing virtual environments: https://docs.python.org/3/library/venv.html. Alternatively, you can use third-party tools such as `conda` from the Anaconda team: https://conda.io/projects/conda/en/latest/user-guide/tasks/manage-environments.html.

To install `netobs`, simply run the following:

```
# From the root of the repository (network-observability-lab)
$ pip install .
```

You can verify the installation by running the `netobs --help` command. This will display a list of available commands and options for the `netobs` utility. The following is an example output:

```
> netobs --help

 Usage: netobs [OPTIONS] COMMAND [ARGS]...

 Run commands for setup and testing

╭─ Options ──────────────────────────────────────────────────────────────────╮
│ --help          Show this message and exit.                                 │
╰────────────────────────────────────────────────────────────────────────────╯
╭─ Commands ─────────────────────────────────────────────────────────────────╮
│ containerlab   Containerlab related commands.                               │
│ docker         Docker and Stacks management related commands.               │
│ lab            Overall Lab management related commands.                      │
│ setup          Lab hosting machine setup related commands.                   │
│ utils          Utilities and scripts related commands.                       │
╰────────────────────────────────────────────────────────────────────────────╯
```

Figure A.2 – The netobs --help command output

Next, let's set up an example network observability scenario by running the `netobs lab deploy --scenario batteries-included` command. This will do the following:

- Use `containerlab` to create the network lab topology, with devices pre-configured and ready for metrics collection and syslog message transmission. See the following figure that shows the output of the `containerlab` section:

```
[00:49:55] Deploying lab environment for scenario: batteries-included
           Network create: network-observability
           Running command: docker network create --driver=bridge  --subnet=198.51.100.0/24 network-observability
c369095805b89ad363a5cb1e1b7a4cee909611982004b7d6d3100dfc908b63c7
           Successfully ran: network create
────────────────────────────── End of task: network create ──────────────────────────────

           Deploying containerlab topology
           Topology file: containerlab/lab.yml
           Running command: sudo containerlab deploy -t containerlab/lab.yml
INFO[0000] Containerlab v0.55.1 started
INFO[0000] Parsing & checking topology file: lab.yml
INFO[0000] Creating lab directory: /root/network-observability-lab/clab-lab
INFO[0000] Creating container: "ceos-02"
INFO[0000] Creating container: "ceos-01"
INFO[0000] Running postdeploy actions for Arista cEOS 'ceos-01' node
INFO[0001] Created link: ceos-01:eth1 <--> ceos-02:eth1
INFO[0001] Created link: ceos-01:eth2 <--> ceos-02:eth2
INFO[0001] Running postdeploy actions for Arista cEOS 'ceos-02' node
INFO[0061] Adding containerlab host entries to /etc/hosts file
INFO[0061] Adding ssh config for containerlab nodes
+---+---------+--------------+------------+------+---------+------------------+--------------+
| # | Name    | Container ID | Image      | Kind | State   | IPv4 Address     | IPv6 Address |
+---+---------+--------------+------------+------+---------+------------------+--------------+
| 1 | ceos-01 | ba1644a86fd3 | ceos:image | ceos | running | 198.51.100.11/24 | N/A          |
| 2 | ceos-02 | d8d60cf6592e | ceos:image | ceos | running | 198.51.100.12/24 | N/A          |
+---+---------+--------------+------------+------+---------+------------------+--------------+
[00:51:00] Successfully ran: Deploying containerlab topology
────────────────────────── End of task: Deploying containerlab topology ──────────────────────────
```

Figure A.3 – Containerlab deployment

- Build and spin up the observability stack for the `batteries-included` scenario. The following figure shows the output of this section:

```
Starting service(s): []
Running command: docker compose --project-name netobs -f chapters/batteries-included/docker-compose.yml
--verbose up -d --remove-orphans
[0000] Docker Desktop integration not enabled
[+] Running 14/16
[+] Running 16/16_alertmanager_data"                              Created                    0.0s
✓ Volume "netobs_alertmanager_data"                              Created                    0.0s
✓ Volume "netobs_loki_data"                                      Created                    0.0s
✓ Volume "netobs_nautobot_postgres_data"                         Created                    0.0s
✓ Volume "netobs_grafana_data"                                   Created                    0.0s
✓ Volume "netobs_prometheus_data"                                Created                    0.0s
✓ Container netobs-nautobot-redis-1                              Started                    4.7s
✓ Container netobs-nautobot-1                                    Started                    5.0s
✓ Container netobs-grafana-1                                     Started                    2.0s
✓ Container netobs-prometheus-1                                  Started                    5.0s
✓ Container netobs-alertmanager-1                                Started                    2.5s
✓ Container netobs-webhook-1                                     Started                    3.7s
✓ Container netobs-loki-1                                        Started                    1.9s
✓ Container netobs-nautobot-postgres-1                           Started                    1.8s
✓ Container netobs-telegraf-02-1                                 Started                    2.1s
✓ Container netobs-logstash-1                                    Started                    1.8s
✓ Container netobs-telegraf-01-1                                 Started                    3.7s


Successfully ran: start stack
────────────────────────── End of task: start stack ──────────────────────────

Lab environment deployed for scenario: batteries-included
```

Figure A.4 – Observability stack deployment

At this point, we have both the network devices and the observability stack up and running. Since we are working with the `batteries-included` scenario, there is a Nautobot instance that needs to be populated with information about our network lab. Completing this step will finalize the integration.

Preparing Nautobot

Nautobot is used throughout the book, and this section will guide you on how to populate Nautobot with data from the network topology used in `Containerlab`:

1. **Verify the Nautobot service**: Ensure that the `nautobot` service is up and running by accessing its web interface at `http://<machine-address>:8080`. Use the credentials provided in your `.env` file (NAUTOBOT_SUPERUSER_NAME and NAUTOBOT_SUPERUSER_PASSWORD). It usually takes a few minutes, but the time depends on the resources of the host.

2. **Check the Nautobot initialization**: If unable to connect, Nautobot might still be initializing. Check its status and logs using the `netobs docker logs nautobot -t 50 -f` command. Initialization logs should show messages such as **Applying migration…** and will complete with **Nautobot initialized!**.

3. **Log into the Nautobot instance**: Once initialized, Nautobot should be running but will be devoid of any data. The Nautobot home page should be similar to the following figure:

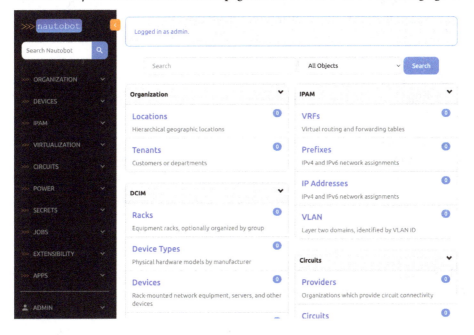

Figure A.5: The Nautobot home page

4. **Populate Nautobot**: To populate Nautobot with information, run the `netobs utils load-nautobot` command. This command reads the variables and content of the network topology and populates it in Nautobot. You can verify this by navigating to **DEVICES | Devices** on the left-hand pane in Nautobot's web interface. A successfully populated Nautobot instance should display the network devices, as shown in the following figure:

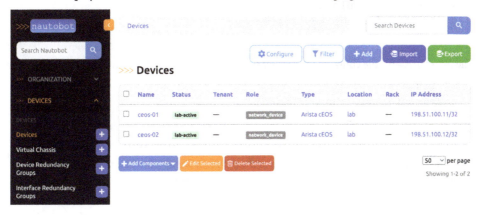

Figure A.6 – The Nautobot Devices page

Nautobot is now ready, with data populated and set to assist in observability tasks.

Let's review the automated provisioning process just in case you prefer this route. This process simplifies the setup and ensures that all components are correctly configured.

Option 2 – automated

The automated process utilizes the `netobs` utility to wrap Ansible playbooks for setting up the DigitalOcean droplet. A *control* machine (typically, your local machine) is needed to execute this command and initiate the process. Follow these steps:

1. **Clone the git repository and install the netobs tool**: The `netobs` tool needs to be installed on your local machine, referred to as the *control* machine. This is necessary because it requires data from the `network-observability-lab` repository to set up a ready-to-go environment.

    Refer to the *Cloning the lab environment repository* subsection for instructions on cloning your forked repository and the *Installing the netobs utility* section for detailed installation steps.

    > **Note**
    >
    > You only need to install the `netobs` tool. There's no need to run the `netobs lab deploy` command, since the local machine will not host the lab environment.

2. **Set up the environment files**: You have seen the setup of the `.env` file in the *Cloning the lab environment repository* subsection. For the automated process, you need a `.setup.env` file to hold environment variables for the DigitalOcean droplet setup. First, create a copy of the provided example file:

    ```bash
    # Navigate to the root location of your forked repository
    $ cd network-observability-lab
    # Copy the example setup environment file
    $ cp example.setup.env .setup.env
    ```

    The `.setup.env` file contains the necessary environment variables with default dummy values:

    ```env
    # Your DigitalOcean API token
    DO_API_TOKEN="do_example123"
    # The fingerprint of your SSH key registered with DigitalOcean
    DO_SSH_
    FINGERPRINT="aa:bb:cc:dd:ee:ff:gg:hh:ii:jj:kk:ll:mm:nn:oo:pp:qq"
    #  The local SSH private key path to connect to DO
    SSH_KEY_PATH="~/.ssh/id_rsa"
    # The local path to the cEOS image
    CEOS_IMAGE_PATH="~/Downloads/cEOS-lab.tar"
    ```

```
# The URL of your forked repository and the branch to be used
# for running the lab scenarios in the DigitalOcean droplet
NETOBS_REPO="https://github.com/<your -user>/network-
observability-lab.git"
NETOBS_BRANCH="main"
```

Next, we will follow the steps to help populate some of these variables:

1. **Download the Arista cEOS images**: Refer to the *Downloading and importing Arista cEOS images* subsection for instructions on downloading the image to your local machine. You do not need to perform the `docker import` step this time. Instead, simply note the path where the compressed `.tar` file is saved and update the `CEOS_IMAGE_PATH` variable in your `.setup.env` file with this path.

2. **Create a DigitalOcean API token**: You need to create a **Personal Access Token** (**PAT**) for DigitalOcean. Follow the instructions in the guide at `https://docs.digitalocean.com/reference/api/create-personal-access-token/` to generate a PAT. Once you have created the PAT, save it in the `DO_API_TOKEN` environment variable in your `.setup.env` file.

3. **Retrieve the SSH key fingerprint**: If you've already created and uploaded your SSH key to DigitalOcean, the next step is to retrieve the SSH fingerprint. This fingerprint is required to provision your VM:

   I.   Navigate to **Settings** | **Security** in your DigitalOcean account.

   II.   Copy the fingerprint value for the SSH key you plan to use.

   Save this fingerprint in the `DO_SSH_FINGERPRINT` environment variable in `.setup.env`. This step ensures that your SSH key is included in the DigitalOcean droplet build process, allowing you access post-creation.

4. **Create a DigitalOcean droplet**: You're now ready to begin setting up the DigitalOcean droplet. Use the `netobs setup deploy` command to start the process. This script will initiate an Ansible playbook that will prompt you to specify the characteristics of your droplet, such as its size and region. Follow the prompts similar to those shown in the following figure, and the provisioning will commence, setting up the environment automatically:

```
> netobs setup deploy
[07:51:28] Running command: ansible-playbook setup/create_droplet.yml -i
           setup/inventory/localhost.yaml
Enter the droplet image [ubuntu-22-04-x64]:
Enter the droplet size [s-4vcpu-8gb]:
Enter the droplet region [fra1]: lon1

PLAY [Stand up netobs-droplet] *************************************************

TASK [Create DigitalOcean Droplets] *******************************************
changed: [localhost]

TASK [Wait for droplets to be ready] ******************************************
ok: [localhost]

TASK [Show Droplet info] ******************************************************
ok: [localhost] => {}

MSG:

Droplet ID is        ID
Public IPv4 is     ip-address
```

Figure A.7 – The droplet deployment

The playbooks executed by `netobs setup deploy` are designed to be idempotent, meaning you can run the command again without adverse effects if you encounter issues (such as DigitalOcean API problems). Just ensure you use the same variables as initially prompted. Once the setup completes, use the `netobs setup show` command to display an SSH command that will allow you to quickly access your droplet:

```
> netobs setup show
[09:33:37] Running command: ansible-playbook setup/list_droplet.yml -i
           setup/inventory/do_hosts.yaml

PLAY [Show Inventory] *********************************************************

TASK [Show SSH command] ******************************************************
ok: [netobs-droplet] => {}

MSG:

ssh -o StrictHostKeyChecking=no -i ~/.ssh/id_rsa root@     ip-address

PLAY RECAP *******************************************************************
netobs-droplet            : ok=1    changed=0    unreachable=0    failed=0    skipped=0    resc
ued=0    ignored=0

[09:33:39] Successfully ran: test
────────────────────────────────── End of task: test ──────────────────────────
```

Figure A.8 – The droplet connection information

At this point, the droplet is fully set up with the Docker, Containerlab, and the `netobs` tool installed. Additionally, your forked repository is cloned and ready to use. By default, the `batteries-included` scenario is running, and Nautobot is configured. Now, let's explore the various ways you can interact with your lab scenarios.

Step 2 – interacting with the lab scenarios

By following either the manual or automated way to provision the VM, you should have the `batteries-included` scenario deployed. This is the default scenario provided in this repository. You can find details of the topology of this scenario in the Git repository: `https://github.com/network-observability/network-observability-lab/tree/main/chapters/batteries-included`.

Now, you can start exploring and interacting with the components of your lab environment. It includes various network devices and observability stack components, each of which is accessible and manageable.

For network devices, you can use SSH or other network management protocols to connect to them. This allows you to modify configurations, run diagnostic commands, and test different network setups.

Here's an example of how to connect to an Arista cEOS network device using SSH:

```
# Connect to ceos-01 via Cli utility command
$ docker exec -it ceos-01 Cli

# Or via SSH (using the default username and password). You can find
the username and password in the .env file.
$ ssh netobs@ceos-01
```

You can interact with observability stack components such as Telegraf, Prometheus, and Grafana primarily through their web interfaces, REST APIs, or command-line interfaces (if available). For example, to access the Prometheus web interface, open a web browser and navigate to the IP address of your droplet and the port on which your Prometheus service is running (`http://<machine-ip-address>:9090`). You can then write queries such as `interface_admin_status` in the query panel, as shown in the following figure:

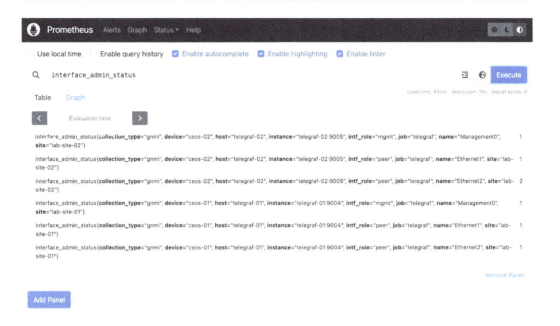

Figure A.9 – The Prometheus web interface

During your interaction with the environment, you might change configurations, experiment with different metrics, create alerts or visualizations, or simulate various network conditions to see how the observability stack responds. This hands-on experience is essential for mastering the skills being taught in this book.

Take your time to explore each component in the observability stack, understand its role, and learn how it contributes to overall network observability. The more you interact with your lab environment, the better you will understand the principles and practices of network observability.

Before proceeding to remove your lab environment, ensure that you save your progress by committing and pushing your changes to your forked repository. This will allow you to easily restore your lab to its current state when needed. For detailed steps on how to remove the lab environment, refer to the upcoming *Removing the lab environment* section.

Step 3 – removing the lab environment

Once you've finished experimenting and learning with your lab environment, you may want to tear it down, especially if you're using resources in a cloud environment where running costs are a consideration.

The process of tearing down your lab environment is referred to as *destroying* the environment. It involves deleting all the components that were created, freeing up the resources they were using, and essentially returning your system to the state it was in before the environment was deployed.

To destroy the lab scenario, you can use the following commands:

```
# Remove the batteries-included scenario
$ netobs lab destroy --scenario <name>
# Remove ALL scenarios setup (useful when you might not remember which
one was up)
$ netobs lab purge
```

To remove the remote VM/DigitalOcean droplet, you can run the `netobs setup destroy` command. This will delete the droplet from your account, which is useful when you have finished the labs for the day.

> **Note**
>
> This action is irreversible. Once you destroy your VM/DigitalOcean droplet, all data and configurations will be permanently lost. If you have any important information or configurations you wish to keep, make sure to back them up before executing the `netobs setup destroy` command.

With this, we conclude the primary actions you can perform in the lab environment, including setup, deployment, and teardown. Feel free to explore the `netobs` tool further, as it is designed to interact with all elements of the lab environment, making management straightforward and hopefully intuitive. The `netobs` tool offers a wide range of commands that allow you to customize, monitor, and troubleshoot your lab scenarios. In the next section, we will discuss the different lab scenarios included in this environment, which can help you practice or explore what can be achieved in your own network.

Step 4 – managing lab scenarios

This book includes various lab scenarios tailored to different requirements for network observability and automation exercises. So far, we have used the `batteries-included` scenario. The `netobs` utility is designed to manage these diverse lab environments effortlessly. The following table presents the different scenarios along with the corresponding commands and brief descriptions:

Scenario	Description
`Batteries-included` lab	Delivers a comprehensive lab setup, inclusive of configs, alerts, dashboards, and an SoT.
Chapter lab	Customized environment with the necessary components and configurations for specific chapter examples.

Table A.4 – Lab scenarios

Apart from the `batteries-included` scenario, which serves as an example lab, each chapter guides you through deploying and interacting with its lab scenario. Typically, you will use the `netobs lab prepare --scenario <chapter>` command, which first runs `netobs lab purge` to remove any lingering services and then executes `netobs lab deploy --scenario <chapter>` to set up the network and observability stack components for that specific chapter.

Summary

The *Appendix A* of this book is your hands-on guide to setting up and managing a network observability lab environment. Here's what we covered:

- **Setting up the lab**: Clear instructions on using `containerlab` and `Docker`, with both manual and automated provisioning on DigitalOcean

- **Forking the repository**: Easy steps to create your own version of a project for experimenting

- **Installing dependencies**: How to install `Docker`, `containerlab`, and `netobs` and set up configuration files

- **Downloading Arista cEOS images**: A guide to getting necessary network device images from Arista's website

- **Preparing Nautobot**: Instructions to integrate Nautobot with network topology data

- **Interacting with lab scenarios**: Manage and interact with network devices and observability components using SSH, web interfaces, and APIs

- **Removing the lab environment**: Commands to safely tear down the lab setup and tips to back up data

- **Managing lab scenarios**: An overview of provided lab scenarios and commands to deploy and manage them, tailored to the book's chapters

This *Appendix A* makes sure you have a clear, tested path to setting up and experimenting with your lab environment, helping you dive into network observability and automation with confidence.

Index

A

access control list (ACL) 10
Advanced Research Projects Agency
Network (ARPANET) 5
agent-based approach 33
versus agentless approach 33
agentless approach 34
versus agent -based approach 33
aggregation operators 201
bottomk/topk operators 202, 203
min/max/avg operator 202
sum operator 202
AI 404
and ML classification 405
alert aggregation 290
strategies 292
alert correlation 290
strategies 292
alert data model 290, 291
fields 290
alert engine architecture 293, 294
alert enrichment 290, 291
fields 291
alerting 285

alerting tips and best practices
AI, role 314
alert challenges, addressing 311, 312
communication process,
improving aspect 312
healthy incident management process 313
Alertmanager 89, 180, 295, 446
architecture 295, 296
lab environment, setting up 296, 297
lab environment, working 297, 298
reference link 446
alerts 286, 287
challenges 289
creating 298
external integrations 308
for Border Gateway Protocol
(BGP) Session Down 299
for interface flapping 299
for Peer Interface Down 299
Grafana 306,-308
Loki rules 303-306
managing, with Keep 310, 311
Prometheus rules 299-303
setting up, considerations 298

Amazon Machine Images (AMIs) 328

annotations, Grafana 275

   configuring, in dashboard settings 276

   creating 277

   creating, manually on panel 276

   HTTP API, using 276

   queries, annotating from log data 278-280

anomaly detection 411

Ansible 326

   URL 447

Ansible Event-Driven Automation (EDA)

   actions 386

   reference link 329, 386

   rules 386

   source 386

anti-lock braking system (ABS) 359

Apache Druid 177

Apache Kafka 161, 162

   example 163-168

   reference link 161

Apache Kafka, components

   cluster replication 161

   data persistence 161

   distributed architecture 161

   fault tolerance and scalability 161

   publish-subscribe model 161

application detail dashboard 346, 355, 356

application owner 343

application performance
   monitoring (APM) 176

Application Programming
   Interfaces (APIs) 17, 52

application summary
   dashboard 346, 354, 355

Argo CD 328

Artificial Intelligence (AI) 27, 360

Artificial Intelligence for IT Operations
   (AIOps) 10, 292, 416, 417

assisted root cause analysis 432, 433

   Alertmanager configuration 439-442

   Alertmanager webhook endpoint 434

   fully assisted RCA 436

   LLM API, asking 435

   LLM prompt, crafting 435

   logs, retrieving from Grafana Loki 434

   Prometheus and Alertmanager,
     configuring 436

   Prometheus recording rules 437-439

association rules 412, 413

attributes 407

automation workflow

   developing, in Prefect 392-400

autonomous system number (ASN) 198, 299

AWS Simple Storage Service (S3) 213

B

backends, for handling tracing data

   Elastic APM 176

   Grafana Tempo 176

   Jaeger 176

   Zipkin 176

batteries-included lab 90

Beats 119

BGP metrics

   retrieving 363-365

BGP Monitoring Protocol 60

BigPanda

   URL 294

Border Gateway Protocol
   (BGP) 25, 197, 344, 365, 405

Borgmon 177

build, versus buy dilemma
cost analysis 323
decision making 325
features and flexibility,
comparing 324, 325
functional requirements 321, 322
in-house capabilities and
resources, evaluating 322
points, comparing 321
risks, assessing 324
Business Calendar 243
business goal 19
business intelligence (BI) 240
business terms
significance 19

C

caching 157
Cacti
reference link 73
URL 319
capacity planning 334
reasons 335
change management (CM) 383
Chaos Monkey idea 24
ChatGPT 415
URL 10
Chef 328
Chronograf 239
classification algorithms 407
Classless Inter-Domain Routing (CIDR) 222
ClickHouse 177
closed-loop automation 384, 387
decision logic 387
execution 388

feedback 388
observe 387
**Cloud Native Computing
Foundation (CNCF)**
reference link 62, 177
cloud networking 18
CloudVision Portal (CVP) 320
clustering algorithms 410, 411
columnar-table databases 176
**command-line interface
(CLI) app 16, 30, 43, 93, 361**
example 369-371
high-bw-links command 374-376
parsing 43-45
setup 371-373
site-health command 377-383
commercial off-the-shelf tools 319, 320
**Common Vulnerabilities and
Exposures (CVE) 442**
compactor 214
config mode changes 278
**configuration management database
(CMDB) 145, 262**
Consul 214
containerized EOS (cEOS) 89
Containerlab
URL 100, 446
controller-based systems 320
correlation engines 292
Create, Read, Update, Delete (CRUD) 52
curse of dimensionality 412
**customer relationship management
(CRM) 310**

D

D3.js
URL 240
Dapper 62
dashboard components, Grafana
data source 242
panels 243
plugins 242
queries 242
transformations 242
dashboarding network systems 340
dashboards 237
application detail dashboard 346, 355, 356
application summary
dashboard 346, 354, 355
architecting 341
detailed dashboards, to consider 356
device detail dashboard 345-352
device summary dashboard 344, 349-351
global overview dashboard 344-348
interface detail dashboard 345-354
site summary dashboard 344-349
types 343, 344
data
use cases, for accessing
programmatically 360
data collection 33
agent-based, versus agentless approach 33
lab environment, setting up 35, 36
network methods 34, 35
data collectors 91-93
characteristics 93, 94
Datadog
URL 320
data enrichment, at query time 158
external data sources, querying 159, 160
info metric 158

data enrichment injection 145
dynamic data enrichment 147-150
static data enrichment 145-147
data enrichment insights 144
enhancing 143
data model 128, 129
network interface example 129, 130
data normalization 126-128
with Logstash filters 135-138
with Telegraf processors 132-135
data pipeline
for observability 80
message broker, need for 160
scaling 160
data pipelines, for observability 80
ETL, unpacking 81, 82
versatility 80, 81
Data Plane Development Kit (DPDK) 3
data visualization
principles 238, 239, 249
deep learning 404, 414
URL 414
deep packet inspection (DPI) 56, 57
device detail dashboard 345, 351, 352
device logs
retrieving, from Loki API 367-369
device summary dashboard 344-351
DigitalOcean 328, 447
dimensionality reduction 412
dimensionality reduction algorithms 412
DistiBERT 415
Distributed Denial of Service (DDoS) 405
distribution 126
distributor 212
division operation 227
docker
URL 446

Docker 328

downsampling 173

droplet 451

dynamic data enrichment 147-150
    enriched data, injecting 154, 155
    incoming metrics, parsing 152-154
    interface data, retrieving via Nautobot's
        GraphQL interface 150-152
    mastering 157
    Telegraf processor configuration 156, 157

dynamic discovery, for scrape targets
    file-based service discovery 190
    HTTP service discovery 190

E

Elastic APM 176

Elastic Common Schema (ECS) 128
    reference link 128

Elasticsearch 119, 176

Elasticsearch (Elastic Observability) 175

Elastic Stack 119

ELK Stack 119

error budgets 24

Ethernet VPN (EVPN) 16

event-driven automation 384-386
    with Prefect 388, 389

event-driven platform
    characteristics 385, 386
    functions 385, 386
    use cases 386

event log 40-43

events 287

eXclusive-OR (XOR) operations 173

exec input plugins 112-119

exporters 185

eXpress Data Path (XDP) 68

extended Berkeley Packet Filter
    (eBPF) 3, 66-69
    networking 66
    security 66

Extract, Load, Transform (ELT) 81
    unpacking, in data pipelines 81, 82

extract, transform, and load (ETL) 93, 126

F

failure scenarios 157

FastAPI
    reference link 420, 427

fault alerts 341

features 407

filter plugins
    reference link 137

flapping 388

FlowCharting 243

flows 54, 55, 56

fluentd/Fluent Bit/Docker driver 215

Flux 328

forecasting 336

fork 449

G

Generative AI 404

geography-based map (GeoMap) 344

GitOps (Git Operations) 327

global overview dashboard 344-348

gNMI input plugin 109-112
    reference link 84
    subscription configuration 111

Goal Question Metric (GQM) approach 25

Golang 95

Google Cloud Storage (GCS) 211, 213

Gorilla compression 173

Grafana 159, 239, 240, 241, 446

accessing 245

architecture 242

dashboard components 242

dashboard variables 266

data sources, connecting 246-248

data sources, exploring 246-248

lab environment, setting up 243, 244

reference ink 89

reference link 446

Grafana Agent 215

Grafana Alerts 306-308

Grafana dashboard

annotations 275-277

creating 249-251

data, overriding 265, 266

details, finalizing 280-283

Device Logs panel 273-275

metrics, inside table panels 271-273

metrics panels 252

saving 260-262

state timeline panels 269-271

table panels, with GraphQL queries 262

transformation, applying to data 263, 264

variables, adding 266-269

Grafana Labs 174

Grafana Loki 177, 211, 445

alerting rules 229

architecture 211

considerations, for using as log
      data database 234, 235

data, reading from 216, 217

data, writing into 215

LogQL queries, running 217-221

logs, retrieving 434

logstash-output-loki plugin 216

metric queries 225

recording rules 229-233

reference link 445

rules 228

working 211

Grafana Loki architecture 211, 212

capabilities, enhancing 214

Loki data model 214, 215

read path 213

storage 213

write path 212

Grafana Loki metric queries 225

derived metrics 226-228

log metrics 225, 226

logs, transforming into 225-228

Grafana plugin library

reference link 243

Grafana Tempo 176

Grafana Worldmap Panel plugin

reference link 348

GraphQL queries 150, 395

execution 395

parsing 395

Graylog 239

Grok Debugger

reference link 139

grok filter plugin 139-142

**gRPC Network Management Interface
      (gNMI)** 17, 198, 320

H

HashiCorp
  reference link 87
hash ring 214
health status
  defining 337, 338
high availability/disaster
    recovery (HA/DR) 86
high-bw-links command 374-376
Huggingface
  URL 415

I

ICMP (Internet Control Message
    Protocol) response 356
If This Then That (IFTTT) approach 27
IGP (Interior Gateway Protocol) 344
inbound interface metrics
  retrieving 365-367
incident 287
incident management 288
incident management workflow 288, 289
individual object identifiers (OIDs) 102
InfluxData 239
InfluxDB 173-176, 239
InfluxDB line protocol 106, 107, 128
Influx Line Protocol 182
info metric 158
Information Technology Infrastructure
    Library (ITIL) 286
Infrastructure as a Service (IaaS) 18, 327
infrastructure as code (IaC) 327
ingester 213

instant vector 193
interface description 104
interface detail dashboard 345, 352-354
interface flaps 278
Interface Message Processor (IMP) 5
interface name field 143
Internet of Things (IoT) 32
Internet Protocol Flow Information
    eXport (IPFIX) 6, 54
Internet Service Provider (ISP) 77
IP Fabric 320
IS-IS (Intermediate System to
    Intermediate System) 356
IT service management
    (ITSM) 84, 240, 286, 340

J

Jaeger 62, 176, 239
Jinja2 templates 326
Juniper MIST 320
just-in-time (JIT) 67

K

Kafka 318, 329, 448
Keep 291
  reference link 291, 309
  setting up 309, 310
  used, for managing alerts 310, 311
key performance indicators (KPIs) 20, 284
  defining 20
Kibana 119, 239
Kubernetes
  reference link 83

L

label filter expressions 223
labels 369
lab environment 446, 447
  architecture, overview 100
  DigitalOcean account setup 451
  Git repository setup 449
  hardware requirements 447
  lab scenarios 90
  lab scenarios interaction 461, 462
  lab scenarios, managing 463
  removing 462
  setting up 88-99, 131, 132, 360-363
  software requirements 447, 448
  VM provisioning 450
Langchain
  URL 415
large language models (LLMs) 10, 415
LibreNMS
  URL 319
line filter expression 222
Link Layer Discovery Protocol (LLDP) 217
Linux networking 17
Linux operating systems (OS) 14
Linux VM (virtual machine) 446
Local Area Networks (LANs) 15
log 40-43
log pipeline 222
log pipeline expressions,
    LogQL queries 222
  considerations 224
  filter expressions 222, 223
  format expressions 223, 224
LogQL 175

LogQL queries 221
  log pipeline 222
  log stream selector 222
  running 217-221
log queries 216
Logstash 119, 211, 445
  reference link 83, 445
Logstash, ad data collector 119
  architecture 119, 120
  syslog input 120-123
Logstash filters
  grok filter plugin 139-143
  used, for data normalization 135-138
Logstash pipeline 119
Logstash plugin 215
log stream selector 222
Loki API
  device logs, retrieving from 367-369
  using 367
Loki rules 303-306

M

Machine Learning (ML) 27, 85, 336, 404
  adopting 417
Management Information Base
    (MIB) 6, 25, 36
Matplotlib
  URL 240
Mattermost 84
Maximum Transmission Unit (MTU) 39
mean deviation (mdev) 339
mean time to change (MTTC) 339
mean time to repair (MTTR) 280, 339
Memcached 214

message 369

message broker 160

Message Queueing Telemetry
    Transport (MQTT) 168
  reference link 168

metadata
  benefits 143

metric queries 216

metrics and logs 287

Metrics Data Model
  reference link 129

metrics panels, Grafana dashboard 252
  Avg CPU metrics panel 257-259
  Avg Mem metrics panel 257-259
  Bandwidth Usage panel 259, 260
  state and latency metrics panel 253-255
  uptime metrics panel 255-257

Microsoft Teams 84

Midwest Internet Cooperative Exchange
  URL 319

Mimir 174

Minimal Viable Product (MVP) 84

ML algorithms 406
  reinforcement learning 413
  supervised learning 407
  unsupervised learning 410

model-driven telemetry 45-52

MQTT 318

MRTG
  reference link 73

multidomain 6

N

Nagios
  URL 319

NATS.io 318

natural language processing (NLP) 414

Nautobot 132, 445, 456
  reference link 89, 132, 445

Nautobot instance
  reference link 244

Nautobot's GraphQL interface
  reference link 150

NetDevOps 9

NetFlow 6

Netflow/IPFIX 31

netobs utility 454
  installing 454-456

network 333

network architectures 14-16

network as a service
  dashboarding network systems 340
  fault alerts 341
  notification of changes 340
  treating 338, 340

network automation 16, 17, 359

Network Automation with Nautobot
  reference link 132

network control monitoring protocols 60, 61

network engineering team member 342

network engineers
  tools for 93

network flows 54
Network Function Virtualization (NFV) 16
networking 14
    cultural changes 18
    technological changes 14-18
Network Management Systems
        (NMS) 32, 286
network manager 342
network monitoring 3, 30
    challenges 30
    evolution 5
network observability 3
    benefits 11
    defining 4, 5
    trends and requirements 7
network observability, expectations
    at scale 27
    faster 27
    full visibility, of network state 26
    heterogeneous and enriched data 25
    more accurate 27
    proactive role, in network automation 26
network observability framework
    control plane 32
    external data 32
    forwarding plane 32
    management plane 32
    planes 31, 32
Network Observability Lab
        GitHub repository
    reference link 449
network observability pillars
    actionable data 8-10
    assisted analysis 8-10
    data quality 8
    scalability and interoperability 8, 9

network operating systems (NOS) 34
network operational data 19
    transforming, into data 24
    transforming, into information 19
network operational data transformation
    business term, significance 19, 20
    KPIs, defining 20
network-related personas 341-343
Network Source of Truth (NSOT) 89
network telemetry 31
Neural Autonomic Transport
        System (NATS) 168
    reference link 168
neural networks 404, 413

O

Object Identifier (OID) 36
observability 360
observability data
    business value 333, 334
observability databases 172
    features 172
    logs 175
    metrics 174
    packet flow data 176
    requirements, matching with 174
    time series databases 173, 174
    traces 175, 176
observability data, examples
    capacity planning 334, 335
    forecasting 336
    health status 337, 338
    percentiles 335, 336
observability pipeline architecture
    stack options 318

observability platform

alerting 75

components 73, 74

databases 75

data collection 74

data processing and distribution 74

deployment methodologies and
    orchestration 326-329

orchestrating 325, 326

orchestration 75

source of truth 75

visualization 75

observability stack

challenges and best practices 82

data pipelines 80

designing 76-78

importance 76

well-designed platform, components 79

**observability stack, challenges
    and best practices**

cost management 85, 86

customization 84

data collection and retrieval 84

extensibility 84

extensibility and customization 85

flexibility 84

key strategies 86, 87

reliability 83, 84

scalability 82, 83

Observium

URL 319

OpenConfig

reference link 129

OpenShift

reference link 87

Open Shortest Path First (OSPF) 225, 406

**open source network monitoring
    systems 319**

commercial off-the-shelf tools 319, 320

controller-based systems 320

time series, versus snapshot
    observability 320

OpenTelemetry 63-66, 77, 128

reference link 62

OpenTelemetry Logs Data Model

reference link 129

Open WebUI 415

operational changes validation 420

pandas library, leveraging 422-424

Prometheus metrics, retrieving 421, 422

using, in real-time service 427-432

values, forecasting with Prophet
    library 424-426

**Operations, Administration, and
    Management (OAM) 30**

Opsgenie

URL 294

organization leader 342

OSPF (Open Shortest Path First) 356

outputs.file plugin 105, 106

overrides 264

P

packet capture 56, 57

packet flow data 176

PagerDuty

URL 295

pandas library

leveraging, for data preparation 422-424

URL 422

parser expression 224

PDU (Power Distribution Unit) 356

percentiles 335
  95th percentile 335, 336

persistence tips and best practices 233
  automation 235
  performance and scale 234, 235

Personal Access Token (PAT) 459

personas 341

pipeline 93, 119

Plotly
  URL 240

Podman 328

Point of Presence (PoP) 25

Power 240

Prefect
  automation workflow, developing 392-400
  event-driven automation 388, 389
  URL 388

Prefect Cloud
  setting up 390-392
  URL 388, 390

Prefect secret 395

Prefect task 395

processing 126

Prometheus 77, 174, 239, 445
  reference ink 89, 445

Prometheus API
  BGP metrics, retrieving 363-365
  inbound interface metrics,
      retrieving 365-367
  using 363

Prometheus data types
  instant vector 193
  range vector 193, 194
  scalars 194

Prometheus functions 204
  aggregation over time functions 204
  basic mathematical functions 204
  range vector functions 204
  special functions 204

Prometheus functions, examples
  average CPU usage over time 206, 207
  interface operational status,
      predicting 207, 208
  network interface traffic 205

Prometheus info metrics 183
  benefits 184
  considerations 184

Prometheus metric types
  counters 183
  gauges 183
  histograms 183
  summaries 183

Prometheus Query Language (PromQL) 363

Prometheus rules 208, 299, 300-303
  recording rules 208-210, **437**

Prometheus TSDB 177
  architecture 178
  data model 181
  dynamic discovery, for scrape
      targets 190, 191
  lab environment, setting up 185, 186
  reading from 192
  rules 208
  scrape jobs 188-190
  server 178
  Telegraf output, configuring 187, 188
  writing to 185

Prometheus TSDB data model 181
  Influx line protocol, migrating to 182, 183
  info metrics 183, 184
  metric labels 181

metric name 181
metric types 183
sample or values 182
Prometheus TSDB server
alerts 180, 181
data, exploring 181
data retrieval 179
HTTP server 180
TSDB storage 179
PromQL operators
aggregation operators 201-204
arithmetic operators 196
binary operators 196
comparison operators 196
logical/set operators 197
vector matching 199
PromQL (Prometheus Query
  Language) 175, 192, 211, 242
data, fine-tuning with selectors 194
data operations 195
data types 193
queries, running 192, 193
Promscale 175
Promtail 215
Prophet library
used, for forecasting values 424-426
Pulumi 328
Puppet 328
pushgateway 179
Python Command-Line Interface
  (CLI) tool 454

Q

Quality of Service (QoS) 26, 339
querier 213
query frontend 213

R

range vector 193
real-world AIOps 418
assisted root cause analysis 432, 433
lab requirements 419, 420
operational changes, validating 420, 421
regression algorithms 408-410
Regular Expressions (regex) 42, 224
reinforcement learning 413
REST APIs 52, 53, 54
retrieval-augmented generation (RAG) 417
root cause analysis (RCA) 27, 405
round robin database (RRD) 73
RRDtool 6
database 319
ruler 228
ruler component 296
ruler functions 214
RYG (Red/Yellow/Green) state 344

S

Salt 328
SAN (Storage Area Network) 334
scalars 194
scraping 185
Seaborn
URL 240
secret block 391
security operations 342
selectors 194
querying, with time 195
Service-Level Agreement
  (SLA) 20, 23, 338, 340
service level definitions, Google SRE guide
reference link 20
Service-Level Indicator (SLI) 20, 21, 338
examples 338, 339

Service-Level Objective (SLO) 20-23, 338
    examples 339
sFlow 6
sidecar pattern 66
Simple Network Management Protocol
        (SNMP) 25, 198, 319
site-health command 377-383
site summary dashboard 344-349
Slack 84
small language models (SLMs) 415
snapshot observability
    versus time series 320
SNMP 30, 36-40
    characteristics 6
Software as a Service (SaaS) 18
Software-Defined Networking (SDN) 17
Software-Defined WAN (SD-WAN) 16
Solarwinds
    URL 320
source of truth
        (SoT) 184, 291, 297, 370, 445
SOW (statement of work) 340
Splunk 175
    URL 319
standard output (stdout) 105, 133
standout tools, log management realm
    Elasticsearch 175
    Grafana Loki 175
    Splunk 175
static data enrichment 145-147
streaming telemetry 46
supervised learning 407
    classification algorithms 407, 408
    regression algorithms 408-410
SuzieQ 320
synthetic monitoring 57-60
synthetic monitoring input plugins 107-109
Syslog 6, 30

T

Tableau 240
technological changes, networking 14
    architectures 14, 15
    cloud networking 18
    Linux networking 17
    network automation 16, 17
    virtualization 16
Telegraf 95, 445
    architecture 95, 96
    configuration 96, 97
    configuration, tips 98
    exec input plugins 112-119
    gNMI input plugin 109-112
    lab environment, setting up 99-102
    reference link 83, 445
    SNMP input plugin 102
    synthetic monitoring input plugins 107-109
Telegraf data normalization 130
Telegraf, InfluxDB, Chronograf,
        Kapacitor (TICK) 127
Telegraf output
    configuring, for Prometheus 186, 188
Telegraf plugins
    reference link 130
Telegraf processors
    used, for data normalization 132-135
Telegraf, SNMP input plugin 103-105
    agent settings 102
    file output plugin 105, 106
    InfluxDB line protocol 106, 107
tenant 213
Terraform 328
Thanos 174
three-tier architecture 14

time series
    cardinality 173
    challenges 174
    dimensionality 173
    versus snapshot observability 320
time series database
        (TSDBs) 6, 173, 242, 296, 363
timestamp 369
Tom's Obvious Minimal
        Language (TOML) 96
total cost of ownership (TCO) 323
training data
    issues 418
transform stage 126
typer
    URL 371

U

UNIX timestamp format
    reference link 428
unsupervised learning 410
    anomaly detection 411
    association rules 412
    clustering algorithms 410, 411
    dimensionality reduction 412

V

value-added resellers (VARs) 324
vector matching 199
    many-to-one/one-to-many
        matching 200, 201
    one-to-one matching 200
Vector Packet Processing (VPP) 34
vendor lock-in 324
VictoriaMetrics 177

Virtual Extensible LAN (VXLAN) 16
virtualization 16
Virtual Machines (VMs) 87
visualization 237
    best practices 283, 284
    in network observability 238
    tips 283, 284
visualization tools, for observability 239
    Chronograf 239
    Grafana 239
    Graylog 239
    Jaeger 239
    Kibana 239
    Prometheus 239
VM provisioning 450
    automated process 458-460
    manually process 451
VM provisioning, manual process 451
    Arista cEOS images, downloading
        and importing 453
    containerlab installation 452
    docker installation 452
    lab environment repository, cloning 452, 453
    Nautobot, preparing 456, 457
    netobs utility, installing 454-456
VMware
    reference link 87
VPN (Virtual Private Network)
        connections 356

W

Web Content Accessibility Guidelines
    reference link 238
Wide Area Network (WAN) 16, 334
Windows Subsystem Linux 2 (WSL2) 447
write-ahead log (WAL) 180, 213

X

XPath filtering 51

Y

Yet Another Markup Language (YAML) 188
Yet Another Next Generation (YANG) 17

Z

Zabbix
  URL 319
Zipkin 62, 176
ZooKeeper 162

packtpub.com

Subscribe to our online digital library for full access to over 7,000 books and videos, as well as industry leading tools to help you plan your personal development and advance your career. For more information, please visit our website.

Why subscribe?

- Spend less time learning and more time coding with practical eBooks and Videos from over 4,000 industry professionals

- Improve your learning with Skill Plans built especially for you

- Get a free eBook or video every month

- Fully searchable for easy access to vital information

- Copy and paste, print, and bookmark content

Did you know that Packt offers eBook versions of every book published, with PDF and ePub files available? You can upgrade to the eBook version at packtpub.com and as a print book customer, you are entitled to a discount on the eBook copy. Get in touch with us at customercare@packtpub.com for more details.

At www.packtpub.com, you can also read a collection of free technical articles, sign up for a range of free newsletters, and receive exclusive discounts and offers on Packt books and eBooks.

Other Books You May Enjoy

If you enjoyed this book, you may be interested in these other books by Packt:

Automating Security Detection Engineering

Dennis Chow

ISBN: 978-1-83763-641-9

- Understand the architecture of Detection as Code implementations
- Develop custom test functions using Python and Terraform
- Leverage common tools like GitHub and Python 3.x to create detection-focused CI/CD pipelines
- Integrate cutting-edge technology and operational patterns to further refine program efficacy
- Apply monitoring techniques to continuously assess use case health
- Create, structure, and commit detections to a code repository

Python for Security and Networking

José Ortega

ISBN: 978-1-83763-755-3

- Program your own tools in Python that can be used in a Network Security process
- Automate tasks of analysis and extraction of information from servers
- Detect server vulnerabilities and analyze security in web applications
- Automate security and pentesting tasks by creating scripts with Python
- Utilize the ssh-audit tool to check the security in SSH servers
- Explore WriteHat as a pentesting reports tool written in Python
- Automate the process of detecting vulnerabilities in applications with tools like Fuxploider

Packt is searching for authors like you

If you're interested in becoming an author for Packt, please visit `authors.packtpub.com` and apply today. We have worked with thousands of developers and tech professionals, just like you, to help them share their insight with the global tech community. You can make a general application, apply for a specific hot topic that we are recruiting an author for, or submit your own idea.

Share Your Thoughts

Now you've finished *Modern Network Observability*, we'd love to hear your thoughts! Scan the QR code below to go straight to the Amazon review page for this book and share your feedback or leave a review on the site that you purchased it from.

`https://packt.link/r/1835081061`

Your review is important to us and the tech community and will help us make sure we're delivering excellent quality content.

Download a free PDF copy of this book

Thanks for purchasing this book!

Do you like to read on the go but are unable to carry your print books everywhere?

Is your eBook purchase not compatible with the device of your choice?

Don't worry, now with every Packt book you get a DRM-free PDF version of that book at no cost.

Read anywhere, any place, on any device. Search, copy, and paste code from your favorite technical books directly into your application.

The perks don't stop there, you can get exclusive access to discounts, newsletters, and great free content in your inbox daily

Follow these simple steps to get the benefits:

1. Scan the QR code or visit the link below

https://packt.link/free-ebook/978-1-83508-106-8

2. Submit your proof of purchase
3. That's it! We'll send your free PDF and other benefits to your email directly

www.ingramcontent.com/pod-product-compliance
Lightning Source LLC
Chambersburg PA
CBHW060640060326
40690CB00020B/4467